Users in Water Management

Users in Water Management

The Andhra Model and its Replicability in India

Edited by

RAKESH HOOJA
GANESH PANGARE
K.V. RAJU

Published under the auspices of

INDIAN NETWORK OF PARTICIPATORY IRRIGATION MANAGEMENT
New Delhi

RAWAT PUBLICATIONS
Jaipur and New Delhi

ISBN 81-7033-715-1

Published by
Prem Rawat for ***Rawat Publications***
Satyam Apartments, Sector 3, Jawahar Nagar, Jaipur - 302 004 (India)
Phone: 0141 651748 / 657006 Fax: 0141 651748
e-mail : info@rawatbooks.com
website: rawatbooks.com

Delhi Office
G-4, 4832/24, Ansari Road, Daryaganj, New Delhi 110 002
Phone: 011-3263290

Typeset by Rawat Computers, Jaipur
Printed at Chaman Enterprises, New Delhi

Contents

IV
Training and Gender

V
Beyond Andhra: Replicability and Sustainability in India

Appendices

Preface

Large public irrigation systems coupled with the green revolution have contributed to a large extent in initially stabilizing and then expanding food production, whereby India's growing population has remained self sufficient in food. However, increasing populations requiring food and fiber in larger quantities and increased competition for water amongst various uses of water, and amongst various regions, are gradually ensuring that the issue of improved and efficient land water management assumes major significance.

A major contribution to improved land water management can come from large scale adoption of Participatory Irrigation Management (PIM) involving a much greater role for active Water Users' Associations (WUAs) working in tandem with governmental agencies which have undergone reorientation as regards attitudes, roles, objectives, organizational structures and working procedures. Many factors have led to governments, aid agencies, water professionals, developmental experts and NGOs taking varied measures aimed at propagating some form or the other of PIM.

In this context, the decision of the Indian state of Andhra Pradesh to go in for legislation followed by massive efforts for the simultaneous and uniform application of a particular model of PIM in irrigation systems throughout the state, has been attracting much attention. It is very different from approaches in many other States which are based on pilot projects, voluntarism, and much more flexibility and variety as regards the structures and roles to be performed by WUAs. Some Indian states are now even trying to emulate the Andhra model of PIM.

Thus, an examination of the Andhra model -- how it is working in the field in Andhra and what lessons its experience holds for other Indian states -- appears to us to be very timely. The idea of putting together a volume of papers which look at PIM in India with special reference to the Andhra model and which is relevant for policy makers and academicians as well as practitioners, first came to us since IndiaNPIM (Indian Network on Participatory Irrigation Management) had commissioned six papers on various aspects of PIM in Andhra Pradesh for an International Researchers' Conference on "The Long Road to Political Commitment: A Socio-Political Perspective in the Process of Irrigation Reform" held at Hyderabad from 11th to 14th December 1999, in conjunction with the Fifth International Seminar on "PIM-Implementing Management Transfer-Lessons from Andhra Pradesh" from 14th to 21st December 1999.

Accordingly, by mid-2000, we had started contacting people to provide us with additional papers to supplement the original six, whereby we could place the Andhra experiment within the all-India context, see what were the special features of the Andhra model which were likely to contribute to its sustainability within Andhra Pradesh, to compare the Andhra model with that in other states, as also to see to what extent the Andhra approach and experiences may be replicable in other India states.

While putting together this volume we have viewed it as the intellectual successor of the earlier IndiaNPIM publication, i.e., L.K. Joshi and Rakesh Hooja (eds.) *Participatory Irrigation Management: Paradigm for the 21st Century (2000)*, whose 23 papers had looked at conceptual and prescriptive issues of PIM with particular focus on the Indian context, at all-India and multi-state overviews of the status of and prospects of PIM, and at specific PIM-related experiments and experiences in some Indian states (this last section had also contained three papers on Andhra Pradesh). Our belief that PIM (of course coupled with related and necessary technical measures) will be the irrigation related paradigm most necessary for the initial decades of the twenty-first century remains unshaken. Accordingly, this book is intended to carry the story, and our hopes from PIM, forward from the point where we had left it in the Joshi-Hooja volume, which we earnestly hope the reader would be able to go through as a publication complementing the current book (which of course is, however, complete in itself).

The papers in this volume by Raju, Sithapathi Rao, Someshwar and Vijaya Shyamala, Venkateshwaralu, Narasimha Reddy, Jasveen Jairath, and Vijaya Shyamala and Sithapathi Rao are basically those received by us for the International Researchers' Conference in 1999 (except that the number of tables in Jasveen Jairath's paper have been reduced). Guru Prem's paper had been presented at the Fifth International Conference as had the note included in the Appendices by Douglas Vermillion which was circulated at the Fifth International Conference (and which Gert Diemer and Raymond Peter of INPIM (International Network on Participatory Irrigation Management) had indicated could be suitably replicated by IndiaNPIM). Peter Mollinga's paper started life as written for circulation at the International Training of Trainers Programme on PIM organized by IndiaNPIM at Hyderabad from 19th to 25th January 2001 and was later revised as IndiaNPIM's Working Paper/Monograph Series No. 1, which the author consented to our including it in this volume. The contribution by Lele, Patil and Sane about NGOs came to us for using in IndiaNPIM Newsletter and we thought of simultaneously using it in this book. Jaya Raj's case study was originally presented at a Seminar sponsored by IndiaNPIM, Government of Tamilnadu and the World Bank at Madurai in the summer of 2001. Ganesh Pangare used a short stint at East-West Centre in early 2001 to write a paper with the intention of using it in the introductory section of this volume. Raymond Peter, Parthsarthy and Turabul Hassan respectively have provided papers specially for this book, as did Rakesh Hooja (who added his introductory overview in the Indian context only after all the other papers had been received and placed in the various sections into which the book has been divided).

The papers have been arranged so that the reader can start with an overview of PIM-related issues and concerns in the Indian context including an idea of approaches being followed in some states other than Andhra. Thereafter, the attempt at large-scale introduction of PIM is discussed in a number of papers which examine the Andhra model, the present status of PIM in that state, problems being encountered and future prospects. It has often been commented that strong political backing is a major factor in the success of the Andhra experiment. Accordingly, papers that discuss the political dimension of the Andhra irrigation reforms have been grouped together in Section Three. After a brief look at training and capacity building and the role of women in the Andhra PIM context, the book moves on to assessing the possibility of the Andhra model being replicated in other states of the country as also its sustainability and effectiveness within Andhra Pradesh, and offers suggestions which may help in the speedy propagation of PIM in the country as a whole.

Three appendices in the form of copies of the APFMIS Act of 1997, the Karnataka amendment reflecting an alternative approach and a note on monitoring and evaluation in the context of PIM in Andhra round up the volume.

The improved management of irrigation and water delivery systems is, in our view, possible only if PIM gains widespread acceptance. PIM, though not a panacea for all water management problems, can go a long way towards ensuring efficient running of canals, optimum on-field water application and reuse or proper drainage of the excess water from irrigation systems and farmer's fields – in short ensuring efficient land-water management. Accordingly, the message of PIM needs to be taken to all sections of society, who need to be made aware of what may be possible, what has already been attempted, and what could be possible ways forward. This book is one more humble attempt at helping to spread the message of PIM.

Editors

Contributors

RAKESH HOOJA	Joint Secretary, Department of Jammu & Kashmir, Ministry of Home Affairs, Room No. 126, North Block, New Delhi - 110001 E-mail: rakeshhooja@yahoo.com
GANESH PANGARE	Chief Executive Officer, IndiaNPIM, Room No. 318, Central Soil and Material Research Station, Olof Palme Marg, Hauz Khas, New Delhi - 110016 E-mail: gpangare@hotmail.com
J. RAYMOND PETER	Executive Director, International Network on Participatory Irrigation Management, Suite 340, 600 Pennsylvania Avenue SE, Washington DC 2003, USA E-mail: rpeter@inpim.org
K.V.RAJU	Senior Fellow, Centre for Interdisciplinary Studies on Environment and Development, Institute of Social & Economic Change, Nagarabhavi, Bangalore - 560072 E-mail: kvraju@isec.kar.nic.in
C. SITHAPATHI RAO	Director, Institute of Resource Development and Social Management, 10-1-123/A/3/1, P.G. College Road, Saifabad, Hyderabad - 500004
K. SOMESHWAR	Research Associate, Institute of Resource Development and Social Management, 10-1-123/A/3/1, P.G. College Road, Saifabad, Hyderabad - 500004

C. VIJAYA SHYAMALA	Research Associate, Institute of Resource Development and Social Management, 10-1-123/A/3/1, P.G. College Road, Saifabad, Hyderabad - 500004
M. JAYARAJ	Faculty Member, WALAMTARI, Himayatsagar, Hyderabad - 500030
D. VENKATESWARLU	Researcher, House No. 68, Doyens Township, Serilingampalli, Hyderabad
D. NARASIMHA REDDY	House No. 3-33-33, L.V. Reddy Colony, Lingampalli, Hyderabad - 500019
JASVEEN JAIRATH	Consultant, Centre for Economic & Social Studies, Aurora Colony, F-I, "Edens Banjara", Avenue No. 8, Street No. 7, Banjara Hills, Hyderabad - 500034
B. GURU PREM	Director, WALAMTARI, Himayatsagar, Hyderabad - 500030
PETER P. MOLLINGA	Visiting Professor, Administrative Staff College of India, Bella Vista, Hyderabad - 500082
R. PARTHASARATHY	Associate Professor, Gujarat Institute of Development Research, Gota, Ahmedabad - 380060
S.N. LELE	Society for People's Participation in Ecosystem Management, 16, Kale Park, Someshwarwadi, Pashan, Pune - 411008 E-mail: soppecom@pn3.vsnl.net.in
R.K. PATIL	Society for People's Participation in Ecosystem Management, 16, Kale Park, Someshwarwadi, Pashan, Pune - 411008 E-mail: soppecom@pn3.vsnl.net.in
S.B. SANE	Society for People's Participation in Ecosystem Management, 16, Kale Park, Someshwarwadi, Pashan, Pune - 411008 E-mail: soppecom@pn3.vsnl.net.in
S. TURABUL HASSAN	Chairman, Institute of Resource Development and Social Management, 10-1-123/A/3/1, P.G. College Road, Saifabad, Hyderabad - 500004
DOUGLAS L. VERMILLION	International Water Management Institute, PO Box 2075, Colombo, Sri Lanka

I

Introduction:
Propagating Participatory Irrigation Management in India

1

Participatory Irrigation Management in the Indian Context

RAKESH HOOJA

Conventional wisdom used to be that only the state was capable of properly handling large modern projects requiring centralised control, heavy capital investment, complicated technical inputs and the legal mandate to distribute water and collect water charges. However, we all know of government-operated irrigation systems that are poorly maintained with steadily deteriorating irrigation infrastructure. And some of these systems have shown dramatic improvement especially in terms of improvement in water delivery services when their management was transferred to Water Users' Associations (WUAs) who took over operating and maintaining portions of the irrigation system or of the entire system.

A large number of countries today are attempting in various ways to bring about various forms of participatory irrigation management (PIM) in their irrigation sector. Whereas the World Bank (1998) states that PIM refers to involvement of irrigation users in all aspects (including initial planning and design of new irrigation projects or improvements, as well as construction, supervision, financing, decision rules, operation maintenance, monitoring, and evaluation of the system) of irrigation management at all levels (the full physical limits of irrigation system and upto the policy level in the capital city), however a more practical definition of PIM is that it (as opposed to mere farmers' participation in irrigation) involves a WUA taking over management (including operation and maintenance) of at least one level of canals (i.e., the minor canal) above the outlet and also being closely associated in efforts at improved water management. Simultaneously, PIM involves a re-definition and re-focusing (and not merely a reduction) of government's role in irrigation so as to lead to a genuine feeling of partnership or joint management between government and WUAs.

Countries like Turkey, Mexico, Chile, Columbia, Indonesia, Nepal, Sri Lanka, the Philippines, Egypt, China, Morocco, Albania, Jordan, Yemen, Pakistan and India have been making efforts at broad basing PIM (referred to by various names in different countries) often with some stimulus in the form of inducing policy dialogue and support for training and capacity building by INPIM (International Network on Participatory Irrigation Management) set up in 1995 by the World Bank's Economic Development Institute (now renamed World Bank Institute).

In India, the National Water Policy adopted in 1987 has stressed that efforts should be made to involve farmers progressively in various aspects of the management of the irrigation system, particularly in water distribution and collection of water rates. The Union Ministry of Water Resources has (1) organised national conferences on PIM in 1994, 1995, 1997 and 1999 with representatives from state governments, NGOs, and various central agencies as well as experts; (2) sponsored regional and state-level workshops; (3) made PIM the subject matter of working groups for five-year plan formulation since the Ninth Plan preparation exercise; (4) incorporated PIM as an important component for which funds are made available under the centrally sponsored scheme of Command Area Development (CAD) and as a part of advocacy in all its irrigation-related activities; and (5) sponsored the creation of IndiaNPIM (Indian Network on Participatory Irrigation Management, sponsored by the Ministry of Water Resources, Government of India) which is now a regular NGO independent of, yet collaborating with, governments and functions as the Indian national chapter of INPIM as well. Most World Bank-water related projects, especially the Water Resources Consolidation Projects and Water Sector Reforms Projects for some Indian states, contain PIM components even though they may be called by some other name. Various state governments have gone in for a rapid "big bang" approach of uniform and simultaneous introduction of PIM throughout the state (the leading example being Andhra Pradesh, which enacted the Andhra Pradesh Farmers' Management of Irrigation Systems (APFMIS) Act in 1997 and covered all irrigation schemes – major, medium or minor – under 10,292 WUAs and 174 Distributory Committees) or have gone in for some form of a "bottom up" (slow and steady, or slightly faster and steady) approach starting with pilot projects, of motivating and facilitating the creation of WUAs, wherever farmers agree to come together and to form them, which gradually take over some functions on their local irrigation systems (Joshi and Hooja, 2000).

Farmers' willingness to undertake PIM as well as both the initial success and long-run sustainability of PIM dependent upon whether the returns from PIM provide enough incentives (Meinzen-Dick, 1997) to the farmers to offset the substantial costs (monetary and otherwise) of participation and undertaking managerial responsibilities. Physical improvements to the irrigation system in terms of deferred maintenance or rehabilitation works provide only short-term incentives for participation. The farmers need to see possible long-term gains from participation. This would depend to a large extent upon the ability of the state to provide a receptive partner bureaucracy and enabling conditions for farmers to take on a greater role in irrigation management which can lead to improvements in water delivery services and system maintenance and the development of a feeling amongst the farmers that they can have a genuine voice in water allocation decisions which in turn could lead to increased yields and incomes because of more predictable, reliable and flexible water supplies (including more convenient operation of irrigation system) and reduction or better resolution of conflicts.

Before spelling out a possible strategy or model of bottom up, flexible, voluntary PIM, it would be appropriate to take a slight diversion and very briefly look at PIM approaches in some Indian states.

The Andhra Pradesh Farmers' Management of Irrigation Systems (APFMIS) Act, 1997 (which is being reproduced as Appendix I to this book), supplementing the Andhra Pradesh Irrigation Utilisation and Command Area Development Act, 1984, provides for the District Collectors to delineate every command area for each irrigation system on hydraulic basis and declare them as water users' areas for each of which a Water Users' Association (WUA) shall exist. All adults who use the water for irrigation purposes and are paying water charges as recorded in the revenue records are automatically deemed members of the WUA. The Collector is to arrange the elections of managing committees for the WUAs. All presidents of the WUAs are to be members of Distributory Committees (DCs) to be delineated by the government. For DCs also the Collector is to arrange elections for presidents of the managing committees. Similarly, the DCs are to be linked to Project Committees whose managing committees are also to be got elected by the Collectors. The government is also to appoint an Apex Committee.

It appears from the functions prescribed for the various levels of these farmers' associations or committees that various functions presently being performed by the irrigation department are to be performed by them and though they have been empowered to levy and collect fees they are to continuously receive government grants. Other agriculture-related activities have not been provided for at all in the objects and functions for the various committees as per the Act.

This has often been a matter of debate the world over. Many experts have felt that WUAs can be successful and sustainable only if they are multidisciplinary and multifunctional bodies (Meinzen-Dick et al., 1994; ISPAN, 1994; Hooja, 1995; Maloney and Raju, 1994) and are not merely being expected to simply carry out functions which in view of the irrigation department are appropriate for them (also see Ambler's paper in Sivamohan and Scott, 1994). This writer clearly feels that each WUA that evolves should be allowed to work out for itself what functions it wants to perform. Though his study of experiences in PIM indicates that WUAs which go in for agriculture-related functions in addition to water management ones and which come together and undertake some group activity on their own before they receive any assistance or stimulus from government, are likely to be more sustained, active and effective. The Government of India's Ninth Plan Working Group on Participatory Irrigation Management (1996) also advocates that water users branch out into activities other than the distribution of water.

It is understood that in Andhra Pradesh elections to WUAs were first held in irrigation systems all across the state and only thereafter were the elected management committee members provided training. This is different from the approach in some other states where on a pilot basis selected WUAs are being facilitated to evolve on their own and much training and orientation effort precedes the formation of each WUA. Of course, in Andhra Pradesh, the 1997 Act was preceded by pilot projects in PIM whose experience provided a basis for the Act, but formation of each individual WUA was not preceded by motivational work so that farmers themselves decide to come together to voluntarily form the particular WUA.

Some of the stages in Andhra's irrigation reform process and the development of PIM are indicated below:

June 1996	Public consultation and diagnostic white paper
April 1997	Water charges tripled
June-November 1997	Elections to WUAs
May 1998	Irrigation sector policy of state framed
August 1998	Training programmes commenced
January 1999	India's 4th national conference on PIM at Hyderabad
March 1999	First DC presidents conference
March-April 1999	ToT workshops commenced
April 1999	District Collectors' conference
May-July 1999	Maintenance and minimum rehabilitation programmes through WUAs and DCs commenced
June 1999	Agriculture intensification progamme commenced which is part of the Andhra Pradesh Economic Restructuring Project assisted by World Bank which also includes components of irrigation performance improvement, scheme improvement/ modernisation and farmers turnover, and institutional development of WUAs and government departments
December 1999	5th international seminar on PIM at Hyderabad, November 2001
December 2000	Two-day workshop at Walamatari of DC presidents
February 2001	Landmark decision of GoAP apportioning portion of actual water cess collected for O&M to WUAs and DCs

It is to be seen to what extent this "big bang" approach of large-scale creation of WUAs in a uniform manner throughout the state by statutory enactment in Andhra Pradesh shall succeed due to the political will behind it (whereby large-scale training and joint walk through survey-diagnotic analysis of existing irrigation systems are also being carried out on a massive scale). At least Maloney and Raju (1994) are categorical that each WUA should be spontaneous, should not be formed by government orders, should be set up using existing social capital, and should be gradually built up over time.

Madhya Pradesh and Rajasthan have gone in for legislation based on Andhra Pradesh's Act and approach (in fact they are extremely similar to the APFMIS Act). Madhya Pradesh earlier in the mid-1980s constituted Irrigation Panchayats under the Irrigation Act which remained as defunct organisations existing only on paper. Subsequently, in early 1990s, some farmers' management committees were set up under the Cooperative Societies Act. With the new legislation in 1999, 2,416 WUAs have been constituted and elections held in April 2000, with elections to DCs in February 2001. In Rajasthan, even though the Act was passed in 2000, the existing WUAs (about 300) have all been formed as per a bottom up, flexible approach and registered under the Cooperative Societies Act or Registration of Societies Act.

Gujarat, Maharashtra, Orissa, Uttar Pradesh, Haryana, Tamilnadu and Karnataka are also amongst the states using the Cooperative Societies Act or the Societies Registration Act to form and regulate WUAs. Maharashtra actually started experimenting with Water Users Cooperative Societies (WUCS) in 1984 and has as per official statistics 241 functioning WUCS. Gujarat efforts date from a policy resolution of 1996 and 144 WUAs have gone through the turnover process though 328 WUAs have been constituted and only slightly more than 50 are registered under the Cooperative Societies Act. Uttar Pradesh and Maharashtra governments have recently issued orders making it mandatory for WUAs to be formed -- thus officially by abandoning the demand driven or bottom up approach. But whether such bodies would eventually exist only on paper (like thousands of "Chak Committees" created all over the country below the outlet level under the CAD programme in the 1970s or 1980s), or whether they actually become functional and actually undertake PIM is yet to be seen. Orissa, Tamilnadu and Haryana are relying on the Societies Registration Act though Haryana is forming WUAs on watercourses below the canal outlet level and this, though involving participation of farmers in irrigation, cannot lead to PIM of canals (in fact these WUAs are somewhat like Chak Committees in various CAD projects).

Recently, Tamilnadu has been active in constituting WUAs without making it mandatory, and without waiting for legislation to be enacted almost a thousand WUAs have been formed.

Karnataka is one state which has brought about an amendment in December 2000 to amend the Karnataka Irrigation Act, 1965 and the Karnataka Irrigation (Levy of Betterment Contribution and Water Rate) Act, 1957 and to provide for Water User Societies/Federations to be registered under the Cooperative Societies Act. This is an experiment (of using legislation but not going in for an Andhra-type "big bang" approach) worth watching and the amending legislation of December 2000 is being reproduced as Appendix II to this book.

So far Andhra Pradesh appears to be the only state which has given comprehensive thought to the changed role and activities (as well as organisational structure) for the functionaries of the irrigation department and related agencies once PIM takes off at least at the canal minor level (if not throughout all irrigation systems).

Now to briefly set forth a strategy for propagating PIM. This strategy is likely to prove more suitable to what has been called the Asian model (Meinzen-Dick et al., 1994) which tends to have smaller base organisational units and allows direct participation of all members rather than the American model which relies more heavily on specialised formal irrigation organisations that employ professionals instead of face-to-face interactions between all members. The Asian model applies to socially cohesive societies with smaller landholdings, low market penetration and less infrastructure (for irrigation, transport and communication).

Since the "big bang" Andhra model is being described in the various papers in this volume in great detail and also because this author feels that the alternative approach is also likely to prove suitable and acceptable in many states in India (and may indeed lead to more active, effective and sustainable WUAs which are bound together by the internal urge to cooperate and manage water rather than external factors like laws), it is the bottom up approach which will be the focus of the rest of this paper. It is also interesting to note that despite many foreign consultants and donor agencies like the World Bank, favouring the "big bang" approach based on legislation in the style of the Andhra model, the two relevant Ninth Plan Working Group reports prepared by the representatives of the central and various state governments and experts in 1996 contain recommendations in line with the bottom up, flexible and voluntary approach. Both these reports (on CAD and PIM) would be summarised later in this paper.

WUAs based on hydrological units and boundaries (with representation of at least 60 percent farmers and 60 percent land on the channel, if not membership of all the farmers) need to be encouraged. As there are regional differences regarding tradition of irrigated agriculture as well as physical features and administrative arrangements a "cafeteria" approach (see Hooja's paper in Joshi and Hooja, 2000) should be adopted with WUAs being free to adopt internal decision-making systems and, provided certain basic functions basic to PIM, like the management of at least one level of canals above the outlet, are undertaken, being free to decide what functions to undertake initially or to add later. The WUAs should however be encouraged to take up a large number of irrigation and agriculture and allied activities (a detailed list of possible activities that WUAs could consider taking up has been presented slightly later in this paper).

WUAs would have to determine the rights and duties of both members and non-members and encouraged to form federations of WUAs so as to take over higher level distributories and main canals as also encouraged to raise their own resources including charging of user fees not lower than government's irrigation charges. A number of model memoranda of understanding (MoUs) should be developed which each WUA could adapt to determine relations between the particular WUA and the government (irrigation/command area development/agriculture/ground water, drinking water, rural development, land records and land revenue departments) as a whole functioning in a multidisciplinary manner including the modalities of arbitrating disputes between WUA and government. These MoUs should emphasise the fact that the WUA and government are joint partners in water resource development and management.

Of course, every WUA would necessarily be performing the function of providing an interface between the farmers and the government agencies.

In fact, the nature of a particular WUA would depend not only upon the manner in which it gets initiated and the nature of external influences and prime movers behind the formation of the WUA, but also on many other factors. One such factor could be whether the WUA is being set up in a system which is managed by an integrated, multidisciplinary CAD organisation, or whether the WUA would have to separately interact with the irrigation department, the agriculture department, the ground water department, the rural development agency, the land revenue administration and other government departments and agencies working in the command area of the irrigation system. Another factor could be whether or not the warabandi system and a warabandi culture already exists in the irrigation project. The example of Chambal-Rajasthan CAD Project informs us that the achievements can be much more dramatic when a WUA comes up in an area lacking in the tradition of warabandi; in that case of Arnetha WUA the water saving was substantial (a reduction of 32 days of watering from 79 days to 47 days) as was the increase in area cultivated (an increase of 145 acres from 660 acres to 805 acres) in the first year after PIM. However, in an area where warabandi already exists (Haryana, Punjab, parts of Rajasthan, Uttar Pradesh etc.), the initial impact may not be so dramatic. Even in such areas the scope exists for better water management which could lead to more optimal returns for the individual members of the WUA as also for the WUA as a whole, based upon more equitable and dependable water supplies. Similarly, a WUA set up before construction starts can comment on the need for modification in the designs and oversee the construction and then continue with other tasks would behave in a very different manner from a WUA set up after construction, which would be more concerned with operation, maintenance and sustainability of operations.

Experience appears to indicate that the local self-government (or village panchayat) institutions and the WUA cannot be one and the same. The local panchayat has many other non-irrigation functions to perform and its boundaries are not coterminous with the command areas of canal systems or sub-systems. Hence, local panchayat leaders have concern and priorities that differ from, or may ever come in conflict with, those of the WUA. Also, the interaction amongst members of WUAs needs to be much more frequent and intense than that of local people with the village panchayat. WUAs thus have to come up separately.

The government has to create an enabling environment through (a) policy resolutions; (b) specific programmes and activities to be implemented or sponsored by government itself; (c) providing intellectual administrative and implementational leadership; (d) putting into place legal and administrative provisions and procedures including Acts, rules, government orders; (e) undertaking mass awareness building and promotional efforts; (f) providing technical advice and technical backup as well as funds to WUAs for selective activities while continuing to perform many tasks on its own or taking up new tasks which were not being undertaken earlier in the pre-PIM era by government agencies; (g) ensuring attitudinal and behavioural change amongst employees of concerned government departments; (h) creating nodal points for PIM in all concerned government departments with a multidisciplinary PIM unit supported by an advisory group of experts in the water resources or irrigation department; (i) creating multidisciplinary mobile PIM propagation teams to move from project to project to motivate and orient farmers and field level government staff, obtain feedback about changes required at the government level, and generally take steps to ensure that PIM takes off and functions properly; (j) preparing guidelines and field manuals; (k) arranging for various sorts of PIM-related training and programmes; (l) helping WUAs rehabilitate old irrigation systems to bring them up to at least a minimal functional level; (m) involving NGOs and community organisers; and (n) providing incentives, monetary or otherwise, to (i) farmers to undertake PIM, and (ii) government staff to facilitate it, etc.

Regardless of whether or not PIM is being introduced, the government needs to ensure that (i) water charges are raised and a substantial part of charges recovered are ploughed back into direct action for improved and sustained operation and maintenance of canals, and (ii) integrated multidisciplinary/ multidepartmental interaction with farmers and WUAs is undertaken. Both these steps would greatly help in the propagation, acceptance, efficacy, sustainability and success of PIM.

Areas where PIM has already been introduced face what can be referred to as "second-generation problems". Svendsen, Trava and Johnson (in Groenfield and Svendsen, 2000) mention insecure water rights, financial shortfalls and lack of financial and administrative experience as amongst the second-generation problems of WUAs and the need to increase farm productivity to take advantage of possible improvements in quality of irrigation service and accessing support services (credit extension, newer technologies, markets and market information, access to inputs, post harvest facilities) as the second-generation problems of individual farmers. These issues need to be thought through and planned for in advance while initiating PIM in a particular area because even one single failed PIM effort at the level of a small WUA can have more of a negative demonstration affect than a hundred successful WUAs. And, the impression cannot be permitted to develop that PIM faces major second-generation problems which people may start speculating could be due to intrinsic defects in the approach of PIM.

While WUAs should be free to decide ("cafeteria approach") the various functions that they would undertake at initiation (provided the basic PIM function of taking over management of at least one level of canal above the outlet is undertaken) or may add on subsequently, Hooja (2001a: 66) has suggested a possible list of functions for WUAs as follows:

- Acting as an interface between the farmers and the main system management of the irrigation project as well as other concerned government agencies (command area development/rural development/agriculture/ground water/ cooperatives/development/land revenue and irrigation departments).
- Water distribution, irrigation scheduling including both rotational running of canals and warabandi, ensuring that water reaches all members (as also non-members) as per their due shares.
- Operation and maintenance of the irrigation and drainage systems.
- Collection (and assessment) of water charges and other user charges or special charges that the WUAs may levy.
- Resolution of local disputes amongst member and if necessary setting up of an internal system of penalties and other incentives or disincentives.
- Conflict resolution between members and non-members.
- Agriculture extension and farmers' training.
- Recommending of cropping patterns and package of agricultural practices suitable for the WUA farmers.
- Management of inputs and credit supply to members.
- Irrigation extension and propagation of better on-farm water application and better intra-outlet command water management.
- Encouraging and taking up of conjunctive water use, or community lift irrigation.

- Soil testing, water quality testing, water table monitoring and salt-sodic correction (if necessary).
- Provision of drinking water from canals.
- Post-harvest practices (grading, packaging, storage, marketing).
- Design and construction of new works (irrigation works, OFD works, drainage works) as well as rehabilitation of canals and structures(deferred maintenance works).
- Maintenance of commercial, financial and water accounting records.
- Cooperating with other WUAs to form federations of WUAs to take over larger canal sub-systems.
- Any other task as mutually agreed upon by the members (e.g., canal side plantation, construction/maintenance of farm tracks, providing of matching funds on behalf of the command area to the District Rural Development Agency (DRDA) for taking up of rural development work).

In fact, since it is hoped that eventually a federation of WUAs shall emerge for each project encompassing WUAs starting from canal minor level and going all the way up to a WUA for the entire project (including main canal and head works), the functions that, this writer has been advocating for Command Area Development Authorities (c.f. Hooja, 2001a) could also be suggested for the higher level WUAs. The list mentioned above is much broader than the activities suggested for WUAs/federated WUAs in the Government of India Working Group on PIM for Ninth Five-Year Plan (1996) which are also being indicated later in this paper, though it must be admitted that we are still to witness an attempt at actually federating WUAs in any part of India..

Each WUA should be free to choose its own internal structure as well as the manner in which it would like to become a legal entity (i.e., a cooperative society, registered society, company, or where special legislation may be carried out to enable it, special WUA registered as per the special Act). The WUA should be persuaded to adopt a consensual style for decision-making, as also for electing its executive, and also to avoid head-tail inequities by giving representation to each outlet command or chak on the executive body. The rights and duties of members and non-members would need to be got spelt out in advance.

WUAs should be permitted to charge water rates equal to or higher than those prescribed by government. They should also have authority to levy service charges and to charge slightly higher water rates from non-members who would not be contributing in the management and activities of the WUAs. The WUAs should be able to raise resources from various activities, and to borrow and repay bank loans so as to enable them to carry out their activities in a sustained manner. If WUAs could levy volumetric rates then it would ensure more optimal water application and reduce the excess water that needs to be drained out.

The government's role has to be that of a facilitator of PIM, and thus can be divided into policy-related aspects and those related to providing technical or financial, or training and orientation related, assistance and creating mass awareness about PIM.

On the policy front the government needs to create not only an enabling environment but also a much more positive encouraging environment by providing intellectual, academic, administrative, implementational and policy-making leadership.

As WUAs take over many functions and the management of canals, the government's CAD (or irrigation) department would have to gear itself to look more to construction and management of dams and the main canal system or drainage works, providing technical advice, training, introduction of new technologies, major repairs, creation of an organisation for interfacing with WUAs and other government departments and for motivating and catalysing the formation of WUAs, as also undertaking long-term water resource planning.

Statutory and administrative empowerment orders may have to be issued by government to facilitate the adoption of PIM. WUAs should be enabled to charge for services (provided the rates prescribed by them are not less than those of the government), decide about water distribution, impose punishments, and provide rewards and incentives. The duties of the government as regards PIM, and responsibilities towards WUAs, would need to be spelt out in the MoU between government and each specific WUA so as to act as a supplement to general executive orders and legislation regarding PIM. Such MoUs and orders/ enactments would also specify the expectations from the WUAs vis-à-vis government, WUA members, and society as a whole. If integrated inter-disciplinary and multifunctional CAD agencies exist, then one MoU between the agency and WUA would suffice, otherwise a number of supplementary MoUs between the WUA on the one hand and irrigation, agriculture, ground water and CAD departments (and perhaps even DRDA) respectively on the other hand, would be required.

Since the government has to play the role of the interfacing and catalysing agency, predetermined and transparent provisions and procedures need to be evolved for negotiations/arbitration between WUAs and government whenever required. Provisions would thus have to be put into place for WUAs to, if they so desire, appeal to government agencies in certain situations for help in conflict resolution between the WUA and other agencies/individuals or within WUAs, as also for appeal against government's irrigation, agriculture or CAD departments' decisions, if required.

A large number of pilot PIM projects, and action research in PIM, as well as PIM orientation, training and implementation review workshops should be supported to provide a positive demonstration affect in favour of PIM, as also to provide experience in the working of PIM which would then be reflected in subsequent legal enactment, and manuals and guidelines on PIM.

As regards technical help to WUAs, the government would have to arrange for providing advice, undertaking major repairs and major construction, the availability of machines on hire basis if required by WUAs, helping the WUAs in designing of engineering works and in ensuring quality control in works being implemented by WUAs, and help WUAs and their members in adopting improved on-field water application techniques, and in water regulation and distribution techniques as also in other improved agricultural and land-water management practices through irrigation (including drainage) extension and agriculture extension efforts.

On the financial side a one-time managerial/functional grant should be given. With the central and state governments providing an area-based grant, the WUA adding a smaller contribution of its own, and the entire amount being placed in a fixed deposit as the WUAs trust fund whose interest amount may be drawable by the WUAs for use for its managerial activities. All WUAs should be encouraged to continue to make periodic additions to their trust funds.

A one-time assistance may also need to be provided for rehabilitation or renovation of the canal (and drain, wherever applicable) which would be a sort of deferred maintenance necessary to bring the canal and drainage up to a minimum functional status (or level) so that the WUA would be able to operate, manage and maintain it on its own in future. (This would of course not apply to newly constructed canals and drains where PIM can be taken up forthwith). The government should provide assistance for such works provided the WUA has come into being, gets involved in designing such works, takes up construction at the WUA level and agrees to contribute at least 10 percent (and preferably 20%) in cash or kind for the works. The WUA and irrigation agency should conduct joint inspections of canals and drains to determine the works required to bring the system up to a minimal functional level and to determine the WUAs specific contribution before the work is got started. Government assistance should be provided in three instalments of 40, 40 and 20 percent. The utilisation certificate of WUA as well as joint inspection of the work done by the WUA and irrigation agency should suffice to get the advance adjusted, with detailed audit being left to the annual Chartered Accountant's audit of the WUA. The WUA should receive 8 to 10 percent extra funds in lieu of the time and effort being put in by the WUA to manage itself and its works so that it can ensure quality control and proper supervision of the works. In this process the WUA should not be treated like a contractor, but as a user group and stakeholder that is sharing in irrigation management.

The government has to agree that part of the water charges recovered shall be the share of the WUA based on the number of activities that it takes over from the government as well as the level of the irrigation system (minor, branch, distributory, main canal, drains, headworks and dam), whose management the WUA takes over. A formula may even need to be worked out for providing

O&M grants which could be performance-based depending upon ensuring supply of due share of water to all, the O&M efforts put in, and recoveries of water charges affected (if the WUA takes upon itself this task) by the WUA. Simple but objectively verifiable criteria for assessing WUAs' performance are required to be established in advance.

It should also be prescribed beforehand that if the WUA raises additional resources for certain activities other than the basic O&M of the canal, then the government would make available matching grants to the WUA.

In the initial stages the government is obliged to finance (and also assist otherwise) for pilot PIM projects and action research projects to be taken up. NGOs, research institutes and state government/project agencies could take up such projects jointly with WUAs. For such pilot trials (and demonstration projects) WUAs would be given back at least 50 percent of the water charges (presently being levied at nominal rates) from their areas for managing and carrying out the O&M and other WUA functions.

For areas where WUAs take over the management of canals and other PIM activities, the government could accord them a higher priority while sanctioning development schemes of various government departments. At the same time, WUAs should be permitted (and perhaps even encouraged) to implement various government schemes not directly related to PIM and for which the government should also make available funds to them. The government should also coordinate and facilitate the availability of institutional finance in case WUAs want to take loans for their activities.

As regards training, mass awareness building and orientation, keeping in mind the apathy, indifference, or even resistance to PIM at various levels in government departments and even amongst farmers, large-scale mass awareness building and conscientisation about PIM and re-orientation of thinking and approach is required for both farmers and government functionaries, as also for all citizens in general. Training for PIM is too important to be left only to trainers in places like Water and Land Management Institutes (WALMIs). Multi-disciplinary mobile teams consisting of experts in social interaction, irrigation, agriculture, ground water, rural development, extension services, drainage, training, administrative coordination, etc., can to collaborate with government functionaries, farmers and NGOs who are intimately involved in PIM implementation, in imparting PIM training. Besides, the government staff and the mobile multidisciplinary PIM propagation teams being used for implementing PIM programmes, WUA chairmen and community organisers concerned with the PIM pilot projects, should also be jointly used to impart PIM training and orientation. The state-level PIM Cell, responsible for policy making and implementation of PIM, should directly coordinate such training. Thus, a PIM Training and Orientation Centre should be placed under the direct control of the state-level PIM Cell (of which more later in this paper).

PIM training should include (i) efforts at general orientation about PIM and attitudinal reorientation; (ii) knowledge and skills in improved water and agriculture management, including drainage, i.e., technical matters; (iii) PIM-related procedural, administrative and organisational issues (including those related to social interactions and group dynamics as well as participation), incentives available, sources of incentives and funding; (iv) lessons from a variety of case studies about successes and failures of PIM initiatives and projects in India and abroad; and (v) commercial matters and accounting procedures including the use of traditional accounting systems (e.g., 'bahikhata' system) which local farmers may be also to understand.

The training has to be two-way process serving as an instrument of change in the irrigation reform process (as also making available skills, knowledge and orientation) on the one hand, and on the other hand providing a mechanism for feedback from the field whereby policies, procedures, laws, rules, strategies and schemes can be modified and the adoption of PIM facilitated.

PIM training would be required not only for all farmers using surface irrigation, but also for all officers and staff of irrigation, agriculture and CAD departments (and perhaps even for ground water department and DRDA and Panchyati Raj staff) as well as for politicians and senior policy makers, WUA personnel (elected and employees), community organisers, NGOs, trainers, etc.

Extensive use of mass media should be made. In addition to the central government generating and providing mass media packages, at the state level also, video and audio cassettes, TV and radio programmes, newspaper articles, pamphlets, leaflets, posters need to be prepared and disseminated on a continuing basis. Schools and colleges could be tapped through debates and essays. Planning Forums and the National Service Scheme in colleges could also be used to make people aware about PIM -- its need, the problems in its implementation, and the likely benefits. Government agencies staff should also be convinced and motivated to take a proactive approach to the propagation of PIM.

Now, to turn to the governmental machinery that is required to be set up to further PIM.

Even though the main interaction in PIM would be between agencies of the state government and WUAs, however the central government needs to provide leadership, support and encouragement in various ways. A National PIM Coordination Committee should be created with membership from central ministries, states, experts working in government organisations, as well as outside governments, NGOs, etc. Similarly, there is a need for a National PIM Support Group to assist the National PIM Coordination Committee, and the PIM unit of Ministry of Water Resources and the state governments. Perhaps, the IndiaNPIM could play this role. A full-time Joint Secretary should be designated for PIM by the Ministry of Water Resources along with a multidisciplinary team consisting of an engineer, an agronomist, an economist, a social scientist, a (land) revenue administrator, and a ground water expert. This team should suggest policy and procedural interventions to the central government and the states, provide for training and mass awareness, help prepare manuals, guidelines and material for use in the mass media, facilitate research and pilot projects, undertake evaluation and ensure documentation, act as a clearing house of information about PIM, and provide financial assistance to the states.

At the state level, the Secretary for CAD (or if that does not exist the Secretary for Irrigation) should be made the nodal secretary for PIM. A full-time Deputy Secretary-cum-Director, PIM should head the multidisciplinary PIM Cell containing staff from the irrigation, agriculture, ground water and CAD departments as well as social scientists and generalist administrators. The PIM Training and Orientation Centre should work under Director, PIM.

A high-level empowered committee for PIM should be created under the Chief Secretary. This committee should have a sub-group (headed by Secretary, CAD, with Secretaries of Agriculture, Ground Water, Cooperatives and Irrigation, as members) to assist it by taking up policy issues, coordination, annual action plan preparation, monitoring, etc. The sub-group (with added representation from Government of India) could also act as a sanctioning committee for Government of India's assistance for PIM, state government's annual action plan for PIM, and pilot projects. The CAD, irrigation, agriculture, cooperatives and ground water departments should also designate a nodal officer for PIM in their departments.

Between two and four (depending upon the size of a state and its area under surface irrigation) mobile multidisciplinary PIM propagation teams should be constituted in each state. Gradually, the number of such teams could be stepped up.

If at the project level a single integrated and multidisciplinary CAD organisation (as in the case of CAD in Rajasthan) does not exist, then a multidepartmental committee should be set up to deal with the WUAs in an integrated manner. For each of the larger projects of 80,000 hectares or more a multidisciplinary and mobile PIM unit should be created at the project level. Community organisers, who may be government functionaries, NGO personnel or local volunteers, could also be used to spread the ideology of PIM. And, if government servants, they could be agriculture extension staff, or from the rural development service, or from irrigation, agriculture, CAD, cooperatives or ground water departments, or from institutions like WALMIs.

Regional administrators and elected local self-government/council heads like Divisional Commissioners, District Collectors and Zila Pramukhs (who would all need orientation sessions for this) should be involved in motivating and orienting farmers so that the farmers themselves take the initiative in forming WUAs and in approaching the irrigation/CAD departments to enter into MoUs.

Such WUAs as take over the O&M responsibilities of canals as part of their activities should be provided representation in Command Area Development Authorities, Water Regulation Committees, Zonal Agriculture Extension and Research Advisory Committees, etc., and in district and state level Irrigation/or CAD/or Agriculture Advisory Committees.

When agriculture extension agencies organise farmers field days in irrigation commands, it should be done in conjunction with the WUAs. Both irrigation extension and agriculture extension including use of border strip/ridge irrigation, proper irrigation scheduling of canals, salt/sodic correction of soils, weed and fertiliser management, integrated pest management, control over water application to avoid wastage, maintenance of canals, water courses and field drains, conjunctive use of water, use of devices like sprinkler and drips, and field drains, should be propagated by government agencies in association with the WUAs.

Government employees included in the mobile PIM teams, or as community organisers, as well as in state-level PIM cells, would be required to work with extra dedication as champions of PIM. Normally, in government service, no special incentives are given for particular tasks which may be difficult. However, for personnel employed in PIM work, a 'special pay' like that given to certain Secretariat Officers, or a 'project allowance' like that given in remote areas, should be provided. State/district awards for PIM and special mentions in Annual Performance Appraisal Reports (APARs -- also called ACRs or Annual Confidential Reports in some states) should also be provided for.

Thus, after discussing the measures for furthering the adoption of PIM in India, it may be pointed out that a slow and steady approach at propagating PIM would be better, in our opinion, that what some people have referred to as the "big bang" approach (like the PIM effort in Andhra Pradesh where the entire state was covered under PIM in one go with legislative Acts being framed, elections being held and all PIM-related activities being initiated simultaneously throughout the state) or what could be called the high target-oriented approach. Even one failed effort at PIM on any one canal minor can have more of a negative demonstration affect than the positive impact of one hundred successful PIM experiments. Thus, monitoring of the PIM effort should not be merely quantity-based but quality-based. As Meinzen-Dick, Raju and Gulati (in an earlier version in Joshi and Hooja, 2000) point out that organisations alone will not have an effect on irrigation performance unless they are active and it is the extent of collective action that enables us to determine whether they are active or exist only "on paper". Unduly high targets should not be fixed. In the initial three to four years the evolution, or creation, of between 200 and 400 new and effective functioning and active WUAs a year in each state should be sufficient; later the speed could, and should pick up. Though one must admit that many

PIM experts feel that going in for Andhra type "big bang" approach from the very onset has a better chance of success.

Similarly, the state governments should not rush into carrying out legal enactments until experiences within their state of a sufficiently large number of working PIM projects have been absorbed and can be reflected in such legislation. Till then use of administrative orders and detailed governmental nurturing of PIM pilot projects would be required. In fact, we have the experience of Panchayat Raj in India to caution us where over forty years of legislation in most India states with numerous elections to the panchayat bodies and about a decade's experience of panchayats having been given constitutional status has not ensured that all the panchayat bodies are institutions of local self-government rather than agencies for development dependent upon the state governments (Hooja and Hooja, 2001; Institute of Social Sciences, 2000).

It is now proposed to use the rest of this paper piece to look at two reports of working groups (actually they were sub-groups) set up by Government of India (GoI) to prepare initial recommendations for India's Ninth Five-Year Plan with regard to CAD and PIM. Because, both the reports have drawn from an ongoing interaction between the Ministry of Water Resources, Government of India and the Economic Development Institute (EDI) of the World Bank, including a series of seminars and workshops (which in turn have influenced the EDI *Handbook's* contents and recommendations), we have also included comments on *Handbook on Participatory Irrigation Management* in the following paragraphs.

Since GoI's efforts at propagating PIM have been built upon the existing CAD programme, and since the PIM working group report itself suggests that irrigation schemes covered under the centrally sponsored CAD programme or under World Bank aided projects should be taken up first for introduction of PIM, it would be appropriate to start by looking at the CAD working group report.

The CAD working group suggests that there should be a unified administrative set up for Command Area Development Authorities (CADAs) which should have complete technical and administrative control over officers from various departments deputed to it for reasonably long and stable tenures. CADAs should also have unified command from head works down to last level in the field and up to last outfall of drainage network and its activities should span the water supply-utilisation spectrum integrating economic, engineering and socio-economic paradigms. PIM should necessarily be promoted as part of CAD. The present frame of on-farm development (OFD) needs to be broadened to include land levelling, field drainage, rationalisation of unit command size, etc., besides construction and lining of field channels. An efficient on-farm management needs to be intimately linked with OFD and with improved and optimal crop management. Extension workers should be geared to work towards achieving productivity as per international standards. Enhanced subsidy should be given for development of ground water, and conjunctive water use be encouraged. In addition to major and medium projects, minor irrigation projects (500 to 2,000 ha) should also be included under centrally sponsored CAD. Emphasis on sprinkler and drip irrigation should be given to economise on use of water. Consolidation of holdings should be made compulsory for irrigated command areas. For the work of adaptive trials and demonstrations on farmers' fields and training of functionaries, farmers and NGOs, as well as monitoring and evaluation by a multidisciplinary team, a 100 percent grant should be provided. This grant should be shared on 80:20 basis by centre and states and available for (i) survey, planning and construction of OFD, (ii) construction of field channels, (iii) construction of field, and collector drains (collector drains were not covered earlier under central assistance for CAD), (iv) reclamation of waterlogged areas, and surveys of such waterlogged areas, (v) rehabilitation of minor/distributories linked to PIM, (vi) enforcement of warbandi, (vii) partial subsidy to small and marginal farmers and SCs/STs against loans taken by them, (viii) crop compensation, (ix) establishment costs of work items, and (x) one-item functional grant to WUAs.

The shortcomings and constraints which the CAD working group hopes the CAD programme would enable the country to overcome are: (i) lack of reliability of water supply at the canal outlet due to inadequate maintenance of irrigation system, (ii) difficulty in obtaining land for field channels due to lack of consolidation of holdings, and lack of proper alignment of field channels in absence of consolidation, (iii) existence of unlined field channels which can easily be ploughed over leading to conflicts, (iv) lack of warabandi, (vi) lack of maintenance of field channels and field drains and drainage system below field channels, (vii) absence of collector and intermediate drains, (viii) problem of waterlogging, soil salinity and alkalinity, (ix) lack of coordination among different departments at state level concerned with CAD, (xii) lack of farmers' participation, (xiii) lack of marketing infrastructure, (xiv) lack of on-farm water management, (xv) lack of proper crop management, and (xvi) insufficient provision of financial outlays by various state governments. It is interesting to find the EDI *Handbook* quoting the June 1994 national seminar on PIM held at Aurangabad whose list of seven problems parallels many of those mentioned above -- though one problem raised in the 1994 seminar is missing from the Working Group report – that is lack of incentives for saving water. We consider this last problem to be very relevant and one which a genuine partnership between governments and PIM agencies like WUAs could help overcome.

The CAD working group has suggested the following performance indicators for the CAD programme: (a) equitable distribution of water, (b) production and productivity enhancement, (c) water supply augmentation to command additional areas, (d) demand-based average water availability, (e) sustainability of resources base, (f) adequate and balanced level of input use, (g) adoption of appropriate technology, and (h) optimising water use efficiency per unit land and per unit water.

The CAD working group has also recommended the continuation of a one-time functional grant for farmers' participation and the provision of major funding under the CAD programme for PIM, including provision for rehabilitation of canals and tanks, legal backup, training, generation of awareness, and pilot projects.

Now, to turn to the PIM working group report.

This working group has suggested that at Union Ministry of Water Resources level there is need to create a separate multidisciplinary wing headed by the Joint Secretary for PIM. To support the PIM wing a National Support Group should be promoted by the Ministry of Water Resources, by such state governments as are already committed to introducing PIM, NGOs working in the area, and academicians. The National Support Group should provide a forum for exchange of ideas and experiences on PIM, for training, field studies, and suggesting policy options.

At the state government level, a high level committee, preferably presided over by the Chief Secretary and comprising representatives from departments such as CAD, irrigation, agriculture, as well as NGOs, has been recommended so as to formulate policies for implementation of PIM. A monitoring committee may also be set up under Secretary, CAD, or of Irrigation, which should include all concerned heads of departments. Irrigation department in the state is basically engineering and construction oriented. This orientation significantly inhibits the management system's ability to meet the basic objectives of fulfilling requirements of reliable and equitable supply of water. Therefore, there is need to have a separate PIM wing in the Department of CAD (or of Irrigation) with an officer of the rank of head additional head of department as its in-charge. The wing should be a multidisciplinary unit which should include social scientists agronomists, economists, engineers and persons from other concerned disciplines.

At the field or project level also there should be a separate organisation comprising engineers, social scientists, agronomists, etc., who should be carefully picked up, be highly motivated, and be solely concerned with the work of farmers' associations.

The PIM group felt that the suggestion made by Government of Rajasthan should be tried out for setting up mobile units consisting of a team of officers from departments of irrigation, agriculture, CAD, and revenue, and also social scientists, which should move from place to place to build up an atmosphere for PIM, and to motivate the farmers as well.

Mass awareness campaigns and large-scale training of farmers and government functionaries have been recommended. Thus training capacities for PIM have to be build up.

It may also be necessary to employ community organisers for motivating the farmers and later working with them for setting up the WUAs.

The WUAs should be legal entities registered either as societies, or cooperative societies, or companies, or through special provisions for registration enacted in a special law related to CAD and PIM, or through amendments in the Irrigation and Drainage Act. WUAs and the irrigation agencies should sign MoUs which lay down the rights and duties of WUAs and government departments and agencies dealing with irrigated agriculture.

The PIM working group also felt that the WUAs/WUFAs have to be formed based upon hydraulic units. The general consensus which has emerged during various workshops is that a canal minor is the most convenient, viable and sustainable unit. This would have a command of roughly 300 to 500 or even 1,000 hectares. The farmers could be organised at the minor level to begin with and given responsibility for operation and maintenance etc. Thereafter, federations of WUAs at branch distributory level and irrigation scheme level could also emerge.

The objectives of the WUFAs as recommended by the PIM working group on are as follows:

- to improve service deliveries through better operation and maintenance for optimum utilisation consistent with crop needs;
- to achieve optimum utilisation of available resources through improved systems of delivery of water, matching to the extent possible the need of the crops;
- to achieve equity in the distribution of water;
- to increase production per unit of water or per unit of land as the case may be;
- to make optimum use of natural precipitation and ground water in conjunction with flow irrigation;
- to help the farmers to have a greater choice of crops sequence and timing of water supply, period of supply and frequently of supply depending on soils, climate and other infrastructural facilities available in the command such as roads, markets, cold storage, etc., to maximise the income and returns;
- to inculcate a sense of ownership and responsibility among the farmers and to facilitate collection of water charges from the members;
- to develop professional relationship between the personnel of the agency of irrigation and the farmers;
- to promote economy in the use of water and better maintenance of the system;
- to mobilise local resources, which may be in cash, kind or labour towards capital cost and/or O&M; and
- to branch out into activities other than the distribution of water for the raising of resources, so long as they are consistent with the basic responsibilities

of the WUFA (for example, farmers' associations in Mexico have set up highly profitable agro-industries).

The above list may be compared with the indicative list of functions which WUAs could consider taking up that this writer has been advocating and has already set forth earlier in this paper.

The PIM working group had recommended that the primary WUFA could be at an appropriate hydraulic unit such as the minor or small distributory and ideally command an area between 300 and 500 hectares. For projects having commands up to 5,000 hectares a two-tier federation of WUFAs has been suggested, for commands up to 50,000 hectares three-tier federation and for commands above 50,000 hectares a four-tier federation has been suggested.

The federation of WUAs could have the following responsibilities:

- coordinate the activities of lower level organisations and groups;
- monitor the distribution of water by the lower level organisations;
- assist the irrigation department in equitable distribution of water;
- settle disputes which cannot be settled at the lower level;
- identify and prioritise the requirements of maintenance and repair of the system in consultation with the irrigation department;
- monitor the activities of the lower level organisations;
- represent the grievances of the farmers/lower level organisations before the irrigation Department for redressal;
- coordinate the activities of lower level organisations and various departments of government concerned with irrigated agriculture.

The following are some of the advantages that the working group on PIM felt would accrue to farmers, if they adopted PIM:

- farmers' associations get explicit water rights;
- assured delivery of water on volumetric basis to the farmers' associations according to the agreement between the association and the irrigation agency;
- farmers' associations get the right to fix and collect water rates;
- farmers get the freedom of cropping pattern within the limits of allocated water;
- there is freedom for conjunctive use of surface and ground water;
- farmers have access to more information about the availability of water due to the agreement with the irrigation agency;
- farmers can participate in decision-making on operation and maintenance of the main system;
- rehabilitation and physical improvements of the irrigation system is a major gain for the farmers;
- there is cost saving from unwanted physical facilities if farmers participate;
- more efficient, reliable and equitable distribution of available water;
- augmented farm productivity and farmer's income;

- empowerment of farmers and creation of a sense of ownership among them; and
- quick resolution of conflicts over water.

From advantages that farmers can derive from PIM, let us turn to the comparative advantages that farmers possess for managing irrigation. As the EDI *Handbook* reminds us, "they have direct incentives to manage irrigation water in a productive and sustainable manner, they offer an on-the-ground presence that even the most dedicated off-site agency staff cannot equal, and they have an intimate knowledge about their fellow irrigators". Thus, the *Handbook* goes on to indicate that PIM is likely to lead to (a) improved design, construction and O&M, (b) benefit from cost savings to both farmers and government alongwith better service for the farmers, (c) the organisation that the farmers established and the skills they acquire in running the WUAs constitute a form of social capital that can have spin-off effects in other aspects of social and economic life.

The *Handbook* also categorises the possible participatory mechanisms as follows: (1) information sharing, (2) construction, (3) joint assessments, (4) shared decision-making, (5) collaboration, and (6) empowerment. (This categorisation can be profitably compared with the six modes of people's participation set forth in Hollensteiner (1978). As we see it, it is only the last three mechanisms (shared decision-making, collaboration and empowerment) which can be said to constitute PIM – the others only being participation without a significant role in management. These three mechanisms are better described in a somewhat different manner as part of joint management options in irrigation which have been elaborated in Meinzen-Dick et al. (1994) and in Meinzen-Dick in Joshi and Hooja (2000) which sets forth a 6×4 matrix ranging from full irrigation control, to agency O&M with users input, to shared management, to WUA O&M, to WUA ownership with agency regulation, and to full WUA control. Strictly speaking, only the last three options are part of PIM, though in India it is the third option of shared management which is often seen by observers to be one where PIM has been achieved while it is a fact that until the users start actually operating and managing canals as per the fourth option, the benefits of PIM cannot start flowing.

One aspect, which neither of the two GoI working group reports has touched, but which has been highlighted in the EDI *Handbook,* is what has been called rules for fostering long enduring irrigation groups. Borrowing from Ostrom's book (1992), EDI informs us that organisations that endure have clearly defined boundaries (area served, who can belong, functions); member contributions that are proportional to benefits, no taxation and no altering of water allocation rules without representation; auditors of physical conditions and irrigation behaviour being accountable to the irrigator members;

punishments that fit the crimes for the rule violators; users and their officials having low cost and local ways for resolving disputes; the absence of any challenge of the irrigators groups from the side of government; and nested groups of very small and larger organisations each with appropriate responsibilities.

Before leaving the reader to the discussions in the rest of the book about how the Andhra Pradesh model of PIM reforms has faired, how the reforms can be made more sustainable within Andhra and how the model could be replicated in other Indian states, two points of caution must be emphasised: (i) PIM is not a panacea for all the difficulties being faced in managing irrigation. Many issues have purely technical solutions. Many others would require government action. However, without the userism involved in PIM, technical fixes and governmental actions by themselves may not suffice. (ii) It needs to be recognised that financial and infrastructural creation/repair support as well as government orders and legislation may enable WUAs to be initially created, but unless major social, institutional and HRD support is made available and farmers enabled to successfully manage canals on their own, the sustainability of PIM may not be achievable.

Our goal has to be to move towards universalisation of PIM in the irrigation sector within India and elsewhere. A joint government and water users partnership for improved water management must go hand in hand with sustained and judicious investments in water resource development and maintenance to ensure that the water-related global challenges facing us in the present century do not become insoluble.

REFERENCES

Brewer J., S. Kolavalli, A.H. Kalro, G. Naik, S. Ramnarayan, K.V. Raju and R. Sakhtivadivel, 1999. *Irrigation Management Transfer in India: Policies, Processes and Performance,* Oxford & IBH Publishing House, New Delhi.

Cornea, M.M. and R. Meinzen-Dick, 1992. Design for Water Users Associations: Organizational Characteristics. In *Developing and Improving Irrigation and Drainage Systems,* World Bank Technical Paper No. 178, Washington, D.C.

Freeman, D.M., 1989. *Local Organizations for Social Development: Concepts and Cases of Irrigation Organization,* Westview Press, Boulder, Colorado.

Groenfeldt, D., 1998 (and earlier editions). *Handbook on Participatory Irrigation Management,* The Economic Development Institute of the World Bank, Washington, D.C.

Groenfeldt, D. and Mark Svendsen (eds.). *Case Studies in Participatory Irrigation Management*, Economic Development Institute, Washington, D.C.

Goldensohn, M.(also often referred to as ISPAN, 1994), 1994. *Participation and Empowerment: An Assessment of Water Users Associations in Asia and Egypt*, Irrigation Support Project for Asia and the Near East (ISPAN), Washington, D.C.

Gulati, A., R. Meinzen-Dick and K.V. Raju, 1999. *From Top Down to Bottom Up: Institutional Reforms in Indian Canal Irrigation*, draft volume circulated for a National Workshop at Institute of Economic Growth, New Delhi, 2 January, 1999.

Hollensteiner, Mary R., 1978, People Power: Community Participation in the Planning and Implementation of Human Settlements, *Philippine Studies*.

Hooja, Meenakshi and R. Hooja, 2001. Institutions of Self-Government or Agencies for Development: Panchayats in India, *South Asia Journal of Socio-Political Studies*, Vol. 1 No. 2, January-June 2001.

Hooja, R. 1995. *Multidisciplinary Management of Irrigation Schemes and the CAD Approach in India*, Occasional Paper No. 2, HCMRIPA, Jaipur.

Hooja, R., 1996. Objective of PIM, *INPIM Newsletter*, No. 2, March 1996.

Hooja, R., 1997 (a). Promoting Water Users Associations in Participatory Irrigation Management: Some Reflections, *Indian Journal of Public Administration*, Vol. 43, No. 3, October-December 1997.

Hooja, R., 1997 (b). Participatory Irrigation Management: A Report, *Prashasnika*, Vol. 24, No. 2 July-December 1997.

Hooja, R., 2000. Administering Water in India in 2050 AD, *Water and Energy International*, Vol. 57, No. 4, October-December 2000.

Hooja, R. 2001 (a). *Multidisciplinary and Participatory Water Management for Irrigation Command Areas*, IndiaNPIM Working Paper/Monograph No. 2, New Delhi.

Hooja, R., 2001 (b). Guest Editorial, *Water and Energy International*, Vol. 58, No. 2, April-June 2001.

Hooja, R. 2001 (c). Training for PIM in India: A Supplementary Note to IndiaNPIM's Working Paper/Monograph No. 4, prepared for the workshop on "Capacity Building for Integrated Water Resources Management for Human Development in South Asia", organised by SaciWATERs and Centre for Alternatives at Dhaka, 22 to 24 September, 2001.

Hooja, R., forthcoming. *Integrated Irrigation Management: Need for Multidisciplinary Command Area Approach and Improved Farmer-Management Interface*, Arihant Publishers, Jaipur.

Hooja, R., Anjali, M. Bhatia and Ganesh Pangare, 2001. *Training for Participatory Irrigation Management: Present Status and Needs: A Policy Aid Document*, IndiaNPIM Working Paper/Monograph No. 4, New Delhi.

Hooja, R., R.C. Bower and S.N. Mundhra (eds.) 1997. *Irrigation, Agriculture and Social Development*, Himanshu Publication, Udaipur.

Indian Network on Participatory Irrigation Management (IndiaNPIM), 2000. *Proceedings of the Fourth National Conference on Participatory Irrigation Management at Hyderabad 19-25 January, 1999*, New Delhi.

IndiaNPIM 2001. *Irrigation Management Transfer: Sharing Lessons from Global Experience* (reprinted from INPIM, FAO International E-conference on Irrigation

Management Transfer) supplement to *IndiaNPIM Newsletter* Nos. 11 and 12 July 2001.

Institute of Social Science, 2000. *Status of Panchayati Raj in the States and Union Territories of India-2000, Concept, New Delhi.*

ISPAN, 1994. *Participation and Empowerment: An Assessment of Water Users' Associations in Asia and Egypt.*

Joshi, L.K. and R. Hooja (eds.) 2000. *Participatory Irrigation Management: Paradigm for 21st Century* (2 volumes), Rawat Publications, Jaipur.

Maloney, C. and K.V. Raju, 1994. *Managing Irrigation Together: Practice and Policy in India*, Sage Publications, New Delhi.

Meinzen-Dick, 1997. Farmers Participation in Irrigation: 20 Years of Experience and Lessons for the Future, *Irrigation and Drainage Systems*, Vol. II.

Meinzen-Dick, R., M. Mendoza, C. Sadoulet, G. Abiad Shields and A. Subramanian, 1994. *Sustainable Water User Associations: Lessons from a Literature Review*, presented at Water Resources Seminar Landsdowne, Virginia USA and reprinted in A. Subramanian, N.V. Jagannathan and R. Meinzen Dick(eds.), *Users Organisations for Sustainable Water Services*, World Bank Technical Paper No. 354, Washington, D.C.

Meinzen-Dick, R., K.V. Raju and A. Gulati, 2000. What Affects Organisations and Collective Action For Managing Resources ? Evidence from Canal Irrigation Systems in India. EPTD Discussion Paper No. 61, IFPRI, Washington.

Meinzen-Dick, R. Reidinger and A. Monzardo, 1995. *Participation in Irrigation* Environment Department Participation Papers Services, World Bank, Washington, D.C.

Ministry of Water Resources, India, 1996. *Report of the Working Group on Command Area Development Programme*, New Delhi December 1996.

Ministry of Water Resource, India, 1996. *Report of Working Group on Participatory Irrigation Management for Formulation of Plan Proposals for Ninth Five Year Plan*, New Delhi, August 1996.

Oblitas, K. and J.R. Peter with G. Pingle, H.M. Qaddami and J. Pererra, 1999. *Transferring Irrigation Management to Farmers in Andhra Pradesh, India*, World Bank Technical Paper No. 449, Washington, D.C.

Ostrom, E., 1992. *Crafting Institutions for Self-governing Irrigation Systems* Institute for Contemporary Studies, San Francisco, California

Parlin, B.W. and M.W. Lusk (eds.), 1991. *Farmers' Participation and Irrigation Organization*, Westview Press, Colorado.

Raju, K.V. and J. Brewer, 1996. Types of Irrigation Management Transfer Models in India, paper presented at International Association for the Study of Common Property Resources Meeting, Berkley, California, June 1996.

Salman, M.A. Salman, 1997. *The Legal Framework for Water Users Associations: A Comparative Study.* World Bank Technical Paper No. 357, Washington, D.C.

Sampath, R.K. and R.A. Young (eds.), 1990. *Social, Economic and Institutional Issues in Third World Irrigation Management*, Westview Press, Boulder, Colorado.

Sivamohan, M.V.K. and C.A. Scott (eds.) 1994. *India: Irrigation Management Partnerships*, Booklinks, Hyderabad.

Tang, S.Y. and E. Ostron, 1993. *The Governance and Management of Irrigation Systems: An Institutional Perspective,* ODI Irrigation Management Network (Network Paper 23), London.

Uphoff, N., R. Meinzen-Dick and N. St. Julien, 1985. *Improving Policies and Programmes for Farmer Organization and Participation in Irrigation Water Management,* Water Management Synthesis Paper 1, Cornell University.

Uphoff, N., P. Ramamurthi and R. Steiner, 1991. *Managing Irrigation: Analyzing and Improving Performance of Bureaucracies,* Sage Publications, New Delhi.

World Bank, 1998. *Handbook on Participatory Irrigation Management.*

2

Scaling up Participatory Irrigation Management in India: Lessons Learnt from the Andhra Pradesh Model and Future Strategies

GANESH PANGARE

The past decade has witnessed widespread reforms in the irrigation sector across the world. The state of Andhra Pradesh in India is pioneering the irrigation reforms under way in the country. This paper reviews the factors that lead up to the reform process and then goes on to analyse the irrigation reforms taking place in Andhra Pradesh after the introduction of the Andhra Pradesh Farmers' Management of Irrigation Systems Act, 1997. The paper reasons that a political will and an enabling legal environment form the institutional framework conducive for the implementation of irrigation reform process. It highlights the fact that there could be qualitative scaling up through a robust policy initiative by the government. In a political democracy like India this could be one of the approaches that could be tried out in other states wherein elements of "good governance" are introduced.

The reforms in Andhra Pradesh have a lot more to achieve and it is too early to even say whether they are successful or failed, but what is important is that they have shown the way forward. Terms like successes and failures are many a time misleading, and can also become tools in the hands of both proponents and critiques. What is important from a rational perspective is to see what one can learn from these reforms that are in progress. And can this learning be translated into a feedback mechanism for managing the programme better and achieving the goals it set out to achieve. And how lessons can be drawn for replication of the programme in other states of India.

I
Introduction

Today, the world over, examples of successful community-based natural resource management examples can be seen. With very few exceptions these initiatives have remained at the local level, and these projects have largely remained "islands of development". In countries as vast as India, these successes will need to be repeated tens of thousands of times. Clearly to facilitate such a transformation, some scaling up will have to be attempted. NGOs, academics, and even policy makers frequently romanticise the successes of these "experiments" or the virtues of small indigenous practices for natural resource management. What is important however is to bring the "good" elements of these successes into a national or state policy for scaling up. Instead of concentrating more and more resources into an already "successful" project, it makes more sense to transfer the learning to and invest available resources in other projects that cater to underprivileged populations and resource-poor regions.

Over the last decade, governmental and non-governmental programmes have attempted to apply learning from these small-scale "development experiments" to a larger area. This includes programmes formulated by the government. The paper asserts that with a political will that is pro-poor and ready to move away from populist ideology, and a robust policy initiative backed by legal sanctions, benefits from development programmes can reach more people in a short time. Critics may argue that there are qualitative losses in scaling up but the paper argues, by using the case of the participatory irrigation management reforms going on in India, that once reforms assume the shape of a people's (or in this case, a farmers') movement scaling up with quality can occur.

Latest estimates show that 36 percent of India's population live below the poverty line. Poor survival chances, landlessness, malnutrition, environmental pollution and social exclusion, arising out of caste and gender discrimination, characterise the lives of these people. Increased scarcity and degradation of land and water resources makes it difficult for people to meet their daily subsistence needs. Livelihoods dependent on land-based resources are adversely affected. Most of the time, the effect on women is particularly severe because they are the primary collectors and users of natural resources in most parts of rural India. Regeneration of natural resources through effective conservation and management practices is crucial to achieving the goal of food and livelihood security.

Agricultural progress and regeneration cannot take place without water. In spite of large-scale irrigation systems developed in India, there are many problems in maximising the full benefits of these installed irrigation systems. One of the main reasons being that unless people accept responsibility and ownership of the interventions, water management initiatives cannot become sustainable. Farmers who depend on irrigation water for their livelihood have the strongest incentive to manage that water carefully (Groenfeldt 2000). One of the ways in which this issue can be addressed is through the introduction of participatory irrigation management.

Participatory irrigation management, or PIM, refers to the involvement of irrigation users in all aspects and at all levels of irrigation management. "All aspects" includes the initial planning and design of new irrigation projects or improvements, as well as the construction, supervision, financing, decision-making, operation, maintenance, monitoring and evaluation of the system. "All levels" refers to the full physical limits of the irrigation system, up to the policy level. In any management function, including the setting of policies, can and should have a participatory dimension to it (Groendfelt 1998).

This paper argues that PIM cannot succeed without a political will and the readiness of the irrigation bureaucracy to change. In addition, while critics may maintain that there are qualitative losses in scaling up, this paper claims that once reforms assume the shape of a people's movement, or at least a farmers' movement, successes may be institutionalised at broader social scales while the quality of programme benefits is maintained.

II
Background and Conceptual Framework

A quick overview of the growth of the irrigation sector in India after independence will help set the context for the current situation. The total geographical area of India is 329 million hectares (mha). The total cultivable area of the country at present is about 142 mha and the gross cultivated area nearly 185 mha. At the time of independence (1947) the irrigation potential in India was 22 mha and by the end of 2000 it was around 90 mha. Thus full irrigation potential is estimated to be about 90 mha (Joshi 2000). India has the second largest irrigated area in the world after China.

Under the leadership of Pt. Jawaharlal Nehru, the first Prime Minister of India, a great deal of emphasis was placed on centralised economic planning and on construction of large irrigation projects both for hydropower as well as for increasing the foodgrain production. In the 1960s, coupled with construction of the irrigation projects, there was a focus on the use of high yielding varieties of seeds, chemical fertilisers, etc., which ushered in the Green Revolution in India. The production of foodgrains increased from 50 million tons in 1951 to nearly 200 million tons at present.

Today, the environmental, social and second-generation problems associated with the Green Revolution are well documented. Given the significant role of irrigation in bringing about the Green Revolution in India, it is a matter of worry that the irrigation sector is problematic as well. It is important to note here that in the welfare state approach adopted by the various political parties in India this kind of over-governmental zeal and enthusiasm to provide everything to the masses was inevitable. The belief was that government intervention and control over the economy and the assets of the growth engine was to ensure economic welfare of the people. But, today, the stagnant growth rate and decline and deterioration of the system have led the government into rethinking their strategies.

Government-owned and managed large irrigation systems have become so centralised and one-sided that farmers have been almost completely excluded from them. The system was set into a downward spiral of disuse and neglect, and ultimately broke down in many places. Faced with strong criticism, policy makers were introduced to concepts such as people's participation, bottom up approach, decentralisation and devolution. A 1995 government document (GOI 1995) reports that the irrigation sector in India is beset with problems related to investment priorities, poor management, corruption, deteriorating physical infrastructure, lack of adequate financing, social inequity and no incentives to use water optimally. Today, the government has become sensitive to the issues of equity and efficiency and has come to understand, to a large extent, the importance of involving people in the development process.

Problems with the Irrigation Sector in India

The irrigation sector in India is facing following problems:

- For the last 50 years the main focus of the irrigation department has been on construction activities.
- Serious deficiencies in project formulation, planning and implementation have led to major time and cost overruns in almost all the major irrigation projects, further leading to cost escalations and delayed realisation of planned benefits.
- Large gaps in irrigational potential created and actual utilisation is seen largely due to the fact that there is no "culture" of water management in the department, rather, there is an over-emphasis on the "works" component alone.
- Increased tendency of rent seeking (corruption) in the department from the lowest to the highest level of power is evident.
- Many of the systems today are rapidly deteriorating because of poor management and inadequate maintenance, leading over a period of time, to the collapse of the entire system.
- The water delivery service is unreliable, in regard to both quantum and timeliness.
- The water rates for irrigation are one of the lowest in the world. Added to this is the poor recovery of the charges. With most of the state governments

in India on the verge of bankruptcy, there are just no finances with the irrigation department to provide funds for operation and maintenance of the system. The annual budgets of many irrigation departments from various states in India over the last few years have been spent only on administrative costs with negligible spending on operation and maintenance of the system.
- Finally, there is also the social question of inequitable distribution of water in the command of the irrigation project. And there is also a lack of incentive for the farmers to save water and use it optimally.

Irrigation Management Transfer Around the World

Increased farmer participation in irrigation is part of a worldwide trend in devolution in natural resources management (Meinzen-Dick 2000). In many countries the move to transfer responsibilities of managing the irrigation system from government agencies to farmers' organisations is gaining momentum. Mexico, Philippines, Colombia and Turkey are some of the well-known examples where there has been a major effort for transfer of irrigation systems through PIM processes. Mexico has served as a model for other countries considering irrigation management transfer programmes. By the end of 1996, 87 percent of the area under medium and large irrigation districts in the country had been transferred to users' associations to manage (Palacios 2000). Turkey has succeeded in transferring nearly 1 million hectares, or 61 percent of the publicly managed irrigation in the country, to local government units or special purpose irrigation associations created at the local level (Svendsen and Gladys 2000). Irrigation management transfer in Colombia was unusual in the sense that it was initiated by farmers rather than the government. Since the first two schemes were transferred in 1976, 16 out of the 24 medium and large schemes in the country, covering about two-thirds of the 750,000 hectares under such irrigation, have been transferred to local associations to manage (Pinto 2000). Transferring irrigation responsibilities has been seen as a way of reducing pressures on thinly stretched government finances, while improving irrigated agricultural production and ensuring the long-term sustainability of irrigation systems (Brewer 1999).

Irrigation management transfer around the world covers a wide spectrum. Wageningen (1990) and Vermillion (1991) list the following forms of transfer found worldwide:

- Introducing irrigation service fees.
- Fostering competition in service delivery.
- Contracting with the private sector to provide irrigation services.
- Vending irrigation services to those who pay a fee.
- Franchising the right to provide irrigation services to private organisations.
- Subsidising a local organisation to provide irrigation services.
- Joint investments by agency and users.

- Making the management agency financially autonomous.
- Joint management by agency and users.
- Devolution of responsibility or control to users.
- Shedding of government functions.
- Privatisation of assets.

Irrigation Management Transfer or Participatory Irrigation Management in India

PIM is a concept that is not new to India. In fact, farmer managed irrigation systems are recorded as far back as the 3rd century B.C. There were a wide range of systems and practices across the country, some of which are evident even today. Usually, the kings would provide the material for building the dams and canals and then the people would be responsible for managing and maintaining the systems, resolving conflicts etc. In South India, the thousands of small irrigation tanks seen today, were owned, operated and maintained entirely by farmers in the command area either through formal or informal institutional mechanisms. The aim is not to go into the virtues of these farmer-managed systems, but to highlight the fact that the concept of participation is not a recently introduced concept in India, aid driven or otherwise, but a practise that existed and was lost over time. Ironically, as stated earlier, the government-owned and managed large irrigation systems over the last 100 years or so became so centralised and one-sided that there was a near exclusion of the farmers from them -- a process started during the colonial rule and perfected by the Indian bureaucracy after independence.

The Indian Constitution separates various subjects into those that are the responsibility of the central government and those that are the responsibility of the states. Though water, including irrigation, is a state subject, the central government through allocating financial resources as well as centrally sponsored schemes, can influence state policies. The medium and major surface irrigation systems operated by the states account for 36 percent of the net irrigated area in India. The remainder includes area irrigated by private groundwater irrigation (49%), traditional tank irrigation schemes (7%), and other sources of irrigation (8%). Irrigation management transfer in India is thus a matter of transfer of irrigation management responsibility from state government agencies, mainly the irrigation department, to Water Users' Associations (WUAs) or farmers' organisations (Brewer 1999).

Over the last two decades several efforts have been made to introduce PIM. But most of them have been half-hearted, with no real seriousness on the part of the various state governments to transfer systems or responsibilities to farmers' organisations. Sporadic efforts from NGOs remained localised with a few WUAs formed below the minor outlets in some of the major canal systems. In the 1990s, due to financial pressure on various state governments, a need to improve the performance of irrigated agriculture, and to ensure the sustainability of the irrigation systems, many states have woken up to the fact that reforms are inevitable. Today, virtually, all states in India are considering adoption of irrigation management transfer policies, specifically after seeing the results of the irrigation reforms in Andhra Pradesh. Conditionalities attached with World Bank loan agreements are also hastening the process in a few states. The only caveat here being that hopefully because of the loan conditionality, reforms will take place, but then care needs to be taken that they do not merely remain reforms on paper.

Andhra Pradesh is the first and leading state in India to bring about overall institutional reforms in the irrigation sector at the state level (Peter 2000; Raju 2000). The main components of the reform process are:

- Introduction of a suitable policy and legal framework (i.e., Andhra Pradesh Farmers' Management of Irrigation Systems Act, 1997).
- Formation of WUAs across all types of irrigation systems in the state.
- Implementation of large-scale training on the issues related with the irrigation reform programmes both for farmers and the staff of the irrigation department.
- Bringing in significant financial reforms to influence quality performance of users' organisations.

Conceptualising the Andhra Pradesh Experience

Currently, the installed irrigation capacity in India is managed by the government agencies. This capacity is in the form of dams that are constructed, owned and managed by the irrigation department. PIM therefore, cannot succeed without a political will and the readiness of the irrigation bureaucracy to change. There are also a variety of factors that determine the incentives for farmers to manage and maintain the irrigation system and their lands and to invest in increased agriculture production. For reforms to succeed there needs to be genuine collaboration between the farmers' organisations and the government line departments. We also need to identify the conditions for collective action. And if after all this, the promised powers are not devolved to the farmers' organisations, the reforms will fail to achieve their purpose.

Through an analysis of the Andhra Pradesh experience, this paper addresses two issues as under:

- Scaling up of development programmes with quality and through government departments is possible with the right kind of policy framework and a political will to implement these policies.
- Learning from the Andhra Pradesh irrigation reforms can provide valuable insights for other states who are in various stages of formulating and implementing irrigation management reforms.

In Andhra Pradesh irrigation reform process, devolution of power has occurred at the tertiary or secondary level, but at the primary level there is no transfer of power as of now. Also "decentralisation" in its truest sense is not seen, rather, there is more of administrative devolution of powers for management of the systems. The main issue therefore is that in large-scale government-owned irrigation systems, the ownership or primary responsibility of the physical assets (at least) will remain with the government system. What is seen is that the government retains overall ownership and control over the water resources, reservoirs, and main canals. Maintenance and (perhaps) operation of lower level canals are turned over to WUAs.

The Andhra Pradesh experience shows that there is clearly a need for an enabling legal environment. The WUAs formed under the legal regime have got rights, local control and there is integration with higher levels of the system as characterised by self-governing systems. Downward accountability or officials of the irrigation department being accountable to their client "farmers" is the fundamental requirement of any irrigation reform process. Do legal provisions also bring about this accountability?

There is enough evidence to show clearly that the performance of government-controlled systems is consistently low on several dimensions than that of local self-governing systems (Tang 1992). The author would like to argue in this paper that it is necessary to bring in elements of "self-governing" (Ostrom 1992) at the WUA level and government control at the main system level for better management of irrigation systems. What is needed is a mechanism for management/joint decision-making between the two. The argument is that the legal enabling environment has given the right kind of impetus and powers to the WUAs and a reorientation of the irrigation bureaucracy and thus there is a basic change in the institutional framework.

Ostrom, Schroeder and Wynne have pointed out that temporary decentralisation and local organisations established, as part of construction or rehabilitation projects, often did not have a significant long-term impact because the basic institutional framework was not changed. The paper asserts that in the Andhra Pradesh irrigation reform process there has been a conscious effort to bring in institutional changes in the overall irrigation sector. Further resulting into decentralisation of decision-making powers at the WUA level.

Further, while partnership and ownership should be built from the bottom, they can be strengthened from the top -- through interventions and policies that create enabling environments for participation. The widespread move towards more democratic decentralisation has opened new spaces for grass roots participation to merge with state policy (Blackburn 2000).

III
Andhra Pradesh: Pioneering Irrigation Reforms in India

Andhra Pradesh is the first state in India to have made a legal provision in the form of an Act called "The Andhra Pradesh Farmers' Management of Irrigation Systems Act (Act 11 of 1997)" with a view to involving farmers in irrigation management and ultimately achieve irrigation management transfer to farmers in the entire state.

Andhra Pradesh at a Glance

Andhra Pradesh is the fifth largest state in India with a population of about 75 million and a geographical area of 27.68 mha. Of the state's geographical area, 47 percent (12.9 mha) is under cultivation and the net sown area is about 11.04 mha (4.88 mha irrigated). It is primarily an agrarian state with about 70 percent of its population dependent on agriculture. The economy is essentially agriculture-based. The majority of the farmers have small landholdings, averaging less than 2 ha. The agricultural sector generates about 36 percent of the state's gross domestic product and employs some 70 percent of the population. Irrigated agriculture constitutes 40 percent of the state's cropped area and contributes over 60 percent of the state's total agricultural production. Ninety-five percent of rice -- the main crop grown in Andhra Pradesh and the state's principal staple food – is produced on irrigated lands. A wide variety of other irrigated crops are also grown, including cereals, pulses, oilseeds and non-food crops. Irrigated agriculture contributes to about 18 percent of oilseed production, 55 percent of other food crops like chillies, vegetables and fruits, 15 percent of non-food crops such as tobacco, cotton and sugarcane, and about 36 percent of other minor crops (Oblitas 1999; Peter 2000; Raju 2000).

The state's three major rivers, namely, Godavari, Krishna and Pennar, drain 70 percent of the state's land area. The water potential of Andhra Pradesh is estimated to be 7.78 mha (2746 TMC). The major rivers are seasonal with more than 90 percent of the total flows occurring between June and December depending on the rainfall that varies from year to year. The ultimate irrigation potential from all sources is estimated to be 9.50 mha. This includes 7.30 mha from surface water and 2.20 mha from groundwater (Raju 2000).

Build up to the Reforms

In 1984, the Andhra Pradesh government passed the Andhra Pradesh Irrigation and Command Area Development Act, 1984, which authorised the creation of Command Area Development Authorities and Pipe Committees. These Pipe Committees were formed under the CAD programme for the internal distribution of water below the minor outlet and the maintenance of the micro-system network. The Pipe Committees, however, proved to be unsustainable and powerless. The water supply at the minor outlet was not reliable because of a lack of coordination between the Irrigation and Command Area Development Authorities. The Pipe Committees were too small to have any say in the maintenance of the main system. They had neither clear-cut rights and responsibilities nor any means to raise resources. They lasted as long as they had the support of the Command Area Development Department and became non-functional once this support was withdrawn.

In 1995, the government issued another order for the promotion of farmers' participation in the management of irrigation. The highlights of this order were (GOI 1995):

- WUAs will be formed at the minor level to cover an approximate area of 400 to 750 ha.
- WUAs will be autonomous bodies which will function on democratic lines within the purview of the Act. For the effective functioning of these associations, suitable amendments to the Act will be proposed separately.
- WUAs will be fully responsible for the maintenance and operation of the irrigation network within the area of their operation.
- A programme of promoting WUAs in the Sriramsagar and Srisailam Right Branch Canal Projects would be taken up.
- The endeavour of the irrigation department will be to give increasing responsibility of managing irrigation systems to WUAs and for this purpose it will:
  (i) make available assured and reliable water at the head of the minor;
  (ii) undertake rehabilitation and modernisation of the internal water distribution system wherever WUAs are formed; and
  (iii) facilitate fixation and collection of operation and maintenance charges from water users.
- The WUA will be at liberty to regulate distribution of water-to-water users on volumetric or any other basis and to levy and collect penalties if any, fixed by it for violation of the water distribution schemes finalised for the irrigation system under its administrative supervision.
- By improving efficiency of the irrigation system the WUAs will be at liberty to save water and make the water so saved available to users on payment.

Pilot Project

A pilot project to gain experience and to demonstrate the utility and practicability of the WUA, a farmers' organisation in the management of irrigation system was initiated in the early 1990s. A pilot programme was taken up on a small scale covering a minor, with a command area of about 500 ha under Sriramsagar Project. This was followed up with a large-scale pilot programme covering a command area of about 20,000 ha under six different distributaries in the Sriramsagar Project during 1995 to 1997 through two NGOs. The terms of reference for the NGOs were (Peter 2000):

- Identify a model for creating WUAs.
- Evaluate the present system of water supply and suggest modifications.
- Methodology for optimising the use of waters by farmers
- Identify gaps in the existing dissemination of information to the farmers.
- Train the farmers.
- Identify incentives for effective functioning of WUAs.

Farmers in the command area were helped to form the WUA, take up minimum maintenance works that were feasible to them for ensuring free flow of water up to the tail reaches in the system, organise water distribution in the entire system, resolve disputes, if any, internally and adopt their own crop planning. The net result of this experiment was that the awareness amongst the farmers increased. There was an increase in the irrigated area from a pre-project level of about 50 percent to around 80 percent with the same quantity of water as was supplied in the earlier years and without any substantial investment on rehabilitation or repairs of the system.

Some of the important issues that emerged out of the pilot project were:

- The WUAs should have a clear "water right" so that they can not only insist on getting their supplies accordingly but also develop suitable crop planning for the area and work out appropriate operational plan for distribution of water.
- The WUAs should have proper legal status to be easily recognisable by the irrigation authorities.
- The accountability of WUAs to the irrigation department, in terms of utilisation of irrigation water and area irrigated, is to be spelt out in clear terms so that proper water budget could be developed.
- The financial resources of the WUAs need to be built up.
- Clarity of role is needed for the WUAs for taking over the management of the irrigation system.

The Andhra Pradesh Farmers' Management of Irrigation Systems Act (1997)

In 1996, the Andhra Pradesh government came out with a "White Paper" which noted the following key problems in the irrigation sector (Oblitas 1999; Peter 2000):

- Serious decline in net irrigated area. Only 2.3 mha of the 4 mha of potential irrigated area created was actually under irrigation (a gap of 52 percent between potential created and actual irrigated area).
- Under financing of operations and maintenance.
- Poor irrigation system efficiencies and water distribution (including significant wastage of water diverted and inequitable distribution of water within schemes, especially between head and tail ends).

- Low cropping intensities and yields.
- Low farm yields.

The White Paper also gave an appropriate direction to the reform agenda:

- Farmer empowerment and management should form the heart of the reforms.
- The process should be bold and comprehensive rather than incremental.

The irrigation reforms in Andhra Pradesh would not have been possible unless there was a political will to change and introduce some radical steps to revert the crisis in the irrigation sector. This political will came in the form of the dynamic Chief Minister Mr. Chandrababu Naidu, and some of the pro-poor reforms introduced by his party. The irrigation reforms formed part of his broad-based philosophy -- "Janmabhoomi" concept to help the rural poor in Andhra Pradesh. Janmabhoomi (land of birth) is the term given by the Chief Minister to a concept (programme) of new ways for communities and government to work together to improve their economic and social well-being. Central themes include participation by the people in decision-making and implementation, cultural changes in the way government works, with emphasis on participatory approaches, partnerships and serving the poor, a grass roots orientation in all activities, and programmes aiming for self-help and self-reliance of communities. Janmabhoomi covers many sectors, amongst them, village education committees for development and management of primary schools, Vana Samrakshana Samithis (joint forest management groups), women's thrift groups (DWCRA), youth employment groups, and watershed development committees (Oblitas 2000).

Impact and Analysis of Reform Process

A period of three to four years is a very short span to comment on the success or failure of the reform programme, but a look at what has been achieved can throw some light on the positive impact of the reform process and what mid-course corrections need to be taken so as to streamline the programme.

Increase in Net Irrigated Area

One of the key reasons that hastened the reform process was the fact that there was a gradual decline in the net irrigated area in the state. In the late 1980s the net irrigated area reached a plateau, and it fell from 2.9 mha in 1990 to 2.3 mha in 1994. Out of the 4.8 mha of potential irrigated area created, only 48 percent is actually being irrigated.

The data from the Sriramsagar Project demonstrate the increase in reported irrigated area experienced after the hand over to the WUAs in June 1997. Piloting of WUAs in the project area began on about 8,000 ha two years earlier, but was not extended to the whole scheme until June 1997. There was no increase in total water supplied to the scheme after the piloting was expanded. In fact, less water was used then previously. The total water consumed in the project during 1997 was 28,000 m^3, while in 1996, some 4,300 m^3 were used. The figures show more than a doubling of gross irrigated area (kharif and rabi seasons), with two-thirds as much water (Oblitas 1999).

The reason for this increase has been that due to the improvement in the delivery system, i.e., the cleaning of canals and watercourses, water has reached those areas where it never reached earlier. Also there was some monitoring on the efficient utilisation of water, and a conscious effort on the part of the WUAs to minimise water losses. When analysing the increase in irrigated area, one factor that has to be kept in mind is that, under the management of the WUA, farmers are more willing to report irrigated area than previously to the revenue department officials. The reason being that under reporting of irrigated area meant lesser payment of water tax. But after turnover the WUA is reporting the exact irrigated area and which is resulting in a better tax recovery. Even if one discounts this increase in reported area, there has been a general net increase in the irrigated area.

This point can be further strengthened by taking a look at more empirical evidence from one of the Distributory Committees (DCs) in the Sriramsagar Project area.

The Distributory Committee 16, Tadikal (DBM 7B to DBM 15 of Kakatiya Canal) is located in the district of Karimnagar. The localised command under this distributory is 30,860 acres. There are a total of 13 WUAs. The total membership of farmers is 8,989 (Reddy 2000).

The increase in the area irrigated during the last four years (after reforms) can be seen in the table given below:

Table 1
Increase in Kharif Area Irrigated

(in acres)

Year	*Total Command*	*Area Irrigated Wet*	*Irrigated Dry*	*Actual Command Irrigated*
1995	30,860	3020	3498	6518
1996	30,860	6907	2863	9770
1997	30,860	9280	5350	14637
1998	30,860	13147	5402	18549
1999	30,860	18366	6003	24369
2000	30,860	18505	6003	24508

Increase in Water Tax

Increasing water tax has always been a sensitive and touchy issue in Indian politics. Irrigation water tax has not changed for years in many states and the charge is also a token charge. Even with the charge being so nominal there is hardly any recovery of the water tax from the farmers.

Irrigation water charges were raised three-fold in April 1997. Table 2 gives the before and revised water charges.

Table 2
Water Charges

(in Rs.)

Crop Type	*Water Charges per Acre*		*Water Charges per Acre*	
	Category 1		*Category 2*	
	Pre-revised	*Revised*	*Pre-revised*	*Revised*
First or single wet crop	60	200	40	100
Second and third wet crop	60	150	40	100
First irrigated dry crop	40	100	20	60
Second and third irrigated dry crop	40	100	20	60
Long duration crop	120	350	80	350
Aquaculture (per year)	0	500	0	500

Category 1 – Permanent irrigation source.
Category 2 – Irrigation source for at least four months.

Source: Oblitas (1999).

As mentioned earlier, in the Indian political context, this is a very bold and rare step. There were consultations across the state amongst farmer leaders and various political parties over the price hike. It was after the explanation that the increased water tax would in some way filter back to the WUAs for maintenance of the irrigation system, that the increase was accepted. The principal reason is considered to be the "combination of a widespread consultation and public outreach process preceding the increase, and the presentation of the increase not as a single measure but as a part of a package of measures which were seen by the rural communities and the political parties as overall beneficial to the farmers" (Oblitas 1999).

Good Quality of O&M by WUAs

In 1998, in the first phase of the rehabilitation and maintenance works worth $ 28.1 million that were completed, more than 70 percent of the works were done by the farmers' organisations. Overall the works were planned by walk throughs in the command area along with officials of the irrigation department. And works were carried out on a priority basis in the most crucial areas in consultation with the farmers of the region. The other benefits have been that farmers feel more empowered now and a sense of belonging has been promoted. The scrupulous contractor has been got rid of. Actual funds for O&M have increased for the first time (Raju 2000). And, the same works have been executed by the WUAs for a lesser cost than charged by the contractors earlier.

Environmental Benefits

It is too early to comment on the environmental impacts of the reform process. According to field observations made by Vermillion (1999), one of the farmers in the Godavari Delta area reported that levels of salinity in the soil and water had declined after the reform programme due to improvements in water distribution. More water has been reaching the tail end areas and was used for flushing the salts from the soils. Also, due to more canal water being available, farmers in tail ends had stopped or decreased their use of bore well water (which has high levels of salinity). Discussions by the author with some WUA presidents from the delta region also highlighted the fact that after channels were cleaned and drainage improved after the reforms were introduced, the incidence of waterlogging has reduced.

Such findings, if documented and attributed to the reforms, could be information of potential significance.

Emergence of a New Leadership (WUA Presidents)

According to Shri Kishen Reddy, Chairman, Distributory Committee 16, Tadikal, Sriramsagar Project, "a new leadership was developing because of the reform process which is young and dynamic and concerned with the welfare of the farmers". This is a very positive outcome of the reform process, which was never thought about before. In the Indian political scene this is surely like a breath of fresh air. Though there have been critiques that feel the opposite, that the programme has been hijacked by rural elite and politically powerful (Reddy 1999; Mollinga 2000), the author is of the view that in a politically conscious and active democracy in India this is bound to happen. This is rural reality, what must be seen that at the end of the day the new leadership delivers the good and participates in the reform programme in the right spirit. Also, if the WUA is strengthened, the exploitative tendency, if any, of the rural elite also decreases.

IV
Analysis and Conclusion

Political Will for Reforms

As stated in the conceptual framework, the paper argues that political will is the prerequisite for bringing about irrigation reforms. If one looks at the history of PIM in India, then this point will be appreciated that but for a lack of political will to implement the PIM programme it has not been implemented on a large scale in any of the other states in India. Maharashtra and Gujarat, which were the pioneers in introducing the concept of PIM in India, have to date done little more than setting good examples.

Political will is very significant in a democratic set-up that is prevalent in India. As any wide-reaching policy change can be brought about by only the political party in power, if possible taking into confidence the other political parties in the opposition. Till date one of the major reasons for no political party wanted to touch irrigation sector reforms, was that it feared tackling hardcore basic issues linked to irrigation sector reforms like increase in water tax and restructuring the dominant and all-powerful irrigation bureaucracy. The reason for not increasing water tax being that farmers are "poor" and they also constitute the major vote bank for the politicians, and in no way can be antagonised. This actually in the long run has been the main bane of any good agricultural policy evolving in India.

What the Andhra Pradesh reforms have depicted is that if bold positive steps are taken, and which farmers feel, the government is sincere and accountable then, even harsh measures like increasing water tax will become successful. What the farmers are interested in is delivery of goods (water) at the assured times and right quantity. If they see this happening, then they are willing to pay.

Political will again is just not sufficient for a one-time introduction of reforms. Reforms also have to be sustained for which a continuous follow-up and coaxing is necessary. In Andhra Pradesh, there has been a continuous follow-up from the Chief Minister level to ministers and senior bureaucrats through programmes like "Dial Your Chief Minister" to publication of monthly news magazine, holding of state level conferences of the WUA presidents etc. The result is that at the moment there is a political lax seen, the bureaucracy used to its traditional functioning perceives reforms as no more than some political gimmick and the entire spirit of the reform is killed. The argument thus presented is that a constant political will acts like a watchdog over the bureaucracy and puts pressure on it to perform.

The emergence of the new political leadership (WUA/DC presidents) is also a positive sign that this new leadership that is politically conscious will put pressure on the system to perform. Many of the presidents in the future will contest for mainstream political elections for power, which could be beneficial for the sustainability of irrigation reforms.

Enabling Legal Environment

One of the main characteristics of the irrigation process in Andhra Pradesh is that there is clearly stated Act, viz., Andhra Pradesh Farmers' Management of Irrigation Systems, 1997 defining the various processes, roles and responsibilities of the WUAs and the irrigation department. The WUAs are formed as legal entities, and powers devolved to the WUAs are backed by legal rights and obligations.

As mentioned earlier, the Act mainly gives water rights and control of the system to WUAs. It provides functional and administrative autonomy and the WUAs can take their own decisions. And, most importantly, the Act makes the irrigation department accountable to the WUA. The Act also defines the contracts between the irrigation department and the WUA, which helps clarifying roles and responsibilities, and also provides a forum for communication between the two.

The Act makes it mandatory for all farmers to be members of the WUA. In most other cases or states membership is voluntary. One of the major weaknesses in the WUAs promoted by NGOs in other states is that they have not been legally recognised by the irrigation department so no rights or powers are transferred to them defeating the entire purpose for forming the WUAs.

The Act also provides for the formation of the WUA Federation, which in turn helps in the creation of social capital. This is discussed in the section on WUA Federation.

There have been discussions whether there should be a separate Act like the one in Andhra Pradesh or can "proper" changes in the already existing irrigation laws of the state bring about the same desired effect? Evidence from literature shows that a mere tinkering with the already existing laws has not helped much and unless there is a clear-cut legal provision giving rights and devolving powers to the WUAs, mere cosmetic changes make no major difference and it's back to business as usual. The Andhra Pradesh experience highlights the role of a statutory act applicable to the entire state, and all irrigation (surface) systems. It is important to note here that the administrative bureaucracy at least at the lower levels of power understands only a legal language. Is this in the "rule book" or not?

Irrigation Department as Agent of Change

Large-scale irrigation systems have given rise to large and centralised bureaucracies. Over the years the attitude of the irrigation department has become that of providers and thus brought about certain arrogance in the department itself while dealing with farmers, who are looked upon as recipients of the generosity heaped onto them by the department. The irrigation bureaucracy is resistant to any change, like many other administrative departments. One of the major fears for resisting change being that of losing power and privileges. Also, the fear that if the works component is transferred to the WUAs then chances of rent seeking are lost. So, along with a political will, there also has to be an administrative will for change and reforms.

The administrative will does not mean a few committed and highly motivated individuals at the highest echelons of power to change and try to force a change but a gradual attitudinal and behavioural change amongst all the ranks in the department. This is an easier said than done task. The most important factor is for the assistant engineer responsible at the WUA level to believe in the reform process as well as realise that he is an employee of the WUA and not the reverse. The officer of the irrigation department is nominated as the "competent authority" at the WUA level. The officer coordinates with the WUA and gives technical guidance in the preparation of estimates and maintenance of hydraulic particulars and structures. In Andhra Pradesh, the attitude and behavioural change, especially amongst the technical bureaucracy, was brought about by a series of training and awareness programmes. Also, a will and zeal amongst the higher-level officials has had a positive impact on lower-level staff.

What Andhra Pradesh has shown is that unless the irrigation department is made the agent of change, reforms will not succeed. Constructive farmer participation often hinges on the ability to move beyond an adversarial or paternalistic relationship between bureaucracy and farmers to a partnership. This is not easy, for farmers often distrust the irrigation agency. Developing a service orientation among agency staff and a collaborative attitude between agencies and farmers has been essential for successful joint management and irrigation transfer programmes in the Philippines, Sri Lanka, United States, Mexico and other countries (Meinzen-Dick 1997).

Field interactions with the WUAs revealed that there was a perceivable change amongst the irrigation staff and majority of the WUAs were happy with the support of the departmental staff, and had developed an excellent rapport with the engineers. Discussions with engineers disclosed that many of the lower level employees of the department were unaware about the rapid changes taking place. Only senior officials attended training programmes. So there have to be awareness and orientation programmes for the lower level staff if reforms have to succeed and progress in the right spirit of partnership. Some of the suggestions of the engineers were that:

- there should be continuous training for all levels of staff regarding attitudinal and behavioural change;
- equal participation of farmers and engineers in the reform process;
- least or no political interference in the role of engineers;
- creation of an environment for smooth change over, but at the same time ensuring the feeling of security amongst officials; and
- pick up committed officers and staff to promote and guide the programme at the lower level.

Mobilisation of Financial Resources to Sustain Reforms

One main trigger or motivation for governments to introduce irrigation reforms has been the financial resource crunch the governments are faced with. With thinly stretched financial resources and a majority of this going for administrative requirements (salaries etc.) governments today are ready for PIM. Herein lies the catch. Farmers do not want to take charge of systems that are deteriorated or defunct and then assume full responsibility. Farmers fear that after transfer the government will not provide anything and they will have to pay water tax (increased) but will not get the water as the delivery systems are in disuse or need maintenance. The issue here is that there will have to be commitment of financial resources for the reform process, i.e., basic O&M of the system before transfer or simultaneous with the transfer process.

In the case of Andhra Pradesh the key reformers in the irrigation department argue that minimum rehabilitation before transfer is a myth and they achieved this without doing minimum rehabilitation. In the view of the author this is another myth being created. When the Act was passed and WUAs formed no minimum rehabilitation of the system was done. However, a couple of financial incentives were promised as under:

- There was a financial sop if the election to the WUA presidentship was unanimous. The WUA gets a grant of Rs. 50,000 if elections are not contested.
- With assistance from the World Bank, Rs. 118.1 crores ($28.1) were provided in 1998 for minimum rehabilitation works (an investment of approximately Rs. 100 per acre). Previously, the government allocated Rs. 40 per acre for O&M, but approximately Rs. 35 of this was used for irrigation department staff salaries and fixed overhead costs, leaving barely Rs. 5 per acre for actual maintenance work. The Rs. 100 per acre allocation for minimum rehabilitation from World Bank funds was to be used only for actual maintenance works, exclusive of department salaries. From January 1998 to March 1999, 22,887 works were undertaken, of which 70 percent were implemented by WUAs or DCs (Vermillion 1999).

At present, the water tax is collected by the revenue department and goes into the treasury of the state government. It is proposed that in the near future the water tax would be collected by the WUA and the money thus collected would flow back to the WUA in the proposed manner as shown in the Table 3. The WUA would use this money for O&M purposes as well as administrative expenses etc.

Financial resources were provided to WUAs and DCs for minimum rehabilitation works. So the farmers did not have to spend anything from their pocket for rehabilitation works on the system.

Table 3
Proposed Irrigation Revenue Sharing System

Level	*Major*	*Medium*	*Minor*
WUA	50%	60%	90%
DC	20%	30%	—
PC	20%	—	—
Local government	10%	10%	10%

Source: DICAD, GOAP, 1997.

The point one is making here is that availability of financial funds for minimum rehabilitation of the system prior or during reforms is a prerequisite if reforms have to succeed. The reform programme was funded by the Andhra Pradesh government but with support from the World Bank. Though the government defends that even without World Bank aid they would have gone ahead with the reforms, critiques do not buy this statement. The World Bank grant was simultaneous with the introduction of reforms and all works supported by the loan money from the World Bank. The initial piloting of WUAs on the Sriramsagar Project was assisted, including through use of retroactive financing, under the Andhra Pradesh Third Irrigation Project, World Bank project approved in May 1997. The state-wise reform programme is being financed under the irrigation component of the Andhra Pradesh Economic Restructuring Project (APERP), approved in June 1998 and with retroactive financing available from August 1997. The irrigation component of the APERP provides support, totalling $142 million to the Andhra Pradesh government's irrigation sector reform programme (Oblitas 2000).

Thus, the key argument is that if other states want to introduce reforms but have no World Bank funding what do they do? Some of the options could be:

- The government transfers all the O&M funds from the department to the WUAs.
- The minimum rehabilitation programme is scaled up in a phased manner, first starting with the very crucial areas and then scaling up.
- Mechanism of water tax collection, a portion of which then goes back to the WUA for O&M. Once this cycle is in order, sustainability of the programme can be ensured in the way it is envisaged in Andhra Pradesh (but which is still not put into practice).
- Rights to WUAs to raise their own money. Collecting increased water tax over and above the government rates.
- Promoting and motivating farmer contribution to the programme.
- Tapping into existing state and central government schemes meant for these kinds of activities. The main point here being that the author believes that there are enough financial resources with the government but because of the

high rent seeking tendencies over the last two decades and high level of leakages and sub-standard works, most of the problems have arisen. Hopefully, an accountable WUA in collaboration with the government, and with a proper monitoring and evaluation strategy in place should stem this rot.

Devolution of Powers to WUA and Other Issues

The basic foundation of the irrigation reform process is the WUA. The election process to establish the WUAs was highly democratic and fair and was conducted in a participative manner.

Field observations and discussions with farmers and WUA presidents highlight the following positive impacts after the formation of WUAs:

- For the first time farmers feel they have an equal say in management of their affairs.
- Quality of the structures and maintenance has improved.
- Sense of ownership over assets.
- Increase in crop production.

Many of these issues have already been featured in other sections of the paper. The most important issue here is to identify the powers that have actually been devolved to the WUAs. At the primary level powers like maintenance of the system, conflict resolution, record keeping, etc., have been delegated. There is accountability of the irrigation department towards the WUA. But, the WUA presidents and farmers are of the view that unless the Project Committee is formed and farmers have a right in the management of the main system, reforms will be incomplete.

Some of the other problems that have cropped up in the functioning of the WUAs and need immediate attention are discussed below.

WUAs have Become Petty Contractors

The entire reform process, at least in its first phase, has been construction-oriented, with jobs including repairs, system rehabilitation, desilting works, etc. Since the majority of the works have been done through WUAs the following scenarios have emerged:

- In the best-case scenario WUAs have executed the works in a fair and impartial manner, with proper records of all financial transactions maintained, and contracts given in a fair and just procedure.
- In many cases the WUA president being a dominant person, acted as a contractor and took on the works himself resulted in the siphoning off funds or sub-standard works. Because of his position and fear no one in the WUA complained. As the officials have to pass the bills of the sanctioned works, the compliance with the president has increased their rent-seeking behaviour.

- In the third scenario it is seen that some of the government officials have been petty contractors and in compliance with the presidents executed the works, and also mismanaged the funds.

To curb this in the future more awareness campaigns need to be launched regarding the rights as a salient feature of the programme. Secondly, participatory monitoring and evaluation processes have to be set in place and accountability pressure on both the irrigation department and the farmers' organisation has to be increased.

Collection of Water Tax by WUAs

In the near future it is envisaged that the WUAs will collect water tax from the farmers. While this is the most logical step in the reform process, some of the apprehensions in the minds of the WUA presidents are that, they will not have sufficient manpower to collect the water tax. One of the DC presidents said that at present in his jurisdiction there are more than 16 revenue officials who collect water tax. This is an important point and some thought needs to be given as to how this will be operationalised. The government may have to think of a strategy under which the current staff of the department works under the WUA for sometime to collect the water tax, and as the capacity and financial sustainability of the WUA increases the government can withdraw.

Financial Viability of the WUAs

The WUA at present is completely dependent on government funds for its survival. The physical works because of the World Bank project will continue for sometime, which would provide for some finances for the running of the WUA. But what happens when this stops. In the past, in other parts of the country where WUAs were formed, many became defunct due to financial bankruptcy. WUA will have to be financially self-sufficient. Only when this happens can reforms be called a success.

WUA Federation

One of the salient features of the 1997 Act, as mentioned earlier, is that it stipulates the formation of the Federation of the WUAs. The term "farmers' organisation" is used as a generic term in the Act and includes a Water Users' Association at the primary level, a Distributory Committee at the distributory level, a Project Committee at the project level and an Apex Committee at the government level to resolve any disputes relating to the farmers' organisation (details already mentioned in the section on salient features of the Act).

Till date the Distributory Committees have been formed, but the Project Committees and the Apex Committees have not been constituted. Formation of the WUA Federation is one of the best instruments for building social capital. It allows the issues of the WUA at the lower most end of the system to be raised. It is also a good mechanism for conflict resolution in a multi-structured multi-stakeholder highly complex system.

Series of discussions by the author with farmer leaders and presidents of the Distributory Committees raised the issue as to why the government had not held elections for the Project Committees. The presidents feel that they now want a say in the management of the project at the main system level. The government on its part is unsure whether to allow the farmer representative (WUA) such a major say at the project level. They fear that the Project Committee will be all too powerful and may be difficult for the administration to handle. The author is of the view that elections for the Project Committees should be held as soon as possible and also the Apex Committees constituted. This will complete the entire structure and one could say powers have been devolved at all levels. The non-formation of the committees at this level is also making farmers question the sincerity of the government in putting the reforms in place.

Linkages with Other Line Departments and Institutions

The WUA should become the focal point for all further agricultural development activities. Agricultural credit, extension and seed supply should be done through WUAs. As mentioned earlier, this will also give an opportunity to the WUA to make some money and in the long run become financially viable. Departments like cooperatives, horticulture, etc., should also link up with the WUA. What one would like to see is vibrant and robust WUAs not only managing the irrigation system but all aspects of agriculture production and marketing. For this the WUAs may even hire professional staff once they become financially independent. This will also help agriculture to move from a subsistence level at present in most places to an economically profitable enterprise.

Capacity Building

Though there have been serious efforts at training and capacity building by the government from the beginning of the reform process, there is still a need for strengthening this activity (Guru Prem 1999). Training has to be a continuous process to build up the skills and capacities of both the irrigation department personnel and WUA to address new issues thrown up the reform process.

On its part the government has produced many manuals on how to operationalise various parts of the Act. The author is of the view that this is not sufficient. In the field even after four years many farmers still do not know the existence of the Act, or even those who have heard about it do not know much. Secondly, the emphasis has to be on training the WUA functionaries as to the issues related with office management, accounts, etc., something they were never used to doing before.

At the irrigation department level attitude and behavioural change training programmes need to be conducted for the middle and lower level officials. They need to be more exposed to the new concepts, their roles and responsibilities in the new reform environment. This is not happening at present.

WALAMTARI (the Government Water and Land Management Institute) faculty need to be further trained for the finer aspects of the reform process. Some of the faculty have gone through orientation programmes, but this is not sufficient. Good quality Training of Trainers programmes will have to be conducted. If the trainers themselves are ignorant about the reform process how will they be able to spread the message.

NGOs can play a major role in the entire capacity building part of the programme, but sadly in the entire Andhra Pradesh reform programme they have been kept away. It is surprising that the Andhra Pradesh irrigation reforms that were based to a large extent on the learnings of the pilot projects implemented by the NGOs, should be now conspicuous by their absence. The Act has no reference or role for NGOs. The author is of the view that NGOs can play an active role not only in the capacity building of the irrigation department as well as the WUAs but also in identifying weaknesses and problem-solving measures in the field. The irrigation department should involve a few of the experienced NGOs in conducting action research, monitoring and evaluation exercises and also training programmes to strengthen and add value to the entire programme.

Accountability

The reforms in Andhra Pradesh empowered by the enabling legal provision have brought about accountability into the functioning of the irrigation department. Prior to the Act, the department officials always thought about themselves as providers and had no accountability towards the farmers. Today, the department officials play a facilitators role to the WUAs. In fact, the junior level officials of the of the irrigation department like work inspectors, irrigation construction assistants, and the work charge establishment ahs been transferred from the irrigation departments management to that of the farmers' organisations (Oblitas 2000). And these officials are now directly responsible to the farmers' organisations.

In the case of the WUAs also the presidents and office bearers are accountable to their members. The Act has the provision for recall, wherein if the members of the WUA are dissatisfied with the functioning of the president they can recall him from that position with a third majority of the members calling for the recall. This clause in the Act is a revolutionary step in the devolution process of management powers to the WUA. In the past, in other sectors, once a person is elected, he stays there for five years or the duration of office and has no compulsion to prove himself or perform, because he is guaranteed office for his full term. More than 70 presidents have till date been recalled. This has put a social pressure on other presidents to perform because of the social stigma attached to recall. The author is of the view that these elements of the Act have provided or ensured the downward accountability in the reform process – a vital element for the success of devolution.

One of the key issues in any policy formulation is institutionalising people-centred processes and scaling up participatory approaches in large, public bureaucracies for irrigation management.

Perry (in Merrey 1996) suggested three determinants of "proper" irrigation management as under:

- Defined water rights.
- Infrastructure capable of providing the service embodied in the water rights.
- Assigned responsibilities for all aspects o system operation.

Water rights, infrastructure and responsibilities interact and are inter-dependent. The Andhra Pradesh irrigation reforms have shown that a political will and an enabling legal environment give the the institutional framework that is conducive for the implementation of the factors recommended by Perry.

The irrigation department and WUA are the two sides of a coin. What is important to note here is that reforms are meant for the benefit of the farmer. Policy makers have to keep in mind that if reforms have to succeed then the farmer must be convinced that the reforms are for his benefit and that his participation in the programme is in his interest. The farmer is willing to pay for services (water) if he is assured of timely and delivery at the right time.

Top Down vs. Bottom Up Approach?

From the analysis of the paper the author would like to argue that we have created a myth when we say bottom up approach or top down approach. There is no top without a bottom or no bottom without a top. What is needed is a linkage between the two. Some researchers like Parthasarthy (2000) say that the approach in Andhra Pradesh is top down while the approach in Gujarat in bottom up. The author is of the opinion that this is all a play of words and rhetoric. In Gujarat, there are not more than a couple of hundred WUAs, of which almost all are defunct today. So what is bottom up approach here? The issue is that if a national or state policy that has all the ingredients that NGOs thought should be there (like rights to local communities, involvement in decision-making, equity and efficiency issues etc.), then would you still call this policy or approach a top down approach? The answer to this would be no. In a political democracy like India this is one of the best methods that can work, and this method is basically "good governance".

The reforms in Andhra Pradesh, as stated in the beginning, have a lot more to achieve and it is too early to even say whether they are successful or failed, but what is important is that they have shown the way forward. Terms like successes and failures are many a time misleading, and can also become tools in the hands of both proponents and critiques. What is important from a rational perspective is to see what can be learned and can this learning be translated into a feedback mechanism for managing the programme better and achieving the goals it set out to achieve.

REFERENCES

Bandaragoda, D.J., 1999, *Institutional Change and Shared Management of Water Resources in Large Canal Systems: Results of an Action Research Program in Pakistan*. Research Report No. 36, International Water Management Institute, Sri Lanka.

Blackburn, J., Chambers R. and Gaventa J., 2000, *Mainstreaming Participation in Development*, OED Working Paper Series No. 10, The World Bank, Washington, D.C.

Brewer, J. et al., 1999, *Irrigation Management Transfer in India: Policies, Processes and Performance*. Oxford and IBH Publishing Co. Pvt. Ltd. New Delhi, India.

Brewer, Jeffrey, 2000, Irrigation Management Transfer in Indian Canal Systems, in Joshi L.K. and Hooja Rakesh (eds.), 2000, *Participatory Irrigation Management: Paradigm for the 21st Century* (vols 1 & 2), Jaipur & New Delhi, Rawat Publications.

Department of Irrigation and Command Area Development, Government of Andhra Pradesh, 1997, *The Andhra Pradesh Farmer's Management of Irrigation Systems Act* (Act 11 of 1997) Act and Rules.

GOI, Ministry of Water Resources, CAD Wing, 1995, Report of the Working Group on Participatory Irrigation Management Programme for the IX Five Year Plan, New Delhi, India.

Groenfeldt, D., 2000, Introduction: A Global Consensus on Participatory Irrigation Management in Groenfeldt D. and Svendsen M. (eds.), 2000, *Case Studies in Participatory Irrigation Management*. WBI Learning Resources Series, World Bank Institute, World Bank, Washington, D.C.

Groenfeldt, D. and Svendsen M (eds.), 2000, *Case Studies in Participatory Irrigation Management*. WBI Learning Resources Series, World Bank Institute, World Bank, Washington, D.C.

Groenfeldt, D. et al., 1996, Proceedings of the Second International Seminar on Participatory Irrigation Management, Antalya, Turkey, April 10-17, 1996, Economic Development Institute of the World Bank, World Bank, Washington, D.C.

Groenfeldt, David et al., 1998, *Handbook on Participatory Irrigation Management*. The Economic Development Institute of the World Bank, Washington, D.C.

Guru Prem, B., 1999, Capacity Building for PIM in Andhra Pradesh (India). Fifth International Seminar on Participatory Irrigation Management, December 1999, Hyderabad (being included in this book).

IndiaNPIM, May 2000, *Workshop Report, Attitude and Behavioural Change for Participatory Irrigation Management.*

Joshi, L.K. and Hooja Rakesh (eds.), 2000, *Participatory Irrigation Management: Paradigm for the 21st Century* (vols 1 & 2), Jaipur & New Delhi, Rawat Publications.

Joshi, L.K., 2000, Irrigation and Its Management in India: Need for Paradigm Shift, in Joshi L.K. and Hooja Rakesh (eds.), 2000, *Participatory Irrigation Management: Paradigm for the 21st Century* (vols 1 & 2), Jaipur & New Delhi, Rawat Publications.

Meinzen-Dick, R. and A. Knox, 1999, Collective Action, Property Rights, and Devolution if Natural Resource Management: A Conceptual Framework. Paper presented at the Workshop on Devolution of Natural Resource Management, Puerto Azul, Philippines, 21-25 June.

Meinzen-Dick, Ruth, 1997, Farmer Participation in Irrigation: 20 Years of Experience and Lessons for the Future. IFPRI Reprint No.371, Reprinted from *Irrigation and Drainage Systems*, Vol. 11, 1997, Netherlands.

Merrey, Douglas, 1996, Institutional Design Principles for Accountability in Large Irrigation Systems, Research Report No. 8, International Irrigation Management Institute, Sri Lanka.

Mollinga, P. 2001, Power in Motion: A Critical Assessment of Canal Irrigation Reform, With a Focus on India. Paper presented at the International Training of Trainers Course on Participatory Irrigation Management, January 2001, Hyderabad, India.

Mollinga, P. et al. 2000, Participatory Irrigation Management in Andhra Pradesh, India: Policy implementation and transformation in the Tungabhadra Right Bank Low Level Canal (unpublished paper).

Naidu, Chandrababu, 1999, Transferring Irrigation Management to Farmers. Welcome Address at the Fifth International Seminar on Participatory Irrigation Management, Hyderabad, 1999, India.

Oblitas, Keith and Peter Raymond et al., 1999, *Transferring Irrigation Management to Farmers in Andhra Pradesh*, India, World Bank Technical Paper No. 449, World Bank, Washington, D.C.

Palcios, Enrique, 2000, Benefits and Second Generation Problems of Irrigation Management Transfer in Mexico in Groenfeldt D. and Svendsen M. (eds.), 2000, *Case Studies in Participatory Irrigation Management*. WBI Learning Resources Series, World Bank Institute, World Bank, Washington, D.C.

Parthasarathy, R., 1998, *Coordinating Participation to Transfer Irrigation Management: A Game Theoretic Paradigm*. Working Paper No. 105, Gujarat Institute of Development Research, Gujarat, India.

Peter, Raymond, 2000, Management of Irrigation Systems by Farmers in Andhra Pradesh, in Joshi L.K. and Hooja Rakesh (eds.), 2000, *Participatory Irrigation Management: Paradigm for the 21st Century* (vols 1 & 2), Jaipur & New Delhi, Rawat Publications.

Peter, Raymond, 2000, Salient Features of the Andhra Pradesh Farmers Management of Irrigation Systems Act, 1997, in Joshi L.K. and Hooja Rakesh (eds.), 2000, *Participatory Irrigation Management: Paradigm for the 21st Century* (vols 1 & 2), Jaipur & New Delhi,

Rawat Publications. Joshi L.K. and Hooja Rakesh (eds.), 2000, *Participatory Irrigation Management: Paradigm for the 21st Century* (vols 1 & 2), Jaipur & New Delhi, Rawat Publications.

Pinot Quintero Luis, 2000, Benefits and Second Generation Problems of Irrigation Management Transfer in Colombia in Groenfeldt D. and Svendsen M. (eds.), 2000, *Case Studies in Participatory Irrigation Management*. WBI Learning Resources Series, World Bank Institute, World Bank, Washington, D.C.

Raby, N., 2000, Participatory Irrigation Management in the Philippines: National Irrigation Systems in Groenfeldt D. and Svendsen M (eds.), 2000, *Case Studies in Participatory Irrigation Management*. WBI Learning Resources Series, World Bank Institute, World Bank, Washington, D.C.

Rahaman, M.K., 1999, Farmer's Participation: Successful Stories in Andhra Pradesh (mimeo).

Raju, K.V., 2000, Participatory Irrigation Management in Andhra Pradesh: Promise, Practice and a Way Forward. Working Paper No. 65, Institute for Social and Economic Change, Bangalore, India.

Reddy, D.N., 1999, Designer Participation: Politics of Irrigation Management Reforms in Andhra Pradesh, India. Paper presented at the International Researchers' Conference on Irrigation Management Reform, held at Hyderabad, India, 11-14 December 1999 (being included in this book).

Reddy, Kishan, 2000, Progress Report of the Distributory Committee 16, Tadikal (SRSP), District Karimnagar, Andhra Pradesh.

Ruth, Meinzen-Dick, 2000, Towards New Partnerships in Indian Agriculture, in Joshi L.K. and Hooja Rakesh (eds.), 2000, *Participatory Irrigation Management: Paradigm for the 21st Century* (vols 1 & 2), Jaipur & New Delhi, Rawat Publications.

Shah, Tushar et al., 2000, Institutional Alternatives in African Small-Holder Irrigation: Lessons from International Experience in Irrigation Management Transfer. Unpublished paper, International Water Management Institute, Sri Lanka.

Sutton, Rebecca, 1999, The Policy Process: An Overview. Working Paper No. 118, Overseas Development Institute, London, U.K.

Svendsen, M. and Nott Gladys, 2000, Irrigation Management Transfer in Turkey: Process and Outcomes in Groenfeldt D. and Svendsen M. (eds.), 2000, *Case Studies in Participatory Irrigation Management*. WBI Learning Resources Series, World Bank Institute, World Bank, Washington, D.C.

Vermillion, D.L., 1999, The Participatory Irrigation Management Program in Andhra Pradesh, India: Current Issues and the Need for Monitoring and Evaluation. Notes from a field visit (mimeo being included in this volume).

Vermillion, D.L. and Sagardoy J.A., 1999, Guidelines for Transfer of Irrigation Management. FAO Irrigation and Drainage Paper No. 58, Rome, Italy.

Vermillion, D.L., 1997, Impacts of Irrigation Management Transfer: A Review of the Evidence. Research Report No. 11, International Irrigation Management Institute, Sri Lanka.

Vermillion, D.L., and C. Garces-Restrepo, 1998, Impacts of Colombia's current Irrigation Management Transfer Program. Research Report No. 25, International Water Management Institute, Sri Lanka.

II

Andhra Model: Status and Prospects

3

Irrigation Reforms in Andhra Pradesh

J. RAYMOND PETER

In 1997, the state of Andhra Pradesh (AP) in India embarked on an ambitious program of reforms in its irrigation sector. In this large agricultural state, irrigation management has been revolutionized by transferring responsibility for the operation and maintenance of irrigation schemes to groups of farmers. In total, 10,292 Water Users' Associations (WUAs) have been created. The reform required a number of very difficult changes, as the role of irrigation agency is gradually shifting from service provider to facilitator, and reform has not come without resistance and conflict. However, early indications are overwhelmingly positive as many irrigation systems are realizing increased revenues, an increase in irrigated area, and enhanced involvement of farmers in the operation of irrigation.

The WUAs and 174 Distributory Committees (DCs) were created through a democratic process of elections. The Andhra Pradesh Farmers' Management of Irrigation Systems (APFMIS) Act, 1997 gives the state the power to create WUAs and federate them into higher-level committees. Reform has made the irrigation agency accountable to the farmers' organizations, and resulted in the tripling of water charges and linking of money collected to the costs of operating and maintaining irrigation systems.

Change in the irrigation sector is only a part of a series of reforms initiated over the last five years in AP. The programs aim at modernizing the state government and vitalizing its economy by making the providers of services such as irrigation, education, and health accountable to users and other stakeholders. The emphasis is on making the reform process participatory through a process of extensive public consultations, and on achieving quick results.

Reform will go a long way in making the irrigation sector sustainable. The goal of the reform will be achieved when WUAs in AP become sustainable by raising funds for irrigation operation and maintenance on their own. Water sector reforms in the state break new ground for reforms in the water sector in

India. Several states in the country have launched similar programs with AP as a role model.

Background

Andhra Pradesh, one of the southern states in India, has launched a series of actions over the last five years, aimed at empowering stakeholders under the Naidu administration. The reform process is a part of a statewide policy of administrative reform defined by the Chief Minister. As stated above, it aims to modernize AP's government and vitalize its economy by making the providers of services such as irrigation, education, and health accountable to users and other stakeholders.

The Naidu administration has facilitated stakeholder consultations in key areas, focusing on natural resources, employment generation and human resource development to tackle the complex problems of poverty, unemployment, and environmental degradation.

Philosophy of *Janmabhoomi*

The overarching philosophy of the government has been *janmabhoomi*. It is a process and not just another government program to promote self-help and community work. *Janmabhoomi* creates an enabling environment to launch several innovative approaches to ensure people's participation in development through WUAs, *Vana Samrakshana Samithis*[1], DWCRA[2] groups and CMEY[3] groups. The program *Prajala Vaddaku Paalana*[4] is yet another novel initiative to redress the long-felt needs of the people at the village level and bridge the gap between the administration and the citizens. Box 1 below shows the types of stakeholder groups consulted by the AP government.

The potential of the communities has been tapped through their active involvement and participation in *sramadanam*. People's participation forms the

Box 1
Stakeholder Groups Consulted

Key Sector	*Type of Stakeholders*
Natural Resources	Water User Associations
	and Samrakshana Samithis
	Watershed Committees
Employment Generation	DWCRA
	DWCUA
	CMEY
Human Resources Development	School Education Committees
	Mothers Committees

core of the innovative approach adopted by *janmabhoomi* to bring about social transformation and the development of the downtrodden and deprived sections of the society. People organize themselves, identify their needs through *Gram Sabha*[5] and other grassroots organizations, and participate in the design and implementation of solutions. This approach has given a voice to the people, enabling them to express themselves and solve their own problems.

Janmabhoomi provides the appropriate environment for people's participation in decision-making. This, in turn, enhances the confidence of the people. It can be seen as a people-centered development process aimed at rebuilding the villages and towns and establishing an ideal society which embodies and cherishes the concept of hard work, honesty, team work, cooperation, accountability, and transparency, leading to the achievement of excellence in all walks of life.

Reforms in the irrigation sector were among the first in a series of reforms launched by the state. The success of WUAs in the state has prompted the government to launch similar strategies in other sectors. The process of reform is both difficult and complex. It demands a high level of commitment, open minds, and flexible options for change. Introducing change in any organization is often viewed with suspicion and apprehension. The people's vigorous response to the reform program is what makes AP unique.

The Irrigation Sector

Andhra Pradesh is endowed with very rich water resources and is appropriately called a river state. Waters flow in the large rivers Krishna, Godavari, and Pennar, and in several minor rivers as well. The state is broadly divided into three major agro-climatic zones: the Telangana area, comprising the northern part of the state; the coastal region, covering the coast along the Bay of Bengal; and the Rayalseema region covering the southern part of the state. Table 1 outlines the contribution of major river systems to AP's water resources.

Table 1
Water Resources

River System	*Drainage area (thousands of km)*	*Percent of total drainage area*	*Assessed annual yield (mha)*
Godavari	73.2	26.5	4.23
Krishna	74.4	26.9	2.30
Pennar	48.1	17.4	0.28
Nagavalli	4.8	1.8	0.14
Vamsadhara	1.9	0.7	0.04
Other minor rivers draining into the sea	74.2	26.8	0.79
Total	276.7	100.0	7.78

As irrigation has been the primary source of agricultural growth in AP, it has always been accorded top priority by the state. Traditionally, the irrigation sector has been the largest consumer of plan funds next to the power sector. Over the years, the state has been able to create an irrigation potential of 4.84 million hectares (mha) through 15 major irrigation projects, 75 medium irrigation projects, and approximately 12,264 tanks.[6]

Traditionally, individual farmers have irrigated from wells – groundwater irrigates about 2.2 mha. Around 11 mha, or 40 percent of the state's geographic area, is sown to crops. About half of that is under irrigation. Therefore, the performance of irrigated agriculture is critical to AP's growth. Table 2 shows the historical investment in irrigation infrastructure in the state.

Table 2

Plan Expenditure and Irrigation Potential Created

Period	*Amount Spent (million Rs.)*			*Irrigation Potential Created (mha)*		
	Major and medium	*Minor*	*Total*	*Major and medium*	*Minor*	*Total*
Pre-Plan Period (1951)	-	-	-	1.33	1.37	2.70
First Plan (1951-56)	375	35	410	0.08	0.03	0.10
Second Plan (1956-61)	574	44	618	0.18	0.02	0.20
Third Plan (1961-66)	915	186	1,101	0.37	0.05	0.42
Three Annual Plans (1966-69)	609	108	717	0.08	0.04	0.12
Fourth Plan (1969-74)	1,187	182	1,369	0.19	0.06	0.25
Fifth Plan (1974-78)	2,691	388	3,079	0.21	0.09	0.31
Two Annual Plans (1978-80)	2,577	238	2,815	0.15	0.06	0.21
Sixth Plan (1980-85)	7,296	411	7,707	0.31	0.08	0.39
Seventh Plan (1985-90)	13,064	1,314	14,378	0.09	0.07	0.16
Annual Plan (1990-91)	2,828	478	3,305	0.01	0.01	0.02
Annual Plan (1991-92)	3,339	486	3,825	0.01	0.01	0.02
Total	35,455	3,869	39,324	3.00	1.88	4.88
Eighth Plan (1992-97)	22,653	1,869	24,522	0.04	0.04	0.08
Grand Total	58,108	5,738	63,846	3.05	1.92	4.96

Notes: 1. The above figures do not include expenditure on CADA up to the end of the Seventh Plan.
2. The above figures are adopted from *Eighth Five-Year Plan (1990-95)*, AP-F&P (PLG) Department, Volume II, October 1990.

Landholdings

The average landholding size in the state is 1.56 ha, with irrigated farms averaging 0.88 ha. Land distribution in the state is highly skewed. Farms considered small and marginal cover half of the irrigated area. A majority of farmers (79%) work on these marginal farms. Only 6 percent of the irrigated land consists of farms of more than 10 ha.

Irrigation produces nearly two-thirds of the cereal crops grown in the state and more than half of the fruits and vegetables. Seventy percent of the irrigated area is under cereal crops, mostly rice. Other food crops cover 14 percent of the irrigated area; these include chilies, fruits, and vegetables. Oilseeds cover 12 percent, and the remainder is made up of non-food crops such as tobacco, sugarcane, and cotton.

Structure of the Irrigation Administration

The irrigation sector[7] is under the Irrigation and Command Area Development Department (ICADD).[8] The administration of the irrigation sector is carried out at three levels. First, the highest level of government oversight comes from the Minister of Major and Medium Irrigation, the Minister of Minor Irrigation, the Principal Secretary of ICADD, and three to four secretaries, each of whom are assisted by three to four deputy or joint secretaries.

The second level is made up of the heads of several departments: the Chief Engineer of Irrigation and Administration, the Director General of the Water and Land Management Training and Research Institute (WALAMTARI), the Commissioner of the Command Area Development Authority (CADA), and the Director of the Groundwater Board.

The third level is responsible for field operations at the district level. One or more districts are under the control of a Superintending Engineer. A district may have one or more Executive Engineers heading its irrigation division. Traditionally, each irrigation division has three to four sub-divisions and is manned by a Deputy Executive Engineer. The Assistant Engineer sub-divides each sub-division into three or more sections. One or more work inspectors may assist each Assistant Engineer. The *lascars* are the lowest rung in the Irrigation Department. Lascars work as gatekeepers and assist in the distribution of water. Box 2 presents the structure of the Irrigation Department at the state and district levels.

Types of Irrigation Projects

Irrigation projects in AP have been divided into three categories based on the size of the irrigated area.[9] Major irrigation schemes have command areas of more than 10,000 ha, whereas medium irrigation projects have command areas ranging from 2,000 ha to 10,000 ha.

Minor irrigation projects have command areas of less than 2,000 ha and usually include smaller irrigation schemes such as lift irrigation or schemes with water sources like tanks, diversion weirs, and open head channels. Traditionally,

Box 2

Structure of the Irrigation Department

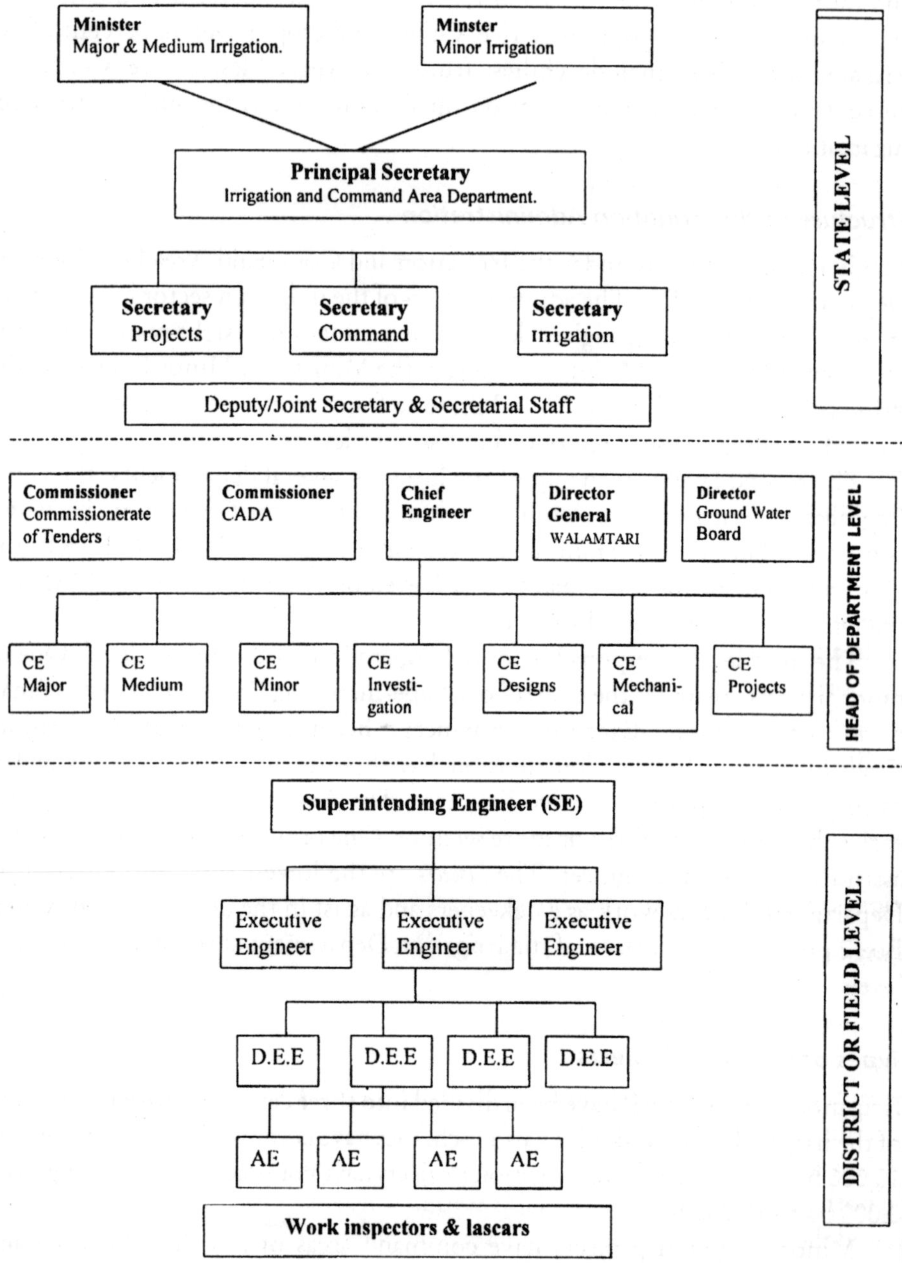

minor irrigation projects and groundwater are under the control of the Minor Irrigation Department, while the major and medium irrigation projects are under the Major Irrigation Department.

Performance of the Irrigation Sector

Despite major plan investments in irrigation by the Andhra Pradesh government which have increased the irrigation potential, most systems are in disrepair and dilapidated due to inadequate maintenance. This has lead to the shrinkage of command areas. Irrigation systems are characterized by low irrigation efficiencies and tail-end deprivation.

An anarchic situation existed in established commands, where head-enders appropriated most of the water for themselves. This situation was exacerbated by a lack of coordination among the various departments, such as Irrigation, Agriculture, and Revenue. The gap between the potential irrigated area created and the actual utilization, popularly known as the "gap area", has been estimated at about half a million hectares.

Lack of established operation and maintenance (O&M) procedures, inadequate funds for O&M, and ad hoc expenditures by the Irrigation Department have compounded the unsatisfactory performance of most systems. Most of the agency's O&M funds were being spent on staff salaries; very little was being spent on effective maintenance. This has led to the siltation of major canals and drains and damage to their lined sections.

In addition, dissatisfied farmers seeking more water, or water deliveries at the appropriate time, have tampered with irrigation structures causing further damage. Such unauthorized irrigation led to a low collection of water charges by the Revenue Department, as the measured water supplied was quite low.[10] The situation in AP is representative of the prevailing conditions in other Indian states. The need for change and reform had become obvious.

Irrigation Management Reform

A major component of the reform in the irrigation sector was aimed at giving a greater role to farmers in irrigation management. To determine a suitable framework for increasing farmer participation, a series of public consultations were conducted throughout the state in most major irrigation commands. This consultative approach marked a dramatic departure from the usual way governments work.

Initially, the consultations were met with severe cynicism and indifference by both farmers and agency staff. Within the Irrigation Department, most viewed the current sordid state of affairs as an outcome of inadequate budgets and increasing political interference. It was felt that adequate infusions of cash for O&M would substantially improve irrigation system performance. Farmers approached the public consultations with a mixture of curiosity and cynicism. While there was great apprehension toward the new approach, things became clear as several actions followed, one reinforcing the other. This sort of a systematic and persistent approach is what makes the program unique.

Public Consultations

Beginning in April 1996, a series of public consultations were held in centrally located places in command areas of major projects. These informal discussions gave a tremendous amount of input to the government for use in designing a suitable plan of action. The first round of public consultations was used to elicit the viewpoints of the farmers and Irrigation Department staff. Subsequent rounds employed a more structured framework for discussion.

The objective of the government was to understand the viewpoints of all concerned. Therefore, participation in the public consultations was not restricted to farmers, but also included politicians, political parties, researchers, and the press. In AP, this was perhaps the first attempt by the government to seek out the viewpoints of beneficiaries and the parties likely to be affected. The consultations have enabled the government to steer the reform process in a transparent manner. From the beginning, there was as no clearly defined blueprint to guide the actions of the government. Rather, the state has attempted to adopt a learning process approach, continually incorporating feedback to guide the reforms.

SRSP Pilot Project

In 1996, two non-governmental organizations (NGOs), viz., IRDAS and SONAR, conducted a pilot project in the Sriramsagar project in Karimnagar district. The results of the pilot project[11] encouraged the government to adopt a concrete plan of action.

A key finding of the project was that substantial improvements in the operation of the irrigation system were realized when farmers in the head, middle, and tail end of the command came together to consult and negotiate with each other. The pilot also revealed that small interventions, such as removal of accumulated silt and weeds and minor repairs to structures, yield dramatic results. Box 3 demonstrates the terms of reference for NGOs working with groups of farmers.

Contrary to popular belief, most works that are considered technical can be undertaken by farmers themselves with no compromise in quality. Some cooperating engineers developed "layman's techniques" for farmers who take up operation and maintenance works. Even thought the pilot study only covered a relatively small area, its results were highly significant to the state government, as

Box 3
The Terms of Reference for NGOs

1. Identify a model for creating WUAs.
2. Evaluate the present system of water supply and suggest modifications.
3. Establish a methodology for optimizing the use of water by farmers.
4. Train the farmers.
5. Identify incentives for effective functioning of WUAs.

they showed a possible strategy that could be adopted throughout the state, namely one of farmer participation in O&M.

Role of the Government of India

The Government of India[12] began an intensive campaign to promote participatory irrigation management through national seminars and workshops in the 1990s. It also funded a study by an NGO – SOPPECAM – to formulate a model legislation. These initiatives provided a platform for debate and discussion on participatory approaches to irrigation management.

In general, there was a broad consensus for the creation of WUAs. These associations would be registered under existing laws such as the Cooperative Act or the Societies Act. A confederation of WUAs would then sign a Memorandum of Understanding (MoU) with the Irrigation Department. They would decide on mutually agreeable terms regarding water supply, maintenance, and collection of water charges. Pilot schemes in the states of Maharashtra and Gujarat were highly successful.

Problems of Scaling Up

The problem facing the AP government was how to scale up participatory irrigation management (PIM) to a statewide level using the lessons learned from pilot projects in AP and other states. A number of issues warranted more detailed debate and analysis, including the wide diversity in irrigation schemes, problems of absentee landlordism, large numbers of unregistered tenants, encroachers and unauthorized cultivators, and the lack of a uniform legislation.

In the absence of a comprehensive legislation[13] on irrigation in the state, it was decided to create new legislation exclusively to enable PIM. Feedback from the public consultations became the main input in drafting the law. The legislation was subjected to several rounds of discussion among the various stakeholders. The Andhra Pradesh Farmers' Management of Irrigation Systems (APFMIS) Act was enacted in 1997. It was the first legislation of its kind in India.

The state government decided to extend PIM to all parts of the state at one stroke. It felt that in order for the reform to be successful the following were important considerations:

(1) ensuring equitable distribution of benefits of the reform;
(2) providing uniform legislation;
(3) completely involving the Irrigation Department and the government; and
(4) spreading the new paradigm across the whole area in the shortest possible time.

Without adequate monitoring, this approach to reform ran the risk of ending in disaster. The consistent actions taken by the state government over a period of time ensured that this would be a remote possibility. A major constraint for implementing such a major program was the diversity of agro-climatic zones and types of irrigation schemes in the state. Despite the difficulties of implementing a single model across a variety of schemes, the state was committed to giving the farmers a new role in managing irrigation, and all projects, big and small, old and new, were included in the reform.

The APFMIS Act

The Andhra Pradesh Farmers' Management of Irrigation Systems (APFMIS) Act of 1997 was a revolutionary legislation. It was the first of its kind in India, seeking to bring a paradigm shift in irrigation management. The act contains broad provisions relating to the types of irrigation schemes, tiers of farmers' organizations (FOs), elections, functions of FOs, resources, and penalties for offenses. Detailed provisions relating to the actual working of WUAs are given in the rules.[14] Box 4 lists some of the most important features of the law.

Box 4
Salient Features of the APFMIS Act

- Transfer of power for the management of state-owned assets.
- Creation of new autonomous institutions as legal entities.
- Areas defined on a hydraulic basis.
- Equity achieved within the structure of a WUA by introducing the concept of territorial constituencies.
- All landholders in possession of land in an irrigation system become WUA members with voting rights.
- One member, one vote.
- Elections by secret ballot.
- Functional and administrative autonomy.
- Freedom to raise resources.
- Resolution of disputes and compounding of offenses.
- Simplified procedures for taking up works.
- Five-year tenure for farmers' organizations.
- Irrigation Department, as competent authority, is made fully accountable to the farmers' organizations.
- Right to recall an elected member after one year.
- Social audit and annual accounts audit.

Process of Formation of Water User Association

The APFMIS Act forms the legal basis for the formation of a WUA in the state of AP. Detailed rules have been separately notified for the delineation, notification and functioning of the WUA. The process of formation of a WUA is relatively simple and entails the following steps:

1. *Delineation of a WUA*: The area proposed to be constituted into a WUA is delineated by the irrigation agency under an irrigation project, either in full or in part, depending on the irrigation system. In the case of a minor irrigation tank, the area could be as low as 60 ha to a maximum of 2,000 ha. In the case of major irrigation scheme it could on average be about 1,500 to 3,000 ha. While delineating, natural and administrative boundaries are preserved as far as possible. Delineation is done on a hydraulic basis.
2. *Notification of a WUA*: The Collector of the district in which the WUA is located notifies the proposed WUA in the District Gazette and calls for objections. After hearing the objections a final notification is made in the District Gazette and the WUA is constituted. Along with the WUA a competent authority to the water user association is notified by law. The competent authority provides technical advice to the WUA and assists in the technical supervision of the works undertaken by it.
3. *Elections to the WUA*: The Commissioner, CADA is the authority who notifies the process of election to the WUA. The Collector of the district concerned issues the election notification. The process of election starts with the following sub-processes:
    (a) *Preparation of electoral rolls*: All the members of a WUA who use water and pay water tax as recorded in the revenue records (Adangal), and who are 18 years and above are the voters of a WUA. The electoral rolls are prepared territorial constituency-wise listed against each survey number (a basis for the landholdings in a village). The voter's list is then issued village-wise.
    (b) *Calling for nominations*: The District Collector issues a schedule for calling for nominations of interested candidates for the post of president and managing committee of the territorial constituency. A time frame for the scrutiny and withdrawal of nominations is specified in the election schedule and the final list of contesting candidates is put up on the notice board.
    (c) *Election process*: On the specified date, the elections are conducted. Each voter casts his vote. The candidate securing the maximum number of votes is declared the winner. In some cases the elections may be unanimous. The consensus candidate is then declared as the winner. The winner is administered an oath and handed over the election certificate.

(d) *Setting up of the WUA office*: The WUA notifies a place as its office and the work begins for a period of five years – the tenure of the WUA. Where a vacancy arises due to the death or resignation of a WUA president or a member, the District Collector fills up the vacancy after following a similar process.

(e) *Election disputes*: Where a dispute arises in the election of a member, the aggrieved could redress in the District Civil Court as an election dispute. All the District Civil Courts have been designated as the election tribunals for the purpose of the election disputes.

4. *Elections to Distributory Committees*: The elections to the Distributory Committees (DCs) are similar to that of the WUAs. The District Collector notifies the date of the election to a DC and calls for nominations. On the same day the election is conducted by a show of hands. The only difference being that the number of persons participating in the poll is restricted to the number of presidents of the WUAs, notified in a DC.
5. *WUA/DC to have Bank Account*: Each WUA or DC has a separate bank account. The president and a member are the signatories to the bank account.

Farmers' Organizations

The term 'Farmers' Organizations' (FOs) covers different types of organizations, namely, the Water User Association (WUA), the Distributory Committee (DC), and the Project Committee (PC). The primary body is the WUA. The number of tiers in a farmers' organization in an irrigation system depends on the type of the irrigation system. Thus, minor irrigation systems, which are typically less than 2,000 ha, have a single-tier system with one WUA. Medium irrigation systems (2,000-10,000 ha) have a two-tier system, made up of WUAs and a PC. Major irrigation systems (more than 10,000 ha) have a three-tier system, made up of WUAs, DCs, and a PC.

A WUA is created by delineating a portion of the command area[15] under an irrigation system. The managing committee of a WUA includes the WUA president and four to ten territorial constituency members. The area of a WUA is subdivided in order to equitably handle water management, maintenance, and governance. The area covered by a WUA generally ranges from 60 to 3,500 hectares. WUAs in the delta regions tend to have larger areas of up to 4,500 ha. In areas with rugged topography, WUAs cover about 1,000 ha. In most projects, however, the size of a WUA is around 600 to 1,000 ha.

Elections to the WUAs are conducted through a democratic process of secret ballots. The election procedure is akin to the *Gram Panchayat* election, so farmers are quite familiar with the election process. The electoral process has several advantages in enabling the election of a leader of choice. The first elections were apolitical. Subsequent elections may however be much more vigorously contested. Box 5 shows the typical structure of a WUA.

Box 5
Structure of a WUA

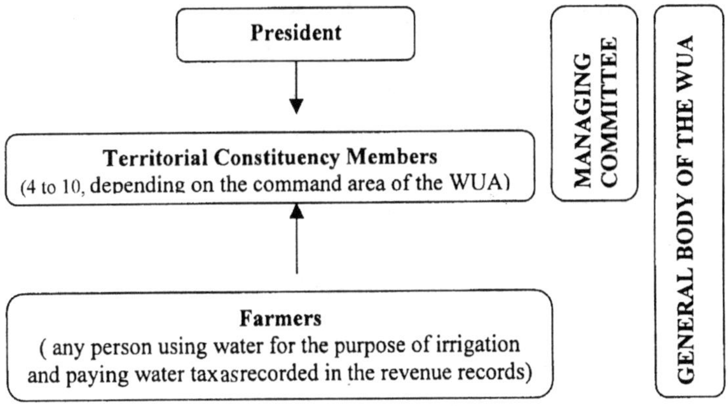

Each farmer has one vote regardless of the extent of his land holdings. This levels out the representation of large and small farmers. An analysis of the elections shows that nearly half of the elections were unanimous and the other half were contested. Unanimous elections were held mostly in the minor and medium irrigation projects, which were largely homogeneous. Elections in the major irrigation projects, which may be made up of a number of factious villages, were intensely contested. The district administration was fully involved in the election process. The total expenditure on elections was Rs. 110 million ($2.5 million).

Table 3 shows the distribution of WUAs in AP. A large portion of the irrigated area (73%) comes under the control of only 2,100 WUAs, or 20 percent of the total number of WUAs. Table 4 depicts the types of individuals elected by farmers to represent their interests in either the WUA or the territorial constituency.

Presently, the state of AP has constituted 10,292 WUAs and 173 DCs. Box 6 shows the typical structure of a DC. The final tier in the irrigation systems,

Box 6
Structure of a Distributory Committee

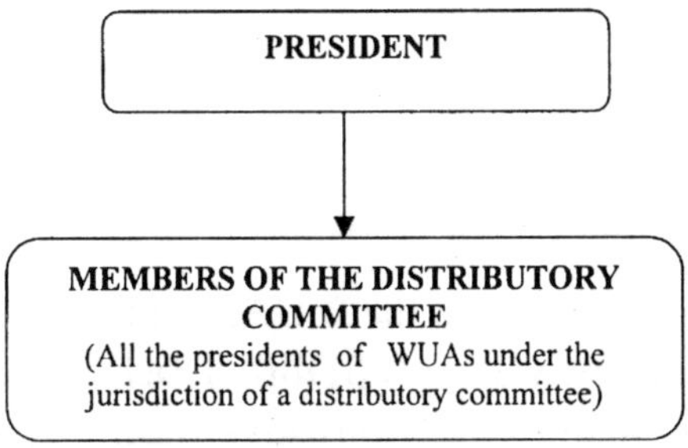

namely, the Project Committee, is yet to be constituted. At present these are manned by the irrigation agency itself. The most important change that has been brought about in the irrigation sector has been the constitution of farmers' organization in an irrigation system. Any O&M work taken up has to be necessarily executed by the WUA or the DC concerned. In the case of a Project Committee the works are to be taken up only after consultation with the DCs. This has brought about a total change in the manner in which irrigation systems are now run in the state.

Table 3
Distribution of Water User Associations

Name of the District	*Major*	*Medium*	*Minor*	*Total*
Adilabad	35	27	221	283
Ananthapur	46	7	305	358
Chittoor	0	51	644	695
Cuddapah	74	8	644	358
East Godavari	106	12	225	343
Guntur	245	8	81	334
Khammam	51	5	181	845
Krishna	189	12	288	237
Kurnool	116	12	153	489
Mahabubnagar	21	31	478	281
Medak	0	12	585	530
Nalgonda	91	45	541	677
Nellore	110	58	695	863
Nizamabad	78	13	267	358
Prakasham	124	5	317	446
Ranga Reddy	0	3	165	268
Srikakulam	37	28	459	524
Vishakhapatnam	28	18	375	421
Vizianagaram	0	22	439	461
Warangal	29	18	683	730
West Godavari	71	6	217	294
Total	1,700	411	8,181	10,292

Source: Commissioner, CADA, Government of Andhra Pradesh.

To give a wider representation, the WUAs and DCs have also been notified in the Irrigation Development Boards (IDBs) of each of the district and CADA Boards. Hitherto, the Members of Legislative Assembly and the Members of

Parliament represented the IDBs. The IDB is chaired by the District Collector and is an advisory body to review the development of irrigation activities in the district. It also recommends to government the closure and opening of canals and the likely area to be included under irrigation in a project.

Table 4

Background of Presidents and Territorial Constituency Members

Background	*WUA Presidents*	*Territorial Constituency Members*
Farmers	8,544	39,606
Business	213	438
Public Representatives	183	151
Professionals	58	118
Service Personnel	40	63
Contractors	35	35
Industry	27	54

A large number of WUAs that have been created are responsible for the minor irrigation tanks which are scattered throughout the state. Historically, tank irrigation has been a part of the village polity and hence there has been no major change in the way tanks are managed.

The major change in irrigation management has been a greater role of the WUAs rather than the irrigation agency. Another major shift, especially in major and medium irrigation projects, has been the realignment of boundaries. New boundaries are not necessarily contiguous with existing village boundaries. It has been necessary to redraw boundaries on a hydraulic basis and prepare new records.

Implementing Reform

The AP reform has been highly successful due to its timely linkage with a World Bank loan under the Andhra Pradesh Economic Restructuring Project (APERP). The $141 million loan is essentially designed to support the WUAs. The APERP plans for minimum rehabilitation of 2.45 mha under various irrigation projects. Minimum rehabilitation is envisaged at an amount of Rs. 1,350/ha ($30/ha). Important features of the APERP are highlighted in Box 7 below.

The AP government made a conscious decision to disburse the amount over a period of three years. This has two distinct advantages: (1) WUAs gain experience in undertaking O&M works; and (2) the Irrigation Department and Finance Department have time to develop new procedures for dealing with user organizations. Disbursing the loan amount in increments also ensures its effective use. The program's scope is vast, and therefore requires a great deal of

Box 7

Salient Features of the Andhra Pradesh Economic Restructuring Project Assisted by the World Bank

The Irrigation Component of the Andhra Pradesh Economic Restructuring Project (APERP-IC) provides support totaling US $142 million to the Government of Andhra Pradesh's irrigation sector reform program. The APERP-IC's main components are:

- *Irrigation Performance Improvement*. This component consists of: (i) minimum rehabilitation of the state's most needing major and medium irrigation schemes (2.45 mha) and (ii) recurrent maintenance on major, medium and minor schemes (3.7 mha) to ensure their sustainable functioning (with declining Bank financial assistance for maintenance ending in March 2000).
- *Scheme Improvement/Modernization and Farmer Turnover (SIFT)*. Piloting of more intensive modernization on selective projects.
- *Agricultural Intensification*. Intensification of agricultural extension services to FOs, training and skills development of agriculture department staff, dissemination of information and productivity-enhancing agriculture technologies, and on-farm demonstrations.
- *Institutional Development of FOs and Government Departments*. Making newly formed FOs fully functional and improving management capabilities through training for farmers, government staff and NGOs.

careful planning. The ultimate goal is complete takeover of O&M activities by farmers' organizations.

Reform of any kind is initially painful because it implies changing from high spending levels to lower, more frugal ones. In this case, the reform is aimed at empowering farmers and reducing the role of the irrigation agency.

Water Charges

The reform in Andhra Pradesh was made possible by two major decisions taken by the state:

- Tripling of water charges and linking of water charges to O&M.
- Linking the WUAs with a program of minimum rehabilitation.

Increasing the water charges by three times is an action, which most governments would shy away from. The increase was made possible in a novel manner. First, water charges were equated with the price of a bag of paddy, and then the administration decided on a policy of giving the revenue collected from water charges back to the farmers to fund O&M activities.

The farmers revolted against the sudden raise in water charges. After a great deal of debate in the State Assembly, it was decided to reduce the water charges to Rs. 497 per hectare, three times the original figure. Table 5 shows the increase in water charges following reform.

Table 5

Water Charges per acre by Crop Type Before and After Revision

(Rs./ha)

Crop Type	Water Charges per Acre		Water Charges per Acre	
	Category I+		Category II++	
	Pre-revised*	Revised**	Pre-revised	Revised
First or single wet crop	60	200	40	100
Second and third wet crop	60	150	40	100
First irrigated dry crop	40	100	20	60
Second and third irrigated dry crop	40	100	20	60
Long duration crop	120	350	80	350
Aquaculture (per year)	0	500	0	500

+ Permanent irrigation sources; ++ Irrigation sources for at least 4 months.
* Effective from July 1, 1986; ** Effective from July 1, 1996.

Source: Government of Andhra Pradesh.

The AP government has established a policy whereby WUAs will collect the water charges in their areas of operation. Finally, the government will hand over collection responsibilities to the WUAs. This will necessarily be a slow process since WUAs need to build up the capability and confidence to assess, levy, and collect water charges. Water charges being collected are to be distributed to the various tiers of farmers' organizations in the ratios given in Table 6 below:

Table 6

Distribution of Water Charges Among the Different Tiers of Farmer Organizations and the Local Body (Gram Panchayat)

(in Rs.)

Level	Major Irrigation Schemes	Medium Irrigation Schemes	Minor Irrigation Schemes
Water Tax (Rs./acre)[16]	200	200	100
Irrigation Department	100	100	nil
Water User Association	50	60	90
Distributory Committee	20	n.a	n.a
Project Committee	20	30	n.a
Gram Panchayat	10	10	10

Minimum Rehabilitation

The minimum rehabilitation program was executed through the farmers' organizations. Farmers were exposed to a new working environment – they had to negotiate for machinery at cheaper rates, persuade the village to take up maintenance works, and maintain records to enable payment. A "mobilization advance" was made available for farmers to start the work. Subsequent payments were given on actual taking up of work. Maintenance works have been taken up by WUAs during the last three fiscal years, viz., 1998, 1999, and 2000.

So far, 49,000 works have been taken up at a cost of Rs. 4 billion ($94 million). Dramatic results were achieved during the first year. An additional area of 207,288 ha was opened to irrigation. Farmers in all commands have been able to get an additional production of 5-10 bags of paddy. Farmers now conduct participatory walkthroughs along with the irrigation agency, prioritize works, and take up works on their own.

In general, the cost of works done by WUAs is 20 percent lower than those carried out by contractors. Most works are done at the estimated rate. This is in sharp contrast to the ad hoc manner in which the irrigation agency paid for O&M. The biggest advantage of the reform program has been that maintenance works are taken up throughout the irrigated command area by all WUAs at the same time. The irrigation agency has maintained responsibility for works on the main canals and headworks, however, as project committees have not yet been formed. Box 8 lists some of the major achievements of WUAs under the reform program.

Box 8

Achievements of WUAs

- Bridging the gap to the extent of 207,000 ha.
- Earlier transplantation in 200,000 ha.
- Relief from submersion in delta areas due to improvement of drains.
- Timely receipt of water in tail ends. Some of the areas have reported receiving water for the first time in the last 15 years.
- Maintenance works and employment generation throughout the state.
- Reduction in the cost of works.
- Empowerment of farmers to manage irrigation systems.
- Closer supervision of works.
- Beginning of a sense of ownership by farmers.
- Choice in prioritizing works.
- Execution of works by themselves.
- Closer interaction of farmers with the Irrigation Department (ID).
- Reorientation of ID role.
- Unification of diverse groups around water.

Assessing the Reform Program

The reform program launched in Andhra Pradesh is both bold and innovative. It attempts to tackle the root cause of poor system performance and unsustainable operation by addressing key issues: institutional structure, incentives, accountability, transparency, and sustainability. The reform is pragmatic and establishes horizontal linkages between the WUAs and their counterparts in the irrigation agency. The reforms have been politically acceptable -- the recent reelection of the administration indicates continued support for the reforms.

The reforms are still in progress. In the future, linkages need to be established which will make WUAs independent of government support. At present, the irrigation agency has been maintained intact. Because of this, there has been less resistance from the irrigation agency. However, the agency must define a new role for itself and attempt a suitable institutional restructuring. Ultimately, WUAs will need to pay the salaries of the staff attached to them. This may be made possible when the Andhra Pradesh government continues further in the reform process.

Another major weak link has been the issue of accountability. This is in part due to the vaguely specified responsibilities of the "competent authority" over the WUAs under the Act. The WUAs need to be given more control of the field staff in the service relationship. The links with the Departments of Agriculture and Revenue continue to be weak. Mechanisms to ensure accountability must continue to be evolved. Also, the roles and responsibilities of all agencies concerned must be further defined. In the long term, there is a need to transfer all O&M functions to the Farmer Organizations. Box 9 gives an idea of the steps that AP has followed in going about the reform process.

The reforms have substantially shifted the responsibility for maintenance functions. However, new ground has to be broken in the field of water management by the users themselves. Farmers have the strongest incentive to provide the services effectively and efficiently. There is a further need to ensure that transactions by the leaders of the Farmer Organizations are made openly and in full view of the farmer-members.

Transparency is critical if the organizations are to progress further. To a large extent, transactions are captured in the accounting system. There is a need to put in place simple, standardized procedures for accounting and finance. The state

Box 9

The Eight-Step Process of PIM

Stage One:	Generating political support at the highest level -this is essential to provide the required motivation
Stage Two:	Creating a favorable environment
Stage Three:	Developing legal framework
Stage Four:	Formation of Farmer Organizations -WUAs and Distributory Committees
Stage Five:	Implementing the program with clarity of roles of the Farmer Organizations, the Irrigation Department, and other government agencies
Stage Six:	Capacity building of WUA, the Irrigation Department and other government agencies
Stage Seven:	Ensuring transparency, accountability in the working of FOs and social audit
Stage Eight:	Monitoring and mid-course evaluation

government has already developed a finance module in consultation with chartered accountants.

The AP Model will be complete when the WUAs are federated into project committees. Continued support and training are required. The awareness base must be expanded to include the farmers. While the awareness of the WUA presidents and managing committee members is quite good, the awareness of the farming committee needs to be made the focus of the training program. Farmer Organizations will become more responsive only when the farming community starts demanding more responsive and equitable service. This constant dialogue would also enable resolution of disputes. Rather than having the irrigation agency undertake capacity building measures alone, it will be far more effective to (i) foster interactions among FOs themselves, (ii) enlist the assistance of NGOs, and (iii) make use of local training institutes.

Replicability of the Andhra Pradesh PIM Model

Irrigation management reform in Andhra Pradesh represents a mixture of top-down and bottom-up processes, each reinforcing the other, organized in a relatively short period of time. Like all participatory programs, there are imperfections in the system that can be remedied over time. A massive reform program such as this one runs the risk of failure if it is not backed up by adequate finances and followed up with a systematic program over at least five years.

Most irrigation systems are deteriorated or fully dilapidated. In this condition, there is little incentive for farmers to take them over. The Participatory Irrigation Management (PIM) program must address the issue of rehabilitation of irrigation systems. In the case of AP, rehabilitation has been done by the Farmer Organizations themselves.

Employing the FOs to perform the minimum rehabilitation has two major advantages. First, FOs are exposed to the procedures of undertaking maintenance and rehabilitation works. Second, in carrying out the works, the Irrigation Department (ID) and FOs are in constant interaction. The new relationships and procedures that are formed early on give considerable time for adjustments and mid-course correction. The state government has issued more than 100 orders and memos to make this happen. Unbundling procedures that have evolved over the last two hundred years is no easy task to achieve!

The Andhra Pradesh government has sought to implement the PIM approach throughout the state quickly and uniformly. Box 10 shows the irrigation reform process in Andhra Pradesh. In a democratic society, bringing about change is sometimes an uphill task. The process of public consultation is a novel and important feature of the AP program. Follow-up, with assistance from the World Bank-assisted APERP project has also greatly contributed to the success of the program. Further refinements have been introduced in the state's *Irrigation Sector Policy Paper*[17] and later reaffirmed in its *Vision 2020*.[18]

Box 10

The Irrigation Reform Process in Andhra Pradesh

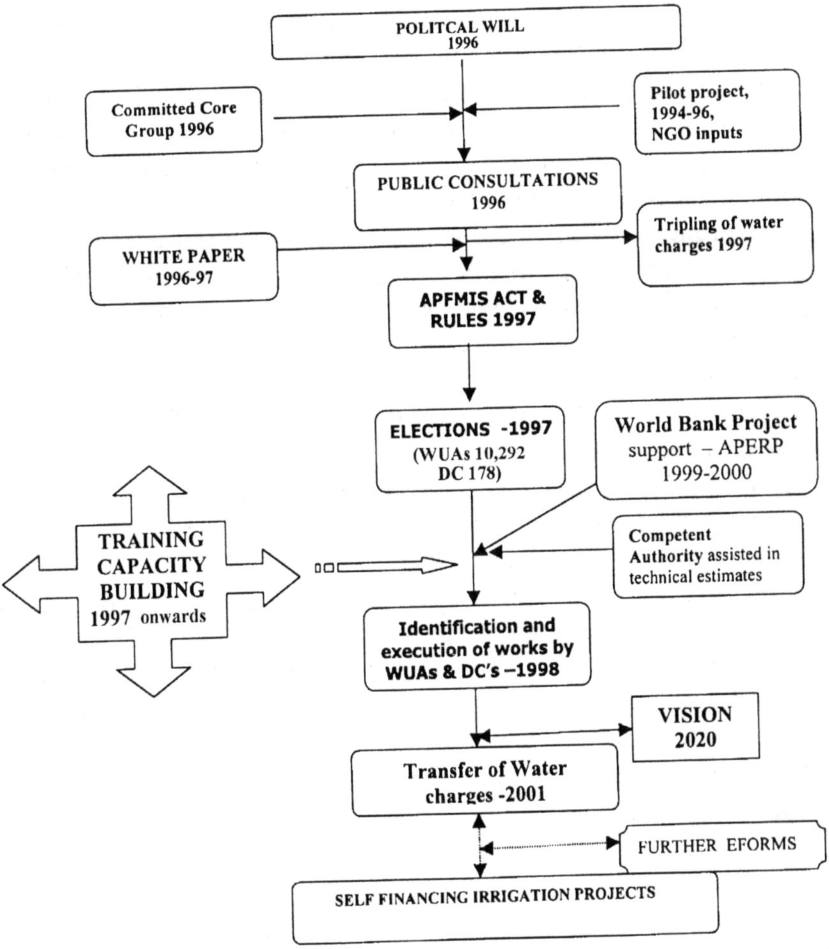

The reform process has been firmly underpinned by a clear legal framework and a committed program of minimum rehabilitation. Reform is not possible without the tremendous energy and support of all concerned: farmers, the Irrigation Department, NGO's, and political leaders. Farmer participation makes sense to both the government and the farmers. Linking water charges to O&M costs has also been vital to the program's acceptance by the farming community. Farmer satisfaction with the program has served to further reinforce the progress of reform. The state has thus become a facilitator in the process.

Summary and Conclusions

The irrigation reforms in Andhra Pradesh break new ground in India. Several states have already launched similar initiatives. The PIM program in Andhra

Pradesh is the first large-scale program of its kind in India and is comparable in scope to programs launched in Mexico and Turkey. The program is quite novel in its approach. An enabling legal framework was created, followed by creation of Farmer Organizations throughout the state. The strength of the program lies in its linkage with timely funding support by the World Bank through the Andhra Pradesh Economic Restructuring Program (APERP). The APERP has further refined the state's strategy by laying down guidelines through the irrigation sector policy. Farmers have a new role to play. The PIM program has enabled a forum for an active dialogue in the irrigation sector. In the context of upper riparian states completing the long pending irrigation projects in the state, the platform could well be used as an excellent tool for debate in water management, crop diversification and improving irrigation efficiency.

For the first time in AP, the focus has been on the linkage of water charges to the operation and maintenance of the system. An explicit linkage justifies increases in water charges.

The PIM program has also empowered farmers. Farmers' Organizations have dramatically changed the socio-economic landscape of irrigated areas. They have given impetus to several new initiatives in the education and health sectors.

The consistent high priority attached to the program by the government has strengthened the PIM program and fostered acceptance by the ID and FOs. The program, launched in 1997, has been able to steer through three intensive O&M cycles with a fair degree of success. The strength of the program lies in its attempts to ensure transparency and accountability and a systematic process of capacity building. The program thus paves the way for a new era in irrigation reform in India.

NOTES

1. *Vana Samrakshana Samithis* – joint forest management groups.
2. DWCRA -- thrift groups of women, for the development of women and children in rural areas; DWCRUA refers to urban areas.
3. CMEY -- Chief Ministers Employment for Youth
4. Prajala Vaddaku Paalana – a program where the administration visits villages to ascertain grievances and take follow-up action.
5. Gram Sabha -- a meeting of villagers under the leadership of the their president (Sarpanch), usually to discuss problems in the village and make decisions. This is part of Panchayati Raj Institutions (PRIs).
6. These tanks are in addition to the tanks vested under the control of the *gram panchayat*. Numbering 60,000, they are mainly used for drinking water, watering cattle and small-scale irrigation.

7. Irrigation in India is a state subject; hence, each state government has a varying departmental structure depending on the extent of irrigated area.
8. The Command Area Development Program was formulated by the Government of India in the 1970s to bridge the gap between the irrigation potential created and actually utilized. The program gave a thrust to on-farm development activities. Andhra Pradesh had major projects such as the Sriramsagar Project, Nagarjunasagar Project, Tungabhadra Project, and the Vamsadhara project, in addition to ten medium irrigation projects under the CADA program.
9. Irrigation schemes are usually classified on the basis of the investments made on them. The Andhra Pradesh Farmers' Management of Irrigation Systems (APFMIS) Act, 1997 has for the first time classified the schemes on the basis of irrigated area.
10. The low collection of water charges is partly due to the misclassification of lands into dry and wet during the localization process, and the low priority accorded to collections by the government after 1984-86, when the land tax and the village officer system were abolished. However, owing to a Supreme Court judgment, the village officer system was reintroduced as the village administrative officer. Nevertheless, the village records substantially degenerated and most land records are outdated.
11. A pilot project was set up subsequent to the visit of a team of officers to Mexico, organized by the Government of India and EDI, World Bank in 1991. Two groups of distributaries in the Sriramsagar project were identified in a total extent of 65,000 ha. Two NGOs, viz., IRDAS and SONAR, were given the task of constituting WUAs on a pilot basis.
12. Irrigation is a state subject in India. The Ministry of Water Resources basically deals with international boundary disputes and inter-state allocation of water. It also funds state governments under the Command Area Development Program to develop the irrigation potential created but not realized. This is done on 50:50 sharing basis.
13. Prior to the APFMIS Act 1997, most provisions relating to violations in water use, giving permissions for the use of water was done under the Boards Standing Orders (BSOs) in AP by the Revenue Department. The BSO provided regulatory and punitive clauses and was administered by the Revenue Department.
14. Some of the important rules under the APFMIS Act are:
    G.O. Ms. 45 dated April 30, 1997 relating to constitution of farmers organizations.
    G.O. Ms. 541 dated December 12, 1997 relating to the functions of the farmers' organizations.
    The Election Tribunal Rules. The Amendments to the APFMIS Act.
15. The command area of a project is the area irrigated by the project. It is popularly referred to as the Ayacut in south Indian states.
16. Water tax in the state is levied under the Andhra Pradesh Water Tax Act. The tax is collected by the Revenue Department which is under the control of the District Collector. The water tax is levied on the basis of a crop/acre/season.
17. The Irrigation Policy articulates the state's vision for the irrigation sector and sets out the reform agenda.
18. Published in 1999, the Vision document reflects the vision of the state in 14 key areas identified as growth engines, and is essential to the government's planning process.

REFERENCES

Government of Andhra Pradesh, 1997. *Statistical Abstract 1994.* Hyderabad: Directorate of Economics and Statistics.

Government of Andhra Pradesh, 1998. *Andhra Pradesh Economic Restructuring Project Irrigation Component, Project Implementation Plan.* Hyderabad: I&CADD.

Government of Andhra Pradesh, 1999. *Vision 2020.* Hyderabad.

Jaya Raj, M., 1999. *A Study on the Impact of the Andhra Pradesh Farmers' Management of Irrigation Systems Act in Andhra Pradesh.* Paper prepared for the Fourth National Conference on Participatory Irrigation Management, National Institute of Rural Development, Hyderabad.

Oblitas, K and Peter, J. Raymond, 1999. *From Political Leadership to Systemic Incentives for Irrigation Sector Reform: The Case of Andhra Pradesh, India.* Paper presented at the World Bank Rural Week, March 1999, Lansdowne, Virginia.

Oblitas, K. and Peter, J. Raymond, 1999. *Transferring Irrigation Management to Farmers in Andhra Pradesh, India.* World Bank Technical Paper No. 449. Washington, D.C.: World Bank.

Perera, J., 1998. *Farmer Organization and Social Change in Andhra Pradesh: A Participatory Rapid Appraisal Evaluation.* Mimeograph.

Peter, J. Raymond, 1998. *Management of Irrigation Systems by Farmers in Andhra Pradesh, India: The Process.* Paper presented at the Fourth International Seminar on Participatory Irrigation Management, Bali, Indonesia.

Peter, J. Raymond, 1998. *Irrigation Management Transfer and the Andhra Pradesh Farmers' Management of Irrigations Systems Act of 1997.* Paper presented at the First National Participatory Management Seminar, IASS, Rampur, Chitwan, Nepal.

Peter, J. Raymond, 1999. *Water User Associations in Andhra Pradesh.* Paper presented at the Fourth National Conference on Participatory Irrigation Management, National Institute of Rural Development, Hyderabad: ICADD.

Salman, M.A. Salman, 1997. *The Legal Framework for Water User Associations: A Comparative Study.* World Bank Technical Paper No. 360. Washington, D.C.: World Bank.

Subramaniam, Ashok, N. Vijay Jagannathan and Ruth Meinzen-Dick, Eds., 1997. *User Organizations for Sustainable Water Services.* World Bank Technical Paper No. 354. Washington, D.C.: World Bank.

Svendsen, Mark and Hupert, Walter. *Incentive Creation for Irrigation Systems Maintenance and Water Delivery: The Case of Recent Reforms in Andhra Pradesh.* Maintain Case Study No. 5, GTZ.

World Bank, 1991. *India -- Irrigation Sector Review.* Washington, D.C.; Agriculture and Water Division, South Asia Department 2.

World Bank, 1997. *India: Andhra Pradesh; Agenda for Economic Reforms.* World Bank, South Asia Country Department 2.

World Bank, 1998. *Andhra Pradesh Economic Restructuring Project, Project Appraisal Document.* Washington, D.C.; South Asia Poverty Reduction and Economic Management Unit.

4

Participatory Irrigation Management in Andhra Pradesh: The Way Forward[1]

K.V. RAJU

The last five years have seen Andhra Pradesh make rapid strides in its economic reforms. The irrigation sector has been positively influenced by institutional reforms largely focussing on: (i) introduction of a suitable policy and legal framework; (ii) formation of water users' associations across (WUAs) all types of irrigation systems in the state; (iii) implementation of large-scale training programmes for farmers and staff of the irrigation department; and (iv) bringing in significant financial reforms to influence quality performance of users' organisations. Today, the state is leading in irrigation reforms in India. This paper, based on field visits, looks at what has made the Andhra Pradesh experiment work, by examining: (a) the promise made through (i) the Andhra Pradesh Farmers' Management of Irrigation Systems Act, 1997 (APFMIS Act); (ii) document on 'Vision 2000: *Swarna* (Golden) Andhra Pradesh'; and (iii) government orders issued over time to encourage formation and functioning of water users' associations; (b) the practice; and (c) key lessons.

I
The Promise

Background

Irrigated Agriculture in Andhra Pradesh

Andhra Pradesh, with a population of 72.7 million (2001 census) and a geographical area of 27.68 mha, is the fifth largest state in the country. About 73 percent of the population lives in rural areas. The labour force constitutes about 45 percent of the total population and about 65 percent of it are engaged in agriculture. Of the state's geographical area, 47 percent is under cultivation and the net area sown is about 11.04 mha (4.88 mha irrigated). The dominant pattern of land ownership in Andhra Pradesh is small private farms with an average of 1.56 ha per holding. Irrigated holdings have an average size of 0.88 ha.

The state's three major rivers, namely, Godavari, Krishna and Pennar drain 70 percent of its land area (see Annex Table A-1). The water potential of Andhra Pradesh is estimated to be 7.78 million hectare meter (ham) or 2,746 TMC (thousand million cubic feet). The major rivers are seasonal with more than 90 percent of the total flows occurring between June and December depending on the rainfall, which greatly varies from year to year. The ultimate irrigation potential from all sources is estimated to be 9.50 mha. It includes 7.30 mha from surface water and 2.20 mha from groundwater. The performance of the irrigation sector in the state is poor, particularly in the case of major and medium irrigation projects. Inadequate budget allocations for the operation and maintenance of irrigation systems have led to poor maintenance and unsatisfactory service over the years. This has resulted in poor irrigation service, low yields and low farm incomes, leading to farmer dissatisfaction and political pressures affecting the allocations for operations and maintenance (O&M) in the state irrigation budget. This vicious circle is observed in most of the Indian states (Gulati, Meinzen-Dick, and Raju, 1999).

The Irrigation and Command Area Development Department (I&CADD) provides water and services to: (a) farmers for irrigation purposes, (b) municipalities and villages for human and domestic uses, (c) state electricity board for power generation, and (d) industries. Water charges for irrigation levied on a per hectare basis, were increased by more than three times from 1997 (see Finance section in Part II). Current O&M budget allocations for the various irrigation schemes are made on the basis of a uniform flat rate of Rs. 100 per hectare, and about Rs. 47 per hectare for special repairs. This figure is approximately 50 percent of the Tenth Finance Commission recommendation and about 30 percent of the actual needs assessed for the purpose of the irrigation component of the Andhra Pradesh Economic Restructuring Project (APERP). Therefore, in effect, the current O&M budget allocations neither cover the full O&M cost nor the full establishment cost estimated at about Rs. 200 per hectare. As a result, no funds are available for the actual O&M work requirements. The deficits in establishment financing are covered to a small extent under special repairs by budget allocations for planned works and O&M works.

Minor Irrigation Tanks

The state has 12,294 tanks under the irrigation and CAD department covering 1.25 mha in 27,379 villages. The overall performance of most minor irrigation tanks is more sensitive to droughts and cyclones due to small size. Their deficient maintenance under the I&CADD has resulted in losses and decline of irrigated areas. Till May 1999, tank-WUAs (8180) constituted 81 percent of the total WUAs in the state. In Rayalaseema region, some 23 tank-WUAs have not used tank water for irrigation, yet they are listed; farmers use their own well water, which is recharged from tank storage. For example, in Chandurti mandal (some 20 kms from Sircilla), few tanks have closed (6-7 years back) their canals to recharge wells during two crop seasons. These WUAs just maintain tanks and not distribution network.

Percolation Tanks

In the Chittoor division area, some 80 percent of the 60 tanks have been converted into percolation tanks over the last 10-15 years, due to fast depletion of groundwater table and siltation of the tank bed. Gradually farmers have shifted to groundwater usage from the earlier practice of tank irrigation. During last 8-10 years, tank sluices have been completely closed to influence continuous recharge of groundwater. Farmers argued that direct tank water usage would empty the storage in one to three months; whereas groundwater usage would help irrigate up to 10 months. This is also because most of these tank command areas are under sugarcane crop.

The major problem with most of these tanks is siltation of the tank bed and thereby a loss of more than half of their storage capacity. The ID staff strongly feels that desiltation is very costly and it is better to opt for construction of new tanks. The percolation tanks are located in the north-western part of the Chittoor revenue division, and spread over 2-3 mandals (a smaller revenue zone below the district/province). But towards the extreme south, east and north-eastern part of the district, tank are still sluice and canal operated surface irrigation systems. Many of these percolation tank command area farmers are debating whether to pay the water fee or not, owing to complete dependence on groundwater.

Policy Framework

The state irrigation policy has kept user involvement in irrigation management as the central theme; all concerted actions revolve round this. The need for reform became essential owing to the following reasons:

- Irrigation sector's low performance despite the massive investment. Traditionally, this sector has been the largest user of the plan funds and even in the Eighth Plan (1992-93 to 1996-97), expenditure amounted to Rs. 25,000 million, or 24 percent of the plan expenditure;
- Infrastructure is in disrepair and irrigated area is declining. From 1991-92 to 1993-94, the gross irrigated area dropped from 4.3 mha to 3.9 mha. In 1998, out of 4.8 mha of net irrigated area created, only 2.8 mha is actually irrigated.

- Low agricultural productivity. Growth in productivity has declined in recent years to less than 2 percent per annum. A major factor has been weak performance of irrigated agriculture. Rice yields an average of only 2.6 tonnes per hectare (see Annex Table A-3).
- Cumulative impact of inadequate maintenance of infrastructure. For instance, expenditure on O&M in 1995-96 was only Rs. 99 per hectare, as against the Tenth Finance Commission (1997) recommendation of Rs. 300 per hectare for major and medium irrigation projects. Further, because of inflation of wage bills, over 75 percent of O&M expenditure went to wages, leaving a negligible amount for actual maintenance works. Low maintenance has been compounded by a purely governmental approach to the sector despite the limited capacity for the government to intervene, especially at the lower levels of the systems, and by an extremely low cost recovery. With the three-fold increase in water charges made effective from the 1996-97 rabi season, revenues, for the first time in many years, exceeded O&M expenditure. As O&M expenditure has remained far too low, revenues remain inadequate to cover their entire needs. Further, to make up for the cumulative neglect, significant additional expenditure is required to rehabilitate the system.
- Rehabilitating and sustaining irrigation and enhancing agricultural productivity are of paramount importance. About 40 percent of the state's gross cropped area is irrigated. Irrigation's contribution to state agricultural production is about 60 percent.

Policy Initiatives

New initiatives in policy reform began with a diagnosis of the situation and subsequent issuance of a white paper on irrigation outlining the performance of the irrigation sector over the years. This was debated in the legislative assembly and the following major actions were taken:

- Three-fold increase in water charges from the 1996-97 rabi season.
- Passing of the APFMIS Act in 1997.
- Creation of WUAs across the state.
- Commencement of a massive campaign to make WUAs functional.
- Constitution of the water charges review committee in December 1997.

Policy Objectives and Strategy

The idea of forming WUAs had been borrowed to a large extent from Mexico and Turkey. Both countries are international 'success stories' in irrigation reform. The major increase in water charges, traditionally a politically sensitive aspect, required extensive public consultation and agreement by all parties. These steps are nevertheless acknowledged as only a start of the difficult actions that must follow. These must be seen in the broader context of Andhra Pradesh's long-term vision for the water resources sector within which irrigation, though important, is only a component. The core elements of the long-term vision encompass two areas: reforms in ownership, financing and management of the state's irrigation systems; and progressive development of comprehensive multi-sectoral water resources management.

(a) *Democratic Decentralisation, Farmer Management, and Financial Autonomy* were considered vital to change the past 'vicious circle'[2] of influences to a 'virtuous circle' where these constraints were tackled systematically to generate mutually reinforcing improvement with in-built incentives to achieve this. The objective was to build rapidly from the WUAs at a minor level to federated WUAs at a distributory, and then project (scheme) level. The (ultimate) objective was to have self-financing and autonomous irrigation schemes managed by WUAs. A farmers' apex committee at the state level would provide a forum for state-wide decision making. The government's role would progressively reduce as provider of technical assistance to manage the headwork of larger systems.

This process was accompanied by institutional and financial reforms and capacity-building to create the respective autonomous entities. The short-term objective was that WUAs and the I&CADD should become financially autonomous for O&M and revenue and expenditure. They would generate their own revenue from water charges and finance O&M, thus giving them financial independence and being cost and quality effective, which was critically lacking in the past. WUA members have to contribute to the interest-earning reserve fund accounts in order to progressively build up the financial capacity to reinvest or undertake improvements. New investment will continue to be financed by the government, but with users or prospective users contributing through appropriate cost-sharing arrangements.

For I&CADD there was radical change in its role as the above progression took place. One of the major transformations was to switch staff orientation to a service provider role to WUAs. Rehabilitation of systems and upgrading maintenance were the responsibilities of the newly constituted Water Charges Review Committee (WCRC). The key requirement was to ensure the necessary direct financial link between revenue and expenditure and the availability of funds for this purpose.

(b) *Sustainable Water Resources Management.* Major principles for development and management of water resources are: (i) comprehensive water resources management; (ii) environmental management; (iii) water as an economic good; and (iv) technology development. To achieve these objectives several steps were suggested.[3]

Strategy Implementation

The immediate priority was to provide a follow-up to the change process already initiated in 1997 by the APFMIS Act (for key provisions, see Annexure 1), the three-fold increase in water charges, and the state-wide elections for WUAs. This was followed in November 1997 by state-wide elections for the second-tier of WUA management structures. If they were to succeed and generate further reforms, these actions required urgent follow-up like restoring and improving the productivity of the existing irrigation systems, assuring their sustainable management by transforming management to farmers through the new WUAs, and putting the sector on a sound financial footing. Concerted actions in the following areas are being launched, such as: (a) creation of a farmer-government partnership, (b) consolidation of irrigation management transfer to farmers, (c) maintenance and rehabilitation, (d) agricultural extension, (e) cost recovery and financial sustainability, (f) expenditure prioritisation, and (g) institutional reforms and capacity building.

Long-term Vision

Andhra Pradesh is the first state in India to put forth a long-term vision through the document 'Vision 2020: Swarna Andhra Pradesh'. Both the irrigation policy and the legal framework were in tune with the long-term vision evolved by the State. The vision document (Government of Andhra Pradesh, 1999) has set ambitious goals and it aims to achieve them through: (a) building capabilities, (b) focusing on high potential sectors of the economy, and (c) transforming governance. As a part of building capabilities, the Vision document highlights involving people to manage services as a key area. It states that an effective way to provide quality and responsive services is to decentralise them and ensure that the people have a role to manage them. Andhra Pradesh's own experiment in water management demonstrates the power of people's participation. Also, the state's own legislation (like APFMIS Act, 1997) in relation to local bodies provides some basis for people's participation.

Investment in the irrigation infrastructure has been a priority. Successive five-year plans have emphasised the need for greater investment in the irrigation sector. Plan outlays on irrigation have increased from Rs. 620 million in the Second Plan to Rs. 31,860 million in the Eighth Plan. This has expanded the net irrigated area from 1,50,000 ha to around 5.9 mha and greatly increased the productivity of land, leading to a substantial impact on agricultural and industrial growth, incomes, and employment. However, to achieve the long-term vision, more was required, such as: (a) realising the maximum irrigation potential; (b) improving the efficiency of the existing irrigation network; and (c) managing water resources better through stakeholder participation. In the absence of funds from the Andhra Pradesh Economic Restructuring Project (APERP) supported by the World Bank, Participatory Irrigation Management (PIM) activities could have had little impact. Good blending of the APFMIS Act and timely assistance from APERP has made a distinct difference in the state. The APERP assisted

WUAs through: (a) minimum rehabilitation of the canal network, (b) regular operation and maintenance, (c) scheme improvement and farmers' turnover (SIFT), (d) agricultural intervention programme (AIP), (e) institutional support for project monitoring unit, and (f) human resources development.

Implementation

Overall, the PIM process followed in Andhra Pradesh was as shown in the Figure 1. In the process of implementation and learning, 110 government orders were

Figure 1

Participatory Irrigation Management Process in Andhra Pradesh

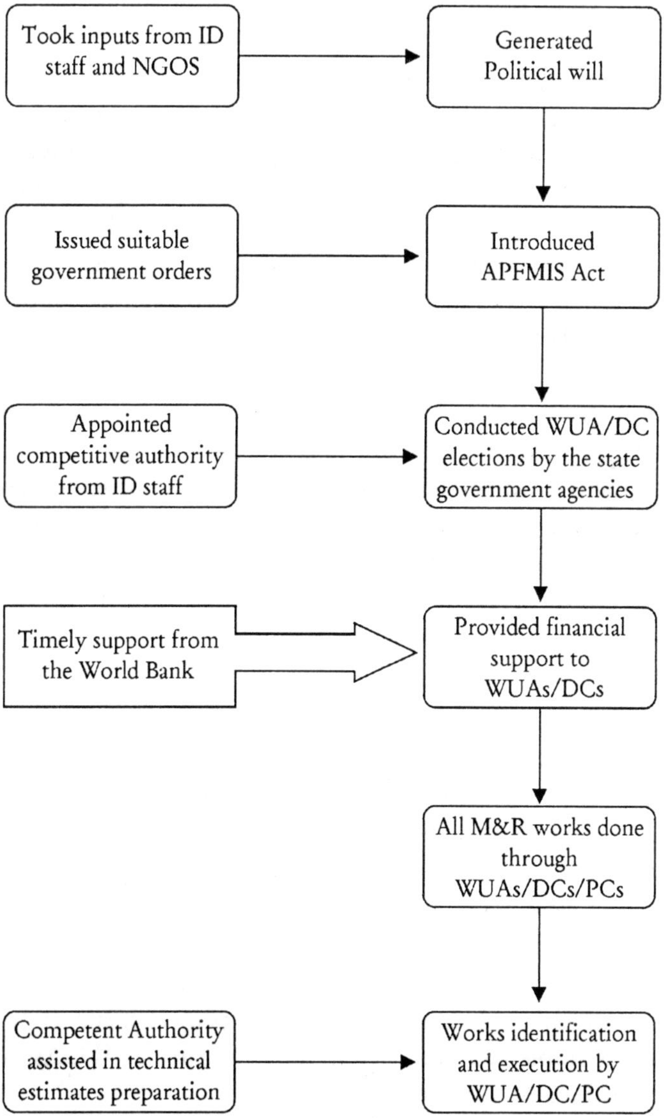

issued by the I&CADD till January 2001, to amend and clarify the APFMIS Act, 1997 and to facilitate better functioning of the WUAs and DCs. To help in physical and financial plan estimates and approvals, junior engineer in case of WUA and executive engineer in case of DC was appointed as Competent Authority (CA). The main focus was construction-oriented – system rehabilitation, repairs, and adding essential structures. Many canals are 10 to 20 years old and have never been maintained. Stress is on cleaning shrubs, desiltation, and strengthening bunds.

Training

The state had launched a massive training programme on WUAs. The programme aimed at covering the entire state by having local training camps for a fixed period (ranging from 2-3 days per batch). A trainer's handbook containing irrigation engineering, agriculture, and revenue subjects was prepared by August 1999 for the ready reference to the trainers.[4] A set of 15 members from each district comprising I&CADD, agriculture and revenue departments of the state were selected for undergoing training as trainers of the WUAs at district level. Outside agencies had collaborated to prepare the training material in local (Telugu) language. The reading material was widely disseminated to participants. As of November 1999, all WUA presidents and TC members had attended the training programme. WALAMTARI (Water and Land Management Training and Research Institute, located in Hyderabad) had become a live partner in the whole training process and dissemination of publications. The institute had shifted their focus from earlier engineering-oriented programmes to WUA training programmes. To bring in a practical perspective to classrooms, WALAMTARI had trained middle and junior level officials of the agriculture, revenue and irrigation department, as trainers under the Training of the Trainers.

Monitoring

From mid-1998, every district had a PIM coordinator, who was primarily a middle level staff of I&CADD and transferred to this position. These coordinators were expected to coordinate all PIM/WUA/DC training programmes, activities, and meetings with district collectors. Currently, they were not dynamic owing to inadequate training and supportive staff. Some 25 PIM coordinators were positioned at district level, mainly owing to problems in career promotion in the I&CADD. The PIM coordinator was responsible for computerisation of WUA details, assigning nine digit code numbers for each WUA (as of now only for major and medium projects), supervising training, planning and its implementation, communicating to all WUAs, and sending letters and publishing in local newspapers.

All ID people have to report to the district coordinator on a regular basis. Because of the weekly (every Thursday) teleconference meetings the district collector along with ID senior staff have to discuss ID, WUA activities with the chief minister. At these teleconferences the Chief Minister receives all newspaper reports (from all district editions of some newspapers) for discussion. This had put all Collectors, in turn all staff of ID on tenterhooks, to regularly keep track of newspaper reports, and answer them.

In some WUAs, the president and/or the executive committee has mismanaged the funds. Though the APFMIS Act provides sufficient insulation, the Irrigation Department has not taken sufficient preventive measures. The Upper Manor Project is a medium irrigation project with command of 13,000 acres. It has seven WUAs, and no DCs or PCs as per the Act. These WUAs got Rs. 50 per acre in the first year and Rs. 100 per acre during the second year. So far, three misappropriation cases have been filed against the presidents for utilising the funds for non-specific purposes.

II
The Practice

In this part the study team had looked at the actual practices of WUAs in the three regions of Telangana, Coastal Andhra, and Rayalaseema[5]. In each region, major, medium and minor irrigation systems were visited and observed on how the WUA functions, its advantages to members and the irrigation department. Also the financial support received and viability of WUAs in the present form have been explored. Many WUAs indicated how the current experiment is playing a pivotal role in the command area development. Some WUAs also expressed their demands and views on the whole exercise initiated by the state government. The number of WUAs visited (given below) varies across the region and the type of irrigation system.

Region	*Number of WUAs visited*			
	Major	*Medium*	*Minor*	*Total*
Telangana	4	1	2	7
Coastal Andhra	7	2	3	12
Rayalaseema	3	2	14	19
Total	14	5	19	38

WUA Formation and Structure

The main role of the government was to promote farmer participation. The structure of users' organisation was given from the top. Each WUA was formed, based on the hydraulic boundary, for 500-8,000 acres, and with four to six Territorial Committee (TC) members in each. Each TC will have an area in the range of 500-1,300 acres. The elected presidents of WUAs will become members of the Distributory Committee (DC) at the distributory level. WUAs and DCs were formed through elections in 1997. Till January 2001, Project Committees (PCs) were not formed. Elections were held for each of these tiers of the organisation as per the government order and all processes were formally handled by the government agencies. Till May 1999, the total number of WUAs formed was 10,292; it included, 1,699 major, 413 medium, and 8,180 minor projects (see Table 1. Also see Annex Table A-4 and A-5 for district-wise details). Elections were conducted for 9,797 WUAs (in minor projects, only for 7,749 WUAs). In the case of 495 WUAs, elections were not held for various reasons. The recall clause in the Act is a large booster for members; so far, seven presidents have been recalled due to misappropriation of funds. Wide publicity of this information in the farmers' newsletters has sent warning signals to others.

Table 1

Distribution of WUAs by Type of Irrigation Systems and Method of Elections

Type of Irrigation	*Unanimously Elected*	*Contested*	*Elections Not Held*	*Total*
Major	649	935	89	1673
Medium	118	179	07	304
Minor	4714	3110	491	8315
Total	5481	4224	587	10,292

Source: Irrigation and Command Area Development Department, 1999.

Since the formation of WUAs, more emphasis has been laid on meeting the president, conducting training courses, short workshops, and keeping him/her as a contact person of the WUA. This has helped the president understand the internal process. In case of many WUAs, TCs and ordinary members have little access to the information flowed from the higher ups and to the CA. With increase in size of funds and powers, WUAs have become more influential. For water allocation, farmers go to the WUA president and not to irrigation department staff. In some places, there are efficient WUAs mainly owing to good senior officials and their efforts (e.g., in SRSP project, Karimnagar, Sri Kakulam, and West Godavari). Inefficient WUAs are mainly because of political interference (e.g., in Cuddapah, Chittoor, and Mehboobnagar). WUAs are categorised (informally by officials) based on the complaints and progress heard from the presidents of WUAs. However, WUAs and DCs face opposition from senior and junior staff of the irrigation department since they have lost money and power.

Court Case of Lascars

A lascar is primarily a system level canal operator and water regulator for approximately 800 ha. For example, in Srikakulam district, there were 574 WUAs and 180 lascars. The services of lascars, originally employed by the state government, have now been transferred to WUAs. They are under the functional control of WUAs and receive salaries from the government. Some 3,500 lascars in the state have opposed this government order and a legal battle is still on. There was also opposition from the engineers' association regarding the transfer of roles and responsibilities to WUAs. The government gave them three choices: (a) a voluntary retirement scheme; (b) work according to government instructions; and (c) punishment of erring staff by government. For now, the staff has agreed to work according to government instructions.

Trendsetters

Srikakulam district claims to be a frontrunner in the WUA movement in the state. WUAs in this district maintain schedules and conduct meetings on time; the irrigation department disburses funds regularly to WUAs, and responds to government orders regularly. Newsletters are published in Telugu regularly. The Superintendent Engineer (S.E.) and his staff take keen interest in all the activities and works. But the middle and lower level officials have complained that their workload has increased. Krishna district is a trendsetter in the delta region, leading both in area coverage and expenditure incurred by WUAs and DCs.

The WUA movement in Andhra Pradesh has gained momentum. With all visible physical benefits, the spirits of the farmers were high as they felt that the system belonged to them. They believed that if users take active part in system maintenance, it brings them better flows and improved yields. Developing the sense of ownership was getting established in many places. WUA members unhesitatingly and with pride said "now water is ours; we have a right on that". In the reform process farmers made no (or in some cases, little) monetary contribution; they were involved in decision-making, work execution; hence, the quality of work was better and farmers felt the structures belong to them. A septuagenarian (a freedom fighter and president of WUA-47 in Manthani village) said "earlier government officials were saying and holding important roles; now that is lost. Users have got a major say in all irrigation related aspects; this is a good trend in the state".

WUA Meetings

The APFMIS act clearly delineates roles and responsibilities at each tier of the organisation. User organisations take pride in saying, "This is the first time, in the state's history, that a chief minister is directly writing to organisations at the village level. So, we feel, we can communicate directly with the chief minister and that is a large achievement for us". Over the last two years (1998-2000), general body meetings have been initiated through CAs. In all, 20 to 50 percent of the members attend these meetings; in a few cases, this has exceeded 60 percent. Otherwise, the WUA/DC president and TC members call small group meetings (6-20 farmers) on their own, as and when required; more specifically, to endorse works identification and execution, and to discuss funds received and expenditure incurred. Informal meetings are held to discuss if voluntary work is required to clean canals, and water regulation across the reaches. For calling a meeting, drum beating is common across the state. For small group meetings, the WUA/DC president informs through oral communication. Unfortunately, WUAs and DCs have not focused on water acquisition, equity in distribution, and efficiency in water use. Neither CAs had stressed these issues nor WUAs have discussed them seriously in their meetings.

WUAs v/s Village Panchayat

There is a growing dissent in some village panchayats regarding the WUA's control on resources, which until now were controlled by the village panchayat. Village panchayat feel that there are too many organisations. The village panchayat in some places, felt that it could have handled water management since farmers belong to the same village. But in many places, with a separate organisation like WUA, farmers feel that they have done better work and have greater freedom to do what is actually required. Water fee can be used for WUA works and functions. In some WUAs (including tank-WUAs) which are close to urban areas and headed by an absentee-landlord, there is non-cooperation from member-farmers. In the case of urban-proximity tank-WUAs, members have secondary interest in tank management.

Physical Works

In Andhra Pradesh, as elsewhere in India, the emphasis till 1970s was on creating infrastructure. From then onwards the state faced a financial crunch leading to lack of management and maintenance. System delivery was weak, and system efficiency was down to 40 percent in major projects. Hence, during May 1998, many works were taken up. Most being repairs to canals and structures and adding some new structures below distributory level. Over the last three years number of works carried out are shown below:

Number of Works Carried Out by Water Users' Associations

Year	*Number of Works*	*Expenditure in Rs. Million*
1998-99	21,406	1,18,562
1999-2000	17,185	16,957
2000-2001	9,289	5,155

Source: Saguneeti Sangam, Volume 2 (12), p.26, May 2001.

By 1999 a new trend had started, i.e., contract being given by local officials (who are CAs) to WUA presidents. Some presidents have taken the work on behalf of WUA and then sub-contracted it to their territorial committee (TC) members, who are basically sub-committee members. In the register of administrative approvals, only TC names figure (e.g., medium irrigation project-UMD, WUA-1-4). Most of these presidents are contractors by profession, and the WUA work came as an opportunity to get involved and gain confidence at the local level to execute these works. Around 30-40 percent of the WUAs presidents in the state have been involved in this kind of processes.

The last two years (1998-2000) have seen the following practice. The WUA president, one/two TC members, and some farmers, through a walk-through survey, identify the required works. These works were then discussed in an executive committee meeting and then passed on to the Competent Authority (CA) for preparation of technical estimates. An agreement is signed between the CA and the WUA president. Funds are released on the basis of Rs. 50 per acre during the first year and Rs. 100 per acre during the second year. For tank-WUAs during the second year it is Rs. 90 per acre (Rs. 10 per acre is given to the village panchayat). Forty percent is paid in advance and the remaining on completion of the works. But in majority of the cases, even after completion of the works, the balance amount was not paid. Whatever funds are released go directly to the WUA account. The ID staff has no role in handling the funds. Proximity to the process of technical and financial estimates, funds disbursements, works execution by local farmers, have all brought more transparency in WUA dealings. Contractors are almost out of the picture; saving on this alone is about 30 percent.

The government has given clear instructions that there will be no new structures, or alteration of original designs. So, major emphasis is on restoring the original designs. Repairs and strengthening of structures, addition of missing structures, desilting canals, installation of water regulation structures wherever damaged or not installed earlier, strengthening canal bunds and construction of drop-structures are some of the major works that WUAs have carried out. In the Godavari and Krishna delta areas, in addition to these works, emphasis was on drainage works. In the tank systems, in addition to the above works, the major focus was on repairing and strengthening of surplus weir and apron, raising and strengthening tank bunds to original designs, and improving sluice structures.

In the process of users identifying and executing the works, lot of attention has been laid on users' needs like cattle pathways, washing steps for women in canal systems, removal of encroachments and clearing shrubs and silt for smooth flow of water till the tail-reach. In most places farmers have contributed their labour for additional working days, as part of their earlier practice of *shramadan* (voluntary work).[6]

Problems of Works Execution

One of the problems in work execution is the rate approved by the government for various works in 1999-2000. The rate for Proclainer machine is Rs. 16.24 per cubic metre and from this 15 percent is deducted as farmer's contribution. This becomes Rs. 13.50 per cubic metre, for which no machine owner is willing to work. Also farmers are not willing to pay this 15 percent. With 40 percent of grants (disbursed as the first installment), work was carried out to the extent possible. Hence, many WUAs and DCs are not taking up any work. The new work-masters like the WUA and DC presidents got trapped in difficulties owing to poor planning of both materials and manpower. Meanwhile, the CA is unable to attend to all works and hence, work is affected in some places.

Water Management

Across the state, unauthorised outlets have retained their status quo. However, WUAs have ensured that water traverses to tail-reaches by canal cleaning, or/and by rotation of minors, as in most problematic areas. Across the state, WUAs have *neeradi* (a water distributor, hired locally on temporary basis by the irrigation department) for water distribution. Earlier the *neeradi* was accountable to an informal group/association of farmers; now, he is answerable to the WUA, as a formal set-up. Some WUAs have hired *neeradis* on their own for the irrigation period and pay them salaries.

Water regulation in the Tungabadra High Level Canal, Stage I (in Ananthpur district) had undergone changes due to the WUAs active role. Owing to the WUAs continuous demand for better water regulation across the reaches, the Irrigation Department devised an alternative system. Accordingly, from 1998, distributories are closed for one day a week (i.e., on Monday). Sluices below the distributory are operated on a rotation basis. During 1999, with the consent of the WUAs, distributories were closed once a week, and thus the approximately 6 TMC water saved was supplied for the second crop in the rabi season to 52,631 ha. For the first time in the project history, such a large area was irrigated without any additional supplies. Field officials felt that gradually this may increase to 72,874 ha.

However, some constraints had remained as reported by the farmers during our interviews:

- Several WUAs and DCs were demanding grants since in some places political leaders had made false promises about liberal grants for office building construction.
- Power supply in rural areas is limited to a total of 5-9 hours a day, supplied in 3-4 slots. This affects commercial crop cultivation and increases the dependency on canal water.[7]

- In many places, the water supply had not reached the designed discharge levels at regulation points, even after some structural repairs and desilting by WUAs.
- WUA and DC presidents were not very clear about their future activities and the potential of these organisations to carry out allied activities.
- Government grants have remained the sole source of funding for most of the WUAs with little (or no) efforts having been made to mobilise own resources.
- Most of the WUAs have not made any serious attempts to mobilise contributions from the farmers to meet the prescribed 15 percent of the estimated costs of works. Hence, physical works have been carried out from the government allotted funds (i.e., 85 percent).

Rayalaseema Region: Acquire and Regulate

In this region surface irrigation sources are limited and groundwater has to be extracted from the deep bottoms. Increasingly, this region has shifted towards mango orchards, citrus fruits and other long gestation and long-term yielding tree products. Farmers prefer this option, because of better insulation from vagaries of rainfall and labour availability along with promised returns over the years. More farmers have opted for low water consumption crops. But, within the authorised command areas, demand for canal irrigation is high owing to paddy cultivation. Most of the tanks in Rayalaseema region, are 100-450 years old. Owing to good physical features, most of them are free from major maintenance problems; bunds are leak-proof, even after 200 years. But, they are constrained by siltation in the tank-bed and reduction in the in-flows.

Coastal Andhra Region: Delta and Drainage

This region consists of two delta areas -- Krishna and Godavari; these irrigation systems are centuries old (with some additions and modifications in recent decades) and well tuned to paddy cultivation. The region has been named as the 'rice-bowl' of India. Over the years, urban-pull and rural-push factors have led to more large landholders shifting their base to urban areas. To add to this, the next generation showed little interest in farming activity. This trend has pushed many farmlands (more than 75 percent) to the hands of sharecroppers. Production sharing varies from 25 to 50 percent, depending on soil quality and water availability. Since, the owners can do little with soil quality, they have taken keen interest to improve water acquisition and distribution. The WUAs then came as a boon to execute essential works and check the system operation and maintenance. Farmers (mainly landowners) have taken enormous interest in WUAs and their activities, mainly to boost up production levels.

Another major problem infested with the delta region is drainage. Being a delta region and close to the sea (ranging from 20--60 km) the gradient is low --

one feet per five kilometres. During the rainy season, a sheet of water covers the whole area with a depth ranging from one to three feet. Inadequate drainage facilities was a major problem. Funds for WUAs, DCs, and PCs came as a boon to this region. During the last two years, all over the delta region, heavy machines were used to desilt main canals, branch canals and distributories. Essential structures were constructed in some places and repaired in many places. This massive work was visible to landowners and sharecroppers. All this had facilitated a smooth flow of water, advancement of transplantation (by two to three weeks) before the arrival of the monsoons, and yield levels were up by 20-40 percent. Now, water flows regularly and drains easily. Hence, there are improved yield levels and farmers are happy too. This has helped to grow a second crop.[8] In the Nagarjuna Sagar Right Bank canal system, owing to a weak database, the project level data could not be collected for seasonwise benefits accrued due to WUA activities. In all these (Krishna, Godavari, and Nagarjuna Sagar) projects, constraint is more of canal regulation and their carrying capacity, rather than water supply levels. The main advantage, as many farmers mentioned was, "after several years, for the first time delta region did not get submerged in the monsoon period; otherwise, by August, one could see all lands are covered by a sheet of water". The study team did not find (during our visit in end August 1999) any submerged paddy fields and most of the transplantation, including in the tail-reaches, was completed before the arrival of monsoon.

Telangana Region: Sustain Current Development

Traditionally, the Telangana region had less experience in canal irrigation; many surface irrigation projects are close to completion stage (some of them have been on-going over the last 20-30 years). The SRSP project, which had created a large irrigation belt, had induced social and economic development in this region; the project was yet to attain its full potential. Hence, the irrigated area (one-third of the designed command area) had surplus water. But the carrying capacity of the distribution system had been declining over time owing to lack of maintenance and repairs. Under those circumstances, funds through WUAs and DCs became handy to carry out works of their choice. This had boosted the local farmer's confidence in WUAs.

Finance

The government was saving about 20 percent on works (which earlier went as rent seeking at various levels) more as cost effectiveness. Same money was spent, but presumably more work receives done. The main idea was not to save money (by avoiding contractors), but to increase the area under irrigation. At the WUA level, contributions were set at only 15 percent, (unlike the 30 percent under *Janmabhoomi* schemes). Unfortunately, only a few WUAs had contributed their

15 percent share. Neither the CA nor the project level irrigation department office had records to indicate the WUAs' contributions.

To begin with, the irrigation department was expecting part of the work (as prioritised from walk-through survey) from the current maintenance grant of Rs. 40 per acre, which was supposed to be given to the WUAs. Based on the criteria of Rs. 40 per acre for maintenance and repairs, the SRSP project received Rs. 63 million per year (out of the state total of Rs. 1,500 million); nearly 85 percent of it was spent on staff costs. It was planned to give this amount to the WUAs in mid-1997, soon after the WUA elections. Meanwhile (after three months of the WUA formation), observing the formation of WUAs and the state's long-term plans, the World Bank expressed interest to support some activities. The Bank was ready to fund, initially, minimum rehabilitation and then operation and maintenance (O&M). The detailed costs worked out to Rs. 1300 per acre in major and medium projects (minor projects were excluded at that time, but for no reason) for minimum rehabilitation. O&M costs worked out to Rs. 500-600 per acre. From this bank grant, the department provided Rs. 50 per acre during the first year and Rs. 100 per acre during the second year to each WUA. Grants were given from 1998 to all WUAs in the state for a total of five years. These grants were meant for the non-SRSP projects under the APERP project of the World Bank.[9]

Proposed Irrigation Revenue Sharing Pattern Across the Schemes

Level	*Major*	*Medium*	*Minor*
Water charges/acre[10] in Rs.	200	200	100
Allocation in Rs.			
WUA	50	60	90
DC	20	-	-
PC	20	30	-
Local government (Gram Panchayat)	10	10	10
Irrigation department	100	100	-

In the reform process, the state budget was not touched. After the APERP project period, WUAs were expected to collect fees, which would be given to all tiers of the organisation, the local government and the irrigation department (see above table). But this sharing pattern was approved only in February 2001. During the first year (1998-99) each WUA and DC got Rs. 50 per acre, and during the second year (1999-2000) Rs. 100 per acre as maintenance grant. In this amount, minor level WUA had 60 percent, DC 20 percent, and PC 20 percent share; since PC was yet to form, Irrigation Department (ID) had taken 20 percent till PC was formed. In this process, Rs. 1,070 million were spent during 1998-99; out of which some Rs. 1,030 million was spent during the 45 days. Every bill of

ID had to be passed by the account officer, before going to WUA. This was a hindrance owing to the delays and rent-seeking attitude at that level. The annual demand raised (based on water fees) was Rs. 1,400 million, while actual collection was only 50-60 percent. However, water fee collection had increased to 65 percent during 1998-99, from 54 percent during 1997-98.

The financial component had played a pivotal role in expediting the whole WUA exercise in Andhra Pradesh. To begin with (during the APFMIS act formulation), the financial map was not clear; the introduction of the APERP project (as part of the World Bank funding for economic reforms in the state) during the later part of 1997 had changed the scenario. The focus shifted to funds allocation for physical rehabilitation. Till 1997, the main problem was lower level of grants for O&M at Rs. 40 per acre, of which Rs. 25 was spent on staff costs. With the remaining money, few essential works were carried out. Now, WUAs receive Rs. 50 per acre during the first year and Rs. 100 per acre during the second year and it is free from staff costs. In the process, farmers were realising benefits from the WUAs' function in the present pattern.[11] Some farmers as well as field officials have agreed that, "the government investment is less than one bag of paddy per acre (i.e., about Rs. 400 per bag of 75 kgs, as per the minimum support price); while the return is 5-10 bags per acre (additional yield per acre). That is the real benefit of formation and funding of WUAs". In some places, funds were also promised for WUA and DC office buildings. Every year WUAs in the major canal irrigation projects receive Rs. 200 per acre (100 for WUA and 100 for DC). During 1999-2000, Krishna delta received approximately Rs. 123.6 million, NSLC Rs. 60 million, medium projects Rs. 3 million, and minor projects Rs. 10.4 million. Under APERP, the Krishna delta got Rs. 40.2 million (during 1999-2000), including project committee works on main canals and drainage works.

One of the key factors attracting the WUAs to function is the size of funds ranging from Rs. 700,000 per WUA in Coastal Andhra to a mere Rs.12,000 for tank-WUA in Rayalaseema region. The former focuses only on operation and maintenance, and repairs of the distribution system in its command; while, the tank-WUA has to take care of both the source (including tank structures) and distribution system. There was an imbalance in the fund allocation pattern, as most of the tank-WUAs in all regions have complained. Indeed, field officials of the minor irrigation department endorsed this view, and suggested modification in the present allocation pattern to encourage tanks rehabilitation.

As of 2000, water fee collection and area assessment was done by the *patwari* (a field staff of the Revenue Department) and in few places WUA president and TCs are involved; together they endorse the figures. About 70 percent of farmers have paid; the remaining 30 percent were mostly large landholders that delay or don't pay quickly. If fee collection was the responsibility of the WUA, it could have been easier, since they can cut off the supplies to the field. Otherwise, there was no record as to who has paid or defaulted. WUAs can raise this issue in the general body meeting and owing to social status, people will pay up; this cannot be done by the revenue *patwari*. On the other hand, WUAs are not focusing on resource mobilisation nor the irrigation department has stressed this point. In some places, farmers were willing to invest on lift irrigation. Priority may be given to such villages where villagers were willing to contribute a major portion of the investment. That would inspire more villages to mobilise their own resources.

Fishing Rights

All tank-WUAs think that fishing rights should be given to them. Currently, fishing rights given to the local fisherman cooperative societies on a 3-5 years lease at much lower rates are unfair. For instance, in Srikakulam's Bajjipuram tank, the fishing rights were sold for only Rs. 2,000, while it could have fetched Rs. 50,000. Many of the fishermen cooperative societies belong outside the WUA boundary and even outside the village boundary. Local WUA feel very strongly about this because these contracts could have boosted their resource mobilisation.

Maintenance of Records

The APFMIS Act, 1997 has suggested maintenance of 12 types of registers at the WUA level; registers ranged from the work register to the WUA property register. The field staff of the Irrigation Department report that most WUAs/DCs being in their early stages, maintain just three or four registers pertaining to the minutes of meetings, payment of bills, physical works, and cash dealings. There were a few WUAs, promoted as models, which have all 12 registers, e.g., Elabotharam-WUA in SRSP and Yenubilli tank-WUA in coastal Andhra. Interestingly, many field staff were not aware of the 12 registers. In few cases, the department staff has got registers printed in the local language. Many WUAs (and field staff) are not clear whether they can spend the grant allocated for administrative expenses for this purpose. Even CA is not clear, in many places, about the nature of registers to be maintained. The training programmes have not incorporated bookkeeping by the executive committee as a main activity, as informed by WUA presidents. All WUAs, except a few, maintained only a cashbook, administration approval, minute's book, and bank passbook. Field staff of the department wrote these registers/records. In a few cases, locally hired persons wrote them. Only a few 'model' WUAs had task-specific printed (as prescribed by the Act) registers. In spite of providing Rs.10,000 for administrative purposes to all WUAs (many WUAs had not even entered this in their cash book till August, 1999), neither they nor the local A.E/DyE.Es had printed task-specific registers.

Area Irrigated Records

Till November 1998, the revenue department with the help of the agriculture department used to assess the irrigated area at the end of each season. Owing to an understanding between the farmer and field staff, reported irrigated area figures and in turn, water fee levels, were less than actual. 'Discounted' figures vary from 20-40 percent less than the actual. By involving WUAs in the assessment process, irrigated area figures had gone up by 20-40 percent owing to correction in the documentation process. During field interviews farmers reported that in several schemes they had no command area maps and information on water fee defaulters; till October 1999, neither the concerned WUA nor ID staff had records to indicate actual levels of water fee collected in their jurisdiction. The revenue staff felt it was not obligatory on their part to share this information with WUAs or ID staff; the village agricultural officer also expressed similar feelings.

Information Diffusion

The major advantage of having WUA was quick information flow from farmers to senior officials and vice versa. Before the formation of WUAs, this process used to take a week to ten days, and some times even more. Reporting WUA's activities in local newspapers and state-level newspapers had contributed significantly to information dissemination. In other states, local newspapers are area-based, and restrict their coverage to the local area (mostly town/city level or district level). In Andhra Pradesh, particularly in recent years, the state level newspapers in Telugu, have started printing editions at the district level. Indeed, these newspaper reports have also become inputs for the Chief Minister's teleconference every Thursday morning.

District Level Tank-WUAs Forum

In Karimnagar, the District Level WUAs President's Forum was started on 20 April, 1998 to provide a link between WUAs and the government in order to solve problems of funds and grants. Till the end of 1999, this forum had 7 meetings. The forum demands were:

- Provide allocation of Rs. 250 per acre instead of Rs. 100 per acre (no basis for this figure).
- *Neeradis* should be under WUAs, rather than under the mandal revenue officer. Eighty percent of WUAs have *neeradis* and their salary should be given by the government.
- Fifty percent of the funds accrued through fishermen cooperative societies should be transferred to WUAs.
- Revenue collected from toddy tappers (from mainly palm trees) through leasing out trees on canals and tank catchment, at Rs. 11 per tree, should be given to WUAs.

- WUAs should receive revenue records, map of command, and voter list.
- The Act should enable WUA to punish or decontrol encroached lands in tank command and catchment areas.
- Presidents are expected to do lot of work, including record-keeping and cash maintenance. WUAs need one accountant per mandal.
- CAs' responsibility is biased. Some CAs have 45 tanks under them; even if he visits one tank per day, he cannot inspect all tanks in a month (this problem is not there in major projects).
- The district pay and accounts officer gives account payee cheque payable at Karimnagar town only, for which banks take 20 days for realisation. Alternative arrangements (like demand draft) at local level should be made.

The Ananthapur District Tank-WUAs Association formed in July 1999 had met once in three months. Its focus was mainly: (a) to fight for auction rights of fish and trees on foreshore and tank bunds area; (b) since many WUAs presidents are illiterates, and not fully aware of the WUAs rights and responsibilities, the district tank-WUA association would like to help them; (c) through this association, the fight for funds from various district offices. However, so far, the association has not discussed problems of siltation and possibilities of increase in storage capacity. Most tanks help recharge groundwater in the downstream, in some cases up to 10-15 km.

Support from Officials

A large number of field staff and senior staff of the Irrigation Department had realised how WUAs and DCs were positively facilitating their functions. The competent authority (CA) sanctions estimates of DC works and approvals up to Rs 100,000 and enters into an agreement for works execution, resolves water acquisition and distribution related constraints, and supervises works given to DC. In most cases, CAs feel that in spite of more work, they are enjoying it, especially since the reform process has facilitated more interaction with farmers. Problems are aggregated at the DC level and become easy to handle. Distributory committee necessitates farmers' participation in operation and maintenance of canal network.[12]

During the field interviews, the senior officials reported that officials like executive engineer (EE), superintending engineer (SE) and chief engineer (CE) should have direct contact with farmers of various reaches. The top officials should frequently visit the sites and should have open dialogue with the users and personally observe the sites. The SE brings out the district-level newsletter on WUAs. This newsletter is quite regular and is released during the first week of every month – reporting the progress made, funds disbursed, and carrying the interviews of one or two farmers. Every month a thousand copies are printed and distributed to all WUAs. The SE himself edits the newsletter (till now, four issues have been published) and receives a monthly supportive grant of Rs. 5,000 from the irrigation department.

Some field officials shared their disappointment; owing to WUAs they have cut into their additional earnings, which may further decline. A deputy executive engineer expressed that owing to too many WUAs everywhere, a new breed of contractors and WUA presidents has emerged, and they charge about 10 percent commission. They would like to have all monetary details with them. Many territorial committees are against such practice. Many presidents do not even discuss these monetary aspects during the general body meetings. On the other hand, engineers are hesitant since they are losing their additional earnings and have to answer to the public. Earlier they were answerable to only the government.

In Ananthpur, the SE of the minor irrigation department, and the SE of the major irrigation department said that before formation of WUAs in major projects, farmers used to have protest march. During emergencies, tractor loads of farmers would barge in to local department office to demand water and some essential physical repairs. By 1999, all that had stopped. Farmers address the WUA president with their problems. Even officials direct farmers to route their complaints through WUA presidents. In effect, for any command area problems, only one/two person's meet or call upon the officials. According to these SEs, major achievements of the reform are: (a) water reaches tail-end; (b) WUAs look after all O&M; and (c) ID is relieved from regular O&M problems. In Rayalaseema region, water regulation was terrible before WUA's formation. From 1998 there was a forum (WUA) to whom ID officials speak about the emerging issues; in turn these WUA presidents brief their members and arrive at a consensus.

III
Benefits

Large sums have been spent through users' associations; ranging from Rs. 5,000 to Rs. 900,000 per WUA. Users have identified works and executed them. They have also carried out their long pending essential physical works. Physical benefits are accrued in a short time, and they are distinctly visible. Across the state, our field observations had indicated that WUAs were happy for the same reason. Wide awareness on WUAs, created by the government and the media, has surely raised the users' aspirations, and thereby pressurised the WUA president and TC members on one hand and the ID staff on the other. It is hard to dispute visible benefits. However, some researchers in the state have a different opinion. The solid benefit across the state is water flowing to the tail-end users; some of them were deprived for years, and now receive water for the first time. Minimum physical rehabilitation works had definitely facilitated better water flow, both across the reaches, and for the length of time, as the study team had observed in various places. This has enhanced crop yield levels by almost 20-30 percent across the state. Every WUA that the study team had visited has vouched on both yield increase and water flow to tail-reach.

Crop Pattern

By and large, the irrigated area had expanded owing to improved water supplies and cleaning of canal network. This had largely benefitted the tail-end areas. All this had increased the production and productivity levels due to better water distribution practices by the WUA. In most areas, shift was towards rice cultivation both in the kharif and rabi season. For example, in Bannuvada village-WUA area, in the kharif season, rice was grown in 3,011 acres; total rabi crop area was also 3,000 acres. There were no open or bore wells owing to loose and sandy soil, and proximity to sea coast. In Duggirala-WUA area, the net profit from kharif rice is Rs. 8,000 to 10,000 per acre, according to farmers. In rabi, the entire 6,727 acres are cultivated; area under rice cultivation has increased from 600 acres before WUA to 2,500 acres after carrying out all physical works. Field official's record shows all 6,727 acres of the command area as wet (paddy) area, plus 1,000 acres non-paddy area. In Emani-WUA in Krishna delta, during kharif all 5,700 acres are cultivated for rice. The command area has some 100 filter points; the groundwater from these wells was used for preparation of seedbeds. Thus, some 200 acres use borewell water. In the rabi season, rice cultivation in 3,700 acres and black gram in 2,000 acres was possible only after the WUA's works.[13]

Even the irrigation department officials are convinced about the benefits. According to one of the executive engineer of the Vamsadhara Project in Coastal Andhra, the following benefits have accrued from WUAs:

- Carrying capacity of canals, both main, distributory and minor canals, has increased by 20-30 percent.
- Irrigated area increased mainly in tail reaches by 10--15 percent.
- Number of complaints drastically reduced.
- The EE and SE have quick access to information. Earlier information was available only from the AE and Deputy EE. Presently, it is from all WUAs.
- Users prioritise works and carry them out with technical approval from the CA.
- Now users have role in works identification, prioritisation, and execution.
- Collective action among users is increasing.
- In the future, WUAs have to collect fees and do all O&M works; but this will emerge gradually.
- WUAs are more beneficial in minor irrigation tanks and medium projects, since source and control can be seen and separated. In major projects (mainly O&M) can be carried out by WUA.

- In the Vamsadhara project, the problem was water carrying capacity and not water availability. WUA solved them since most of it are river diversion schemes and run of the river channels. Canals run continuously from July end to December.

In some places the major achievement was removal of encroachments, particularly in tank-bed areas. Godavari and Krishna delta areas had gained a leading edge in benefitting from the WUA movement and its associated financial allocations. Because of the sheer size of the command area (together, one million acres), and large size WUAs (average is 9,000 acres), work components and benefits were large. Their prime focus on improving drainage facilities and desilting earthen canal distribution network had paid rich dividends. Even officials concur with user pride. Some major benefits in Krishna delta were: reduced canal breaches, building of much awaited bathing and washing stone-steps, construction of cattle ramps, desiltation of minors and distributories, and equity in distribution from tail to head reach. In some distributories, owing to desiltation and canal protection walls, water flows till the tail-reach easily. Hence, water distribution started from tail to head, unlike earlier from head to tail, owing to less water and difficult flow. Paddy crop yield level has increased by 10 bags in the tail reaches (from 15 to 25 bags per acre) and in the head reaches by 5 bags (20–25 to 30 bags per acre).

The top officials said that initial motive was to do away with contractors, involve farmers and spend the whole money on works. Few PIM coordinators in the recently created posts and appointed at the district level, were in full support of WUA activity. Some of them have never worked on WUAs and were close to retirement. These coordinators revealed that owing to PIM they were able to see returns such as expansion of irrigated area, increased yield levels, reduced or almost no protest marches and agitation by farmers, and active involvement of users in irrigation management.

Staff Size Reduced

Since the formation of WUAs, all maintenance and repair works at a circle level (i.e., superintending engineer level) were carried out by them. This had some impact on the ID staff size and their workload. For instance, the SE (major irrigation) in Ananthapur has confirmed that his circle was supposed to have 149 assistant engineers. In reality 60 posts were vacant. The system was still running without any hindrance. The SE said, "this is mainly owing to WUA's involvement and is really beneficial to the department". Even at the middle level, for the deputy engineers, vacant positions were six (sanctioned are 35) and at executive engineers level, vacant posts were 50 percent (sanctioned are eight). In the Sri Ram Sagar Project, reforms led to the following staff scenario till the year 2000:

Staff Size in Sri Ram Sagar Project

Level	*Pre-1998*	*2000*
Chief Engineer	2	1
Circles	7	5
Divisions	36	25
Staff Position		
Sanctioned	2,953	2,078
Working	2,391	1,740

In Chittoor division, the percolation tank-WUAs have focused on improving the inflow channels and strengthening surplus weirs and aprons. Also, they had strengthened tank bunds and cleared shrubs and weeds. To improve storage levels, and in turn their groundwater recharge levels in the command area, most of these tank commands grow sugarcane as a traditional crop.

The commissioner of irrigation and command area development, clearly identified two major benefits due to WUAs and DCs: (a) simplified payment procedures; and (b) introduction of simple monitoring procedures. Further, he listed out some major achievements:

- Empowerment of farmers; they are more articulate now.
- Emergence of young leadership in rural areas
- Teleconferences for effective monitoring (every Thursday of the first three weeks of the month, with all District Collectors and department officials)
- Awareness on both WUAs and the Act is good.
- Farmers and their organisation have a role to maintain the system
- Hardly any complaints on system damages in any project area
- Faction leaders demand collective action in poor areas.

He also identified some major strengths of this experiment: (a) Effective, innovative and quick action, for monitoring and follow-up. Since the state has evolved this process, many irritants have been removed; (b) Regular field inspection; (c) To a large extent rent-seeking levels have been removed. In recent months (early 1999), WUAs have been assigned a nine-digit code for better monitoring and follow-up at the district level and irrigation department level. The state keeps a tab on these inputs; hence, at all levels, it becomes easy to access data and report progress made and problems if any. Formats are being developed for this.

In Andhra Pradesh, strong political will has made a major difference to the whole WUA exercise, followed by continuous bureaucratic support at the top level. This has led to empowerment of farmers and greater awareness among them; they have even been able to question the local officials and field staff.

Economic Benefits

Some of the economic benefits accrued owing to PIM in Andhra Pradesh are:

- Additional 5.21 lakh acres came under irrigation. In the SRSP project, the reported area has increased from 95,000 acres to 237,000 acres in the first year (1997-98), and to 345,000 acres in the second year (1998-99).
- In spite of heavy rains in Coastal Andhra, there was minimum loss owing to drainage works carried out by WUAs. Besides, rice transplantation was advanced by fifteen days.
- Paddy yield level has increased from 2.5 tons per acre to 3.5 tons per acre. Farmers reported that they obtained 45-60 bags per acre (70 kgs/bag) of paddy. Before WUA formation, it was 30-35 bags per acre.
- Total additional production per year was estimated at Rs. 6,140 million

Economic benefits are enormous at the state level[14]; expenditure was more productive-oriented. Wasteful expenses reduced; resource mobilisation at the local level had increased. Rent seeking had reduced. O&M grants can be reduced and the same may be spent on works.

Perceptions of Government Officials

Interestingly, across the state, and across the levels, officials of the I&CADD echoed different tunes on their involvement in the PIM related activities. The following are some key perceptions of officials interviewed by the study team:

- The present course of events would lead to a major shift towards decentralising the perceived power structure of local officials -- both lower and middle; officials feel, at this speed, even senior officials may lose their 'powers'.
- However, there was a growing understanding across the state, and across all levels of officials that the WUA movement had helped in a large way in system operation and maintenance, and to a large extent helped to influence farmers involvement in irrigation management. Officials at all levels, and in all places (more in major irrigation systems), feel their major burden is reduced in operation and maintenance of the system.
- Officials feel that local engineers should have bank cheque signing powers along with the WUA president and TC members. Unfortunately, none of the WUAs were in agreement with this view. They feel that this was another tool for arm-twisting and they pleaded "this should not be encouraged; if any WUA presidents and TC members have mis-utilised the funds, the ID staff can always take stringent action on them".
- Minor irrigation officials argue that present grants (Rs. 50 or 100 per acre) were grossly inadequate for tank systems; in the absence of any DC and PCs for tank systems, WUAs have to take care of both main system (reservoir, sluices, regulation points) and distribution systems. Since recent grants were

proportionate to area size, tank systems received much lower (in many cases, in the range of Rs. 12,000-25,000), while major irrigation system WUAs received in the range of Rs. 300,000-900,000 or even more. Tank systems need different level of grant; they need one time allocation of large amounts to completely rehabilitate the tank systems to their original standards, and then WUAs can receive the present level of grants.

- Some senior irrigation officials felt that during the next five years WUAs should completely operate and maintain the system, regulate water supplies, and collect water fees.
- In the APFMIS Act, according to the SE minor irrigation, Ananthapur, some changes are required: (a) recalling of WUA president should be made more effective; (b) CA should have control over the funds disbursed (because WUA president and TC members are misusing funds, he feels); (c) the Act should provide a clause to take back funds from WUA if they do not execute agreed works.
- In the process of reforms, the government staff, particularly those who are at the level of service delivery and have direct link with the users (like lascars, assistant engineers, deputy executive engineers), feel threatened. Primarily because they lose their undisputed powers and the prestige attached to their positions. There are no more endless requests from the users for their services and thereby a major loss in their additional earnings. Indeed, absence or reduction in these additional earnings affects the upper hierarchy too, owing to lack of demand for such positions and no grease money involved in transfers and appointments. These positions, termed as 'wet' positions, have now become 'dry'. They have to be more accountable and transparent in their dealings.

IV
Key Lessons

This part of the paper focuses on some of the critical issues and key lessons of PIM in Andhra Pradesh. The issues deliberated are based on observations, discussions with the officials of the irrigation department and WUA members, and experiences of other states in India. At the outset, there is no road map as such to drive the WUA movement in the state during the next five or ten years. But, the irrigation department has kept irrigation management transfer as its main goal. Towards that direction, formation of WUAs, DCs and PCs is perceived as most important. "We do not have any blue print; we wanted to build as it grows", says commissioner of I&CADD, who was the key driving force in the irrigation department for the WUA movement.

Both top level and field staff are making efforts to learn while they do things. The chief engineer of the SRSP project would like the following as future plan to give a boost to the WUA movement in the state: (a) linking water fee with revenue; (b) ultimately, making WUAs responsible for collecting water fees; (c) future releases of government funds to be proportionate to fee collection of WUAs; and (d) development of financial management systems for farmer groups at WUA and DC level.

Andhra Pradesh has made a beginning in large-scale PIM. The chief features are:

- For the first time, in the state and also in the country, a suitable legislation has been enacted. The administrators and senior engineers with plenty of field experience crafted it. Users across the state and field staff hail the legislation.
- The state has mounted awareness and training camps in different parts and for different levels.
- Continuous and effective monitoring mechanism both at lower and top levels.
- Strong political will laced with administrative acceptance.

Though the state has made a good beginning, it has to go a long way to claim all round success. Even sustainability becomes crucial in the long run. The state has to gear up the process to stress on the following key issues. Recommendations are also indicated in the appropriate places.

Performance Monitoring Needs to be Strengthened

The key question here is what is the criteria for WUAs' and DCs' performance measurement? The performance evaluation formats (quarterly and annually) should have been devised when WUAs were formed. These formats should be on season-crop basis, covering problems, area covered, production and yield levels in their jurisdiction, activities undertaken, resources mobilisation, and item-wise expenditure incurred. It requires focused planning and frequent appraisals and upgradation to keep up with the growth of WUAs and their functions and allied activities.

Provide Access and Control to WUA Area Records

WUAs and DCs should have access to revenue records related to water fee collection, land acquired on either side of the canal and distributories. They should be able to grow plants and auction them or lease them periodically to mobilise resources. Thus, canal bunds would be protected and organisations will have a regular source of income. This would also help reduce the silt accumulation in the canals. In each WUA, one or two TCs should also be trained in book-keeping; their skills need to be upgraded over time, as WUA/DC/PC activities expand. Training should focus on function-specific aspects. If TC members are not willing to pick up the responsibility, locally hired educated unemployed persons can be trained. WUA/DC/PC should agree to pay their salaries and costs. Copies of all official records[15] should be given to WUAs. This would help remove the encroachments, and improve the storage level and carrying capacity in canals of all types of irrigation systems. WUAs would be able to receive more funds on per acre basis and collect more water fees. This would help strengthen the resource mobilisation capacity of WUAs and reduce disputes relating to authorised and unauthorised areas[16]. The process of rectification and thus revising area records, needs to be done before asking WUAs to collect water fees on their own. It was a common demand across all WUAs in the state.

Restructure the Capacity Building Mechanisms

In any given project area, at any time, there should be a minimum of 60 percent staff trained on PIM related aspects. Their skill needs to be upgraded on regular manner. Organisers have to conduct need assessment on contents, timings, and locations of the training programmes. The staff needs to be reoriented to prepare case studies, design effective training and use of innovative teaching methods for task-specific and level specific programmes. They may need to forge an alliance among NIRD-IRDA-HRD agencies.

The training institute under the control of irrigation department, i.e., WALAMTARI, needs to be restructured, preferably on the lines of recommendations made for WALMI-Gujarat[17] in order to design and offer more demand-driven training programmes, recruit more productive persons on time-bound and incentive based scale, and bring in output orientation to the institute. Currently, in spite of a strong WUA movement in the state, WALAMTARI has a few faculty members with good social science, organisation behavior, and management skills. The institute needs to respond to the changing needs of the state and its activities. It has to achieve the following: (a) play a pivotal role in capacity building for the WUA movement; (b) essentially, track down its road map for the next five to ten years in order to be a productive partner in the WUA movement; (c) be a truly autonomous organisation, free from the irrigation department's control; and (d) upgrade its staff skills[18] since the training materials currently used for WUAs' presidents and CAs' training programmes are designed by outside agencies.

Forge Institutional Alliance

As of now, the WUA movement has been mainly confined to the irrigation department. To utilise the available expertise both within the state and outside, this movement needs to forge an alliance with non-governmental organisations and research institutes. To draw a road map, with clear milestones and financial implications, and potential outputs and inputs, it may be worthwhile to collaborate with mutually beneficial reputed institutes (e.g., International Water

Management Institute, Colombo). All these would enable better performance measurement, appraisals and mid-course corrections, evaluation, and learning from mistakes.

Focus on Organisational Principles

WUAs should be directed from their present orientation towards physical works and meetings. It should shift towards water management and productive functioning of the organisation in terms of resources mobilisation, long-term plans, member education, member participation, water and related activities, value added functions, strengthening rule-making process and rule enforcement mechanisms. The current method does not focus on the organisational design of WUA. The experience with creating user organisations, especially in gravity flow systems, is still formative and yet to yield firm design principles which will ensure that the user organisations will be capable of self-governance and self-management. The WUA movement will be in jeopardy unless it begins with valid assumptions about the design principles that need to be followed to catalyse participatory user organisations. Designing energetic member organisations involves understanding their working through focusing on the interactions between their three constituent sub-systems: members, the governance structure (board or management committee), and the operating system which provides the services.

Extensive research in member organisations in various fields suggests that robust self-governing user organisations achieve high levels of goal-cohesiveness, governance effectiveness, and member-need responsiveness by satisfying four design principles: (a) Member-Centrality of the Goal; (b) Goal-cohesive Governance; (c) Receive the Right Operating System; and (d) Secure and Retain Member Faith and Allegiance.

Develop Model Projects

The APFMIS Act is a trendsetter for the country and also for south Asian countries. The Act remains same for the whole state while the government orders have to make provisions for regional variations. For users, and also field staff, seeing is believing. As of now, the state has no pilot site/project to show as the model from which to learn lessons. The state has to have demonstration sites/projects, wherein the total irrigation management transfer has taken place. Good documentation of lessons learnt will be a good guide during the training programmes and for replication in the state. In the absence of that, both water users and field staff are riding on wish-horses. For demonstrative effect, in each of the three regions (Telangana, Coastal Andhra, and Rayalaseema) one full major project (or a hydraulic boundary of 50,000 acres or above), one medium project, and two minor irrigation projects (or two tanks) should be taken to bring all

WUAs and DCs, and form one PC. Farmers should be involved in selecting the pilot sites for full turnover.

First, bring these sites on to a dynamic function in tune with the act and establish a true organic link across all tiers in their functions, responsibilities, and roles. Each tier, wherever applicable, should be accountable both upward and downward. Gradually, in a phase-wise manner, as the tiers are strengthened, the irrigation department should be withdrawn from all projects (under this experiment). Each tier or concerned organisation level should have office space in the regular ID office and use it for regular liaison purposes; as the ID staff reduces, the WUA/DC/PC should, if they feel, increase their staff. If the user organisations are made to pay staff salaries, they will be in a better position to judge the need-based staff. As today, the lascar is accountable and works under WUAs, similarly AEs and Deputy EEs should work under DCs. Gradually, this should move up to EE and SE level.

The state has to design and attempt three different approaches – one each for minor, medium, and major projects. While the existing Act serves a good purpose at the macro-level, in practice, for effective implementation, problems are mainly technical and social. Feasible approaches have to be evolved and designed, region and type-wise. To test the feasibility of these approaches a period of two crop seasons should be considered. Geographically, the experiment should cover three districts in three regions; in each region it should cover a minor, medium and major projects. During the second year, the approach may be modified and extended to two more districts; and during the third year onwards replicated all over the state.

Expedite to Transform the Technocracy

In the state, the focus is on transforming technocracy. Owing to the Chief Minister's teleconferences and District Collectors' frequent meetings, countless formats have to be sent to the I&CADD secretary regularly. As part of the transforming technocracy, the present set-up of the organisation may be restructured on the following lines. The proposed federal structure will have WUA at the bottom and the irrigation department at the top. The federal structure will:

- Be self-supporting with initial corpus fund from the government;
- Hire professionals at all levels, and will have own staff rules and incentive schemes for high performers;
- Have rights to raise funds and loans at all levels. The federal structure will have four-tier structure – WUAs at the bottom, DCs at the second tier, PCs at the third tier (project level), and the state level federation at the top level.
- Each level will be self-supporting with autonomous powers, clear rights and responsibilities, yet vertically and horizontally well integrated (e.g., for

Figure 2
Suggested Federal Structure

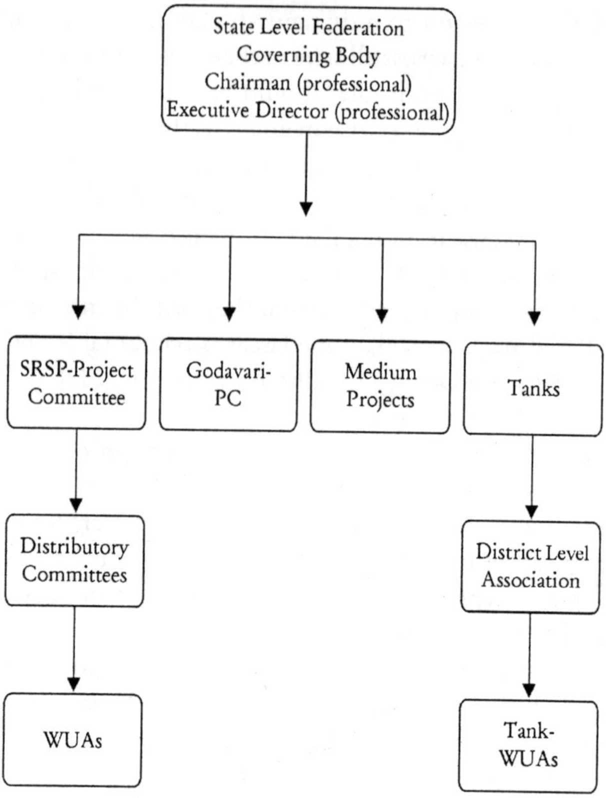

water acquisition, water distribution, system maintenance, and fee collection). The study team feels that the federal structure, as shown in Figure 2, may be useful.

Required Separate Scheme to Tackle Tanks

Management of tanks necessitates an altogether different approach. Tank users' associations have control both on source of water and its entire distribution system. Compared to major and medium irrigation systems, tank systems are pretty old[19]. But present O&M grants at Rs. 100 per acre are inadequate for any tank rehabilitation work. Weed (ipomea and others) growth in the tankbed area had increasingly affected around one-third of the storage capacity, particularly in Coastal Andhra region. Farmers have reportedly said that for permanent removal of weeds, they are willing to share up to 30 percent costs. Tank-WUAs need different financial disbursements. Firstly, they require physical rehabilitation. Prior to that, tank-WUAs should be made vibrant and responsive. This needs a special task force both at the state and regional levlevel2 . A time-bound plan has to be evolved in consultation with the existing tank-WUAs and in some cases (wherever available) district level tank-WUA associations.

Facilitate Additional Resource Mobilisation

WUAs should be encouraged to attain financial independence. Towards this goal, gradually WUAs have to increase their resource mobilisation capacities through (besides water fee collection), auctioning rights on fish, grass, trees, silt, grazing, and able to undertake allied activities. Most tanks have potential to cultivate fish and sell them for Rs. 10,000 to Rs. 200,000 per year. But, until 2000 they were helpless to make use of this opportunity owing to rules. The rules permit fishing by only fishing cooperatives at local level, who pay some nominal amount to the fisheries department. In turn, the fisheries department pays some nominal amount to WUA. According to the Bajjipuram tank-WUA, it can receive Rs. 50,000 per year as fishing rights, while presently it receives only Rs. 2,000 per year from the fisheries department. To improve the resource mobilisation capacity of WUAs, rules need to be revised. The fishing community also does not receive any benefits. As of 1999, middlemen made money.

To have more effective integration of WUAs at all levels they need to be federated into a four-tier structure: primary (at present level); secondary (distributory or DC level); project (PC level); and apex (state level body) level. This would enhance the coordination of various aspects (organisational, financial and manpower). This would also facilitate focusing on essential activities like water acquisition, water regulation and distribution, and moving towards equity within the WUA level and across the reaches in a project. The apex body has to take the full responsibility of all these functions. It should plan and execute its activities on its own, generate funds, and be able to borrow funds and receive grants on its own. The organisation should lay emphasis on member-controlled WUAs, DCs and PCs. The apex body and the four-tier organisation should function as vibrant and truly member-controlled organisations[21].

Need to strengthen local and traditional approaches of labour sharing and community efforts: Before the WUA formation, local informal committees or groups (both in tanks and canal systems) used to collect both men (one/acre or household) and money (Rs10-50/acre or household) for essential works like cleaning canals before the crop season began, desilting, and some essential repairs. After the formation of WUAs, all farmers think that since the government is giving money, let the WUA president and TC members do all the works. Farmers, in general, contribute neither labour nor money. The new initiative of the government has weakened the local initiatives and participation. The new initiatives are defined to be participatory; but the approach is different. The new approach was imposed from outside, and has little member contribution and control, while the traditional approach was locally evolved and member-controlled.

The new approach should have taken local approaches into consideration; it should necessitate that local farmers contribute first, and twice of that amount may be given as a matching grant from the government. Local contributions should have been made compulsory on an acre basis. Rs. 30-50 per acre per year is ideal, as many WUAs have mentioned. Thereby, members would develop stakes in the organisation and its functions. Some WUAs have suggested that the accumulated money may be deposited in a commercial bank and the interest accrued may be spent on activities.

Replicability

The Andhra Pradesh programme on WUAs is a standing example of how an impossible task can be made possible. To surge ahead in a similar and better fashion, any state has to muster the following strengths:

- *Focus on Key Areas*: Emphasis should be laid on getting rid of middleman-contractor system in O&M; users should be responsible for identifying and executing essential works for repair and maintenance. In O&M, irrigation department's role has to be reduced.
- *Building Capabilities*: Training at all levels; focus has to be more on middle and lower level field staff; departmental staff may be trained as trainers; for the users, training should be conducted in all centres; Wide dissemination of specifically prepared training material in local language is essential. Similarly, the Act should also be made available in the local language and needs to be distributed widely all over the state.
- *Good Governance*: Strong politicians at the top level (the chief minister and irrigation minister level) will make a positive difference. Right kind of bureaucrats needs to be identified to head the department/divisions.
- *Effective Monitoring*: Tele-conferences at top-level with all district collectors and senior officials of the department is required. Regular monitoring through specially designed formats by filling and filing to higher level offices and quick consolidation at the top level would help necessary decision-making process.
- *Institutionalisation*: The PIM in Andhra Pradesh was evolved through the government set-up. While in most states of India, WUA movement came through a non-governmental or semi-governmental agency (SOPPECOM in Maharashtra, and in small scale through WALMIs in some states). In Andhra Pradesh, the irrigation department has taken the present initiative; thereby, it has become part of the regular work of the department. It is one of the statewide department's initiative and main activity. Competent Authority for WUAs and DCs are regular staff of the department and currently they are responsible to prepare technical estimates and to authorise work completion.
- *Large-scale Operation*: This has four key features: (a) do it in one-shot, (b) massive scale, (c) learning by doing, (d) modify through government orders

and continuous monitoring. Any thing large is visible and will have wider implications at all levels. Large operation also demand larger human power effort, and related operational and organisational effort; large-scale operations are mounted to achieve larger output in one-shot. If operations are spread out both in space and time, there is scope to learn by doing. If the framework is flexible to achieve the given targets, then one sees variations across regions and levels. Indeed, flexibility over time provides opportunities to modify the process. Senior officials understand that the reform process is moving towards receiving it right and learning by doing. To support the process, government orders/circulars/guidelines are issued regularly.

Impressed by the performance, some states have made a modest beginning to replicate the experiment. Madhya Pradesh has already formulated legislation mostly in tune with the APFMIS Act. Tamil Nadu seems to have started an internal exercise to move in this direction. Officials of the Andhra Pradesh irrigation department strongly feel that the states interested to replicate the experiment should have taken the Andhra Pradesh state personnel during the initial years of WUA formulation. This state could share its rich experience with other states.

Both demonstration sites in Andhra Pradesh and other states have to make well-designed and constant efforts to move towards total irrigation management transfer. To begin with, as shown in Figure 3, responsibilities of the joint government/project committee have to be minimal in the lower level of canal distribution network and gradually increase towards main canal and project level. The above chart indicates kind of responsibilities of WUAs and the joint government/project committee across different levels of the irrigation system.

What is also essential is that both the Planning Commission and the Command Area Development wings of the Government of India should make

Figure 3

Responsibilities of WUAs and government in the process of irrigation management transfer for adoption in demonstration sites in Andhra Pradesh and replication in other states

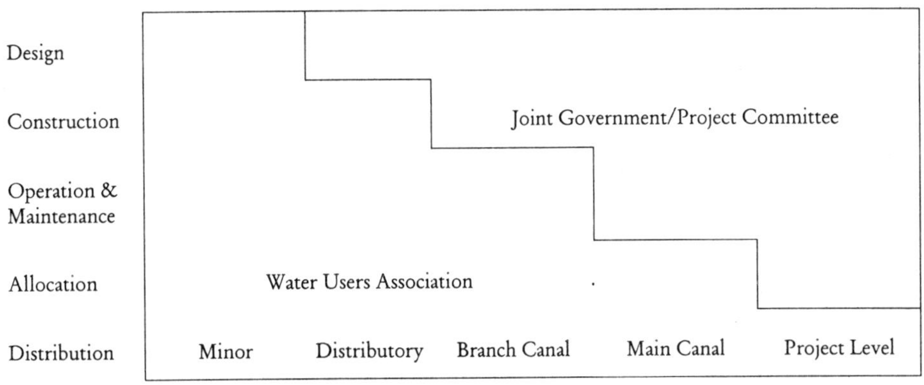

such Acts and internal changes as pre-condition for any assistance, including financial disbursements from the central government. This would help build pressure to expedite the process. In spite of having a separate component on farmers' organisation and turnover in the Water Resources Consolidation Project, three states (Haryana, Orissa, and Tamil Nadu) have made little progress in this direction. These three states had started their exercise much before Andhra Pradesh stepped in. On the other hand, the Indian Network for Participatory Irrigation Management (INPIM), should play a more active role at two levels -- the central level and state level. It should help restructure all state training institutes (like WALMIs) to respond to the emerging needs of the state irrigation sector. The IndianPIM can facilitate at state level: formulation of policy and legal framework formulation, guidelines for monitoring and evaluation of WUAs and irrigation management transfer activities, and for training farmers and officials at various levels.

Table A-1
Water Resources of Andhra Pradesh

River System	*Drainage Area in the State in '000 km*	*% of Drainage Area to the Total Area of the State*	*Assessed Annual Yield in the State (mha)*	*Remarks*
Godavari	73.201	26.45	4.23	Assessed
Krishna	74.382	26.88	2.30	Allocated
Pennar	48.111	17.39	0.28	Assessed
Nagavalli	4.833	1.75	0.14	Assessed
Vamsadhara	1.934	0.70	0.04	Allocated
Other minor rivers draining into the sea	74.239	26.83	0.79	Assessed
Total	276.700	100.00	7.78 (2746TMC)	

Source: NWMP Phase II: Vol.I Project Profile, Andhra Pradesh.

Table A-2
Andhra Pradesh Profile of Plan Expenditure and Irrigation Potential Created

Period	*Amount Spent (Rs. in million)*			*I.P. Created (mha)*		
	Major and Medium	*Minor*	*Total*	*Major and Medium*	*Minor*	*Total*
Pre-Plan Period 1951	-	-	-	1.332	1.372	2.704
First Plan (1951-56)	374.7	35.2	409.9	0.079	0.025	0.104
Second Plan (1956-61)	574.3	43.8	618.1	0.181	0.016	0.197
Third Plan (1961-66)	915.2	186.0	1101.2	0.368	0.050	0.418
Three Annual Plans (1966-69)	608.7	108.1	716.8	0.078	0.037	0.115
Fourth Plan (1969-74)	1187.1	181.5	1368.6	0.190	0.063	0.253
Fifth Plan (1974-78)	2691.1	388.2	3079.3	0.213	0.092	0.305
Two Annual Plans (1978-80)	2576.9	237.9	2814.8	0.154	0.057	0.211
Sixth Plan (1980-85)	7295.9	411.0	7706.9	0.305	0.082	0.387
Seventh Plan (1985-90)	13064.0	1314.0	14378.0	0.089	0.067	0.156
Annual Plan (1990-91)	2827.5	477.7	3305.2	0.006	0.009	0.015
Annual Plan (1991-92)	3339.2	485.8	3825.0	0.009	0.007	0.016
TOTAL	35454.6	3869.2	39323.8	3.004	1.877	4.881
Eighth Plan (1992-1997) anticipated	22653.2	1869.2	24522.4	0.042	0.040	0.082
Grand Total	58107.8	5738.4	63846.2	3.046	1.917	4.963

Note: (1) The above figures do not include expenditure on CADA upto Seventh Plan end. (2) The above figures have been adopted from Eighth Plan (1990-95) AP-F&P (PLG) Department, October 1990, Vol.11, Sectoral Programme.

Table A-3

Average Yield per Hectare, Area and Out-turn of Principal Crops 1993-94

Crop	*Region*	*Estimated Average Yield (in kgs. per hectare)*	*Area under the Crops ('000 hects)*	*Production ('000 tonnes)*
Rice*	Coastal Andhra	2,952	2,249	6,417
	Rayalaseema	2,622	293	752
	Telangana	2,381	1,005	2,393
	Andhra Pradesh	2,759	3,547	9,562
Jowar	Coastal Andhra	945	55	52
(Sorghum)	Rayalaseema	1,037	215	223
	Telangana	763	783	597
	Andhra Pradesh	828	1,053	872
Bajra	Coastal Andhra	943	62	59
(Millet)	Rayalaseema	1,034	32	33
	Telangana	374	49	18
	Andhra Pradesh	778	143	110
Ragi	Coastal Andhra	1,059	79	84
(Finger Millet)	Rayalaseema	1,775	29	51
	Telangana	874	32	28
	Andhra Pradesh	1,164	140	163
Maize	Coastal Andhra	2,497	30	75
(Corn)	Rayalaseema	2,292	2	4
	Telangana	2,562	272	697
	Andhra Pradesh	2,554	304	776
Sugar Cane	Coastal Andhra	72,837	187	8696
	Rayalaseema	79,048	50	2348
	Telangana	75,590	68	4013
	Andhra Pradesh	74,464	305	15057

Source: Directorate of Economics and Statistics, Andhra Pradesh.
* Paddy is 1.5 times the finished product rice.

Table A-4
District/Sectorwise WUAs in Andhra Pradesh
(as on 20 May 1999)

Name of the District	*Total No. of WUAs Notified*			*Total Notified*
	Major	*Medium*	*Minor*	
Adilabad	35	27	221	283
Ananthapur	46	7	305	358
Chittoor	0	51	644	695
Cuddapah	74	8	276	358
East Godavari	106	12	225	343
Guntur	245	8	81	334
Karimnagar	249	10	586	845
Khammam	50	4	183	237
Krishna	189	12	288	489
Kurnool	116	12	153	281
Mahabubnagar	21	31	478	530
Medak	0	12	585	597
Nalgonda	91	45	538	674
Nellore	110	58	695	863
Nizamabad	78	13	267	358
Prakasam	124	5	317	446
Ranga Reddy	0	3	165	168
Srikakulam	37	28	459	524
Vishakhapatnam	28	18	375	421
Vizianagaram	0	25	439	464
Warangal	29	18	683	730
West Godavari	71	6	217	294
Total	1699	413	8180	10292

Table A-5
District/Sectorwise WUAs (as on 23 March 2000)

Name of the District	Total No.of WUAs Notified			Total Notified (3+4+5)	WUAs Formed			Total Elections Conducted (7+8+9)	Elections Not Held due to			Total Elections to be held (11+12 +13)	Total No.of WUAs (10+14)
	Major	Medium	Minor		Major	Medium	Minor		Stay by Court	Stay by Govt.	Other Reasons		
Adilabad*	35	27	221	283	35	27	212	274	3	0	6	9	283
Ananthapur	46	7	305	358	46	7	302	355	3	0	0	3	358
Chittoor	0	51	644	695	0	48	616	664	1	15	15	31	695
Cuddapah	74	8	276	358	74	8	259	341	8	4	5	17	358
East Godavari#	106	12	225	343	106	12	215	333	7	0	3	10	343
Guntur	245	8	81	334	239	8	76	323	3	6	2	11	334
Karimnagar	249	10	586	845	249	10	571	830	4	0	11	15	845
Khammam	51	5	181	237	50	5	180	235	0	2	0	2	237
Krishna	189	12	288	489	166	12	256	434	1	20	34	55	489
Kurnool	116	12	153	281	114	12	145	271	4	2	4	10	281
Mahabubnagar	21	31	478	530	21	29	473	523	1	6	0	7	530
Medak	0	12	585	597	0	12	551	563	2	0	32	34	597
Nalgonda	91	45	541	677	91	44	541	676	0	1	0	1	677
Nellore	110	58	695	863	100	57	612	769	22	72	0	94	863
Nizamabad^	78	13	267	358	78	13	228	319	3	0	36	39	358
Prakasam	124	5	317	446	113	4	291	408	16	11	11	38	446

Contd...

Contd...

Rangareddy	0	3	165	168	0	2	165	167	0	0	1	1	168
Srikakulam	37	28	459	524	37	28	442	507	6	1	10	17	524
Visakhapatnam**	28	18	375	421	28	18	369	415	0	1	5	6	421
Vizianagaram	0	22	439	461	0	21	422	443	0	0	18	18	461
Warangal	29	18	683	730	28	18	623	669	1	0	60	61	730
West Godavari	71	6	217	294	70	6	205	281	1	0	12	13	294
Total	1700	411	8181	10292	1645	401	7754	9800	86	141	265	492	10292

Notes: *Includes 9 WUAs under Vattivagu & Chelamalavagu.
**Includes TANDAVA (Part Only)
#Includes 12 WUAs under TANDAVA
^Includes 7 WUAs under Koulasnala

NOTES

1. This paper is based on a short study, commissioned by the Indian Network for Participatory Irrigation Management. The study was prepared with cooperation of Raymond Peters, Commissioner, CAD and the Ex-Officio Secretary, Irrigation and Command Area Development Department, currently Executive Director of INPIM, the World Bank, and Irrigation Department staff across Andhra Pradesh State and the farmers. The author is grateful to all of them for willingly sharing their experiences and actively participating in discussions. He is also thankful to Dr.Douglas L.Vermillion, Ruth Meinzen-Dick and Shashi Kolavalli for useful comments on the earlier version of this paper.
2. A purely governmental approach, wherein poor system performance has led to farmer dissatisfaction, low fee payment, low cost recovery, and under-funding of O&M (for more details see Ashok Gulati, Ruth Meinzen-Dick, and K.V. Raju, 1999, *From Top Down to Bottom Up: Institutional Reforms in Indian Canal Irrigation*. A collaborative study of IFPRI-NCEAR-IEG-ISEC).
3. The important steps are: (a) establishment of a state multi-sectoral water resources board or committee to guide development of actions; (b) development of multi-sectoral river basin plans and environmental management plans to provide guidance for management and future development; (c) continued development of watershed management practices and integrated conjunctive use of surface and groundwater resources; (d) promotion of technologies for greater water use efficiency, with emphasis on irrigation; (e) progressive development of institutional and human resource capabilities in water resources management; and (f) public awareness campaign on water resources management issues to foster an environment for change.
4. A separate handbook for WUAs and field functionaries on participatory irrigation management was also prepared by IRDAS (Institute of Resource Development and Social Management) and WALAMTARI.
5. (a) Telangana region has 39 percent of the state's population and 42 percent of its geographical area, (b) Coastal Andhra, 43 percent population and 34 percent area, and (c) Rayalaseema region, 18 percent population, and 24 percent area.
6. Owing to these induced works, WUA members of Elabotharam village in SRSP project in the Telangana region feel they have come together and can take more collective action now. In the Narmala village of Upper Manor dam, desiltation of canals was done after 14 years. In Thotapalli tank-WUA, farmers contributed carts and tractors for desilting at a lower rate of Rs. 100 per trip, which is Rs. 30 less than the market rate. The Bajjipuram tank in Coastal Andhra had three surplus weirs at different levels; now there is just one and this helps in having more storage.
7. In spite of cheaper power rates, many farmers have not paid up the bills. Because, the Congress party had promised in their electoral manifesto (of Parliamentary elections, held in October 1999) that they would reduce the current rates and abolish all arrears. These arrears range from Rs. 4,000 to Rs. 35,000 per farmer, with a burden of about Rs. 3,000 million per year on the state exchequer.

8. Duggirala and Emani WUAs' farmers in Krishna delta area have honoured WUA presidents and ID staff for their tremendous efforts in water supply and cleaning canals. They had conducted public functions and honoured these guests by presenting shawl on their shoulders. Now some farmers jokingly say, if they do not perform well, the same farmers will take back the shawl.
9. The World Bank funding for WUAs through Andhra Pradesh Economic Restructuring Programme (APERP) arrived after the whole process had started. Under APERP major items of expenditure (during 1998-99) was a total of Rs. 1,114 crores, and it included: (a) Minimum rehabilitation, Rs. 336 crores; (b) SIFT, Rs. 120 crores; (c) Operation and maintenance, Rs. 600 crores; (d) Management Information System, Rs. 29 crores; and (e) Human Resources Development, Rs. 29 crores.
10. One hectare = 2.47 acre.
11. For e.g., at a division level (in delta area), of Rs. 20 million (at the rate of Rs. 40 per acre), nearly Rs. 13 million goes to salaries. At the sub-division level (at Duggirala in Krishna delta) the Rs. 40 per acre grant was Rs. 6 million, out of which work-charge staff salaries are Rs. 3.5 million, leaving a balance of Rs. 2.5 million for works. In 1999, this sub-division got Rs. 26 million for minimum rehabilitation through WUAs. Clearly a four times increase in funds allocation; and all these without any salary component.
12. For example, according to one of the CA in SRSP project, after DC and WUA formation, irrigated area increased to 3,600 acres from 2,000. The CA said that additional 30 percent was owing to revision of revenue records, and 70 percent owing to better structures and improved maintenance. The SRSP would gain additional 20 percent of irrigated area, if staff could reconcile groundwater irrigated area (now groundwater area is not recorded, properly).
13. The study team also looked at cost of paddy production in Dharmavaram tank-WUA. In this command, generally only one crop is cultivated -- paddy in kharif and no crop during rabi. One acre of paddy produces some 25-30 bags (each bag of 70 kgs); gross value Rs. 15,000; labour costs Rs. 7,000 (5,000+2,000); chemicals and pesticides Rs. 2,500; net profit Rs. 6,000, according to local farmers.
14. But, exact specific figures were not readily available.
15. These records should include storage levels, reservoir size, tank bed area, canal boundary, catchment area, authorised command area with survey numbers, structures, and other irrigation and revenue details.
16. For instance, in Vamsadhara project some 40,000 acres are receiving irrigation by unauthorised lifts through private diesel engines from the canals. This practice was on for several years; yet, official records do not indicate this.
17. For details see "Report of the Small Study Group appointed by the Government of Gujarat to evolve a plan for restructuring the Water and Land Management Institute, Gujarat", April, 1999 (draft).
18. The Institute needs to utilise all those staff trained under USAID assistance, both within and outside the country (currently most of the trained staff had been posted elsewhere in the state).
19. A large number of rainfed tanks in Vishakhapatnam district are of the pre-independence vintage. Most of them have a designed command area of 150 acres and more. In 1953 the Irrigation Department took over these tanks, which hardly

had any repair and maintenance work undertaken. Some of them have silt close to their surplus weir level. Most of the silt comes from nearby hillocks owing to flash floods, high slope, and deforestation in the upper stream and lack of catchment area treatment. What is required is to desilt for sustainable irrigation and increase irrigated area.

20. For example, Karnataka, a neighbouring state, has mounted a separate project to rehabilitate 5,000 minor irrigation tanks on community based approach. The project has taken an integrated approach for the tank system as a whole for forming the stakeholders association, which is quite different from WUAs in major and medium irrigation projects.
21. In India, such member-controlled organisations and their success stories have proved that it is possible to effectively function that way, e.g., three-tier milk societies in many states which has created white revolution in the country; apple growers cooperatives in Himachal Pradesh and Uttaranchal, which has strengthened growers and at the same time, delivered apples round the year all over the country. Locally, Andhra Pradesh can look at the Mulkanoor Multi-purpose Cooperative Society for its effective member-controlled and good servicing.

REFERENCES

Government of Andhra Pradesh. 1999. Vision 2020: Swarna Andhra Pradesh. Hyderabad: Government of Andhra Pradesh.

Government of Andhra Pradesh.1999. *A Compendium of Government of Orders on Farmers' Management of Irrigation Systems in Andhra Pradesh*. Hyderabad: Water and Land Management Training and Research Institute (WALAMTARI).

Government of Andhra Pradesh. 1998. APERP Project Document.

Government of Andhra Pradesh. 1997. *The Andhra Pradesh Farmers' Management of Irrigation Systems Act (Act 11 of 1997): Act and Rules.* Hyderabad: Department of Irrigation and Command Area Development.

Government of Andhra Pradesh. 1998. Andhra Pradesh Economic Restructuring Project: Irrigation Component. Project Implementation Plan. Hyderabad: Engineering Staff College of India, May 1998.

Sreedhar, G. 1997. *Tank Irrigation in Semi-Arid Zones*. Bangalore: Sunrise Publications.

Government of India. 1998. Amendment to Irrigation Act for Participatory Irrigation Management (Revised-Draft). Prepared by SOPPECOM, Pune. New Delhi: Ministry of Water Resources.

Maloney, Clarence, and K.V. Raju. 1994. *Managing Irrigation Together: Practice and Policy in India*. New Delhi: Sage Publications.

Gulati, Ashok, Ruth Meinzen-Dick, K.V. Raju. 1999. From Top Down to Bottom Up: Institutional Reforms in Indian Canal Irrigation. A Collaborative Study of NCEAR, IFPRI, IEG, and ISEC, supported by the Ford Foundation (monograph).

Raju, K.V. and Tushaar Shah 1999. Rehabilitation of Minor Irrigation Tanks in Rajasthan: Project Document (Draft for Discussion). Report prepared for the Swedish International Development Agency, New Delhi.

5

Sustainability of Water Users Associations in Andhra Pradesh

C. SITHAPATHI RAO, K. SOMESHWAR and C. VIJAYA SHYAMALA

Implementation of the policy of participatory irrigation management (PIM) through farmers' organisations, covering all types of irrigation systems in the entire state of Andhra Pradesh (AP) is very complex since it has to cross through various political, economic, and administrative barriers. A series of consultations with farmers, public representatives, administrators and various concerned agencies were held in the state to evolve a strategy for the implementation of PIM.[1]

To provide the legal status, strength and stability to the strategy, a suitable legislation was made in the state through the enactment of the Andhra Pradesh Farmers' Management of Irrigation Systems (APFMIS) Act, 1997.

The introduction to the Act states:[2]

> "whereas the state of Andhra Pradesh is essentially an agricultural state depending on an efficient and equitable supply and distribution of water which is a National Wealth, ensuring optimum utilisation by farmers for improvement of agricultural production is the immediate need;
>
> And whereas scientific and systematic development and maintenance of irrigation infrastructure is considered best possible through farmers' organisation;
>
> And whereas such farmers' organisations have to be given an effective role in the management and maintenance of the irrigation system for effective and reliable supply and distribution of water;
>
> Be it (Act) enacted by the legislative assembly of the state".

The APFMIS Act facilitates the farmers' organisations (FOs) to function effectively in nine different ways:

(1) gives water rights and control of the irrigation system;
(2) provides for functional and administrative autonomy to the FOs;
(3) ensures timely and effective maintenance of the irrigation system within the area of operation;
(4) enables to prepare its own operation plan and water distribution based on the water allocation, to serve the entire command;
(5) crop choices best suited to them;
(6) makes the irrigation department accountable to the FOs;
(7) empowers the FOs to resolve the conflicts within themselves;
(8) accountability to all the farmers in their area; and
(9) transparency in all its activities.

Under the provisions of the Act, three-tier FOs are envisaged. At the primary level there will be the Water Users' Association (WUA) covering a well-defined hydraulic unit with a group of minors. The group of WUAs along a distributory or distributories are federated to form the second tier called the Distributory Committee (DC). The DCs of the entire irrigation system will form the Project Committee (PC). One tier (WUAs) under minor irrigation system, two tiers under medium irrigation projects and three tiers under major projects have been envisaged as per the Act.

After the enactment of the APFMIS Act, the process of formation of WUAs covering all the irrigation projects in the state was initiated. The entire district administration under the District Collector was mobilised and WUAs were formed following the normal democratic processes. In all, around 10,292 WUAs and 173 DCs have been formed covering a total command area of 4.79 mha.

Irrigation Projects	*No of Projects*	*No. of WUAs*	*No. of DCs*	*Total Command (mha)*
Major	13	1729	173	3.20
Medium	63	435	-	0.39
Minor	8128	8128	-	1.20
Total		10292	173	4.79

The implementation of the PIM programme in AP covering all irrigation projects simultaneously is referred to as "big bang" approach[3]. It is also said that "this is the greatest transfer of power since 1947 (and) the state for the first time in history is abandoning control of irrigation and thereby over the beneficiaries and with this we will see solid structure, based on democratic management of vital resources (water) spread all over the state of AP".[4] These pronouncements and expectations have made the entire process of PIM in AP a big movement of the millennium and obviously all eyes are focussed on the working of the WUAs and their sustainability.

Partnership Management for Sustainability

The farmers' organisations constituted with the "big bang" approach, need to be initially given all the support for strengthening them on a sustainable basis. Thus, the PIM programme in the state is a partnership management which involves the government agency (irrigation department), stakeholders (farmers' organisations) and supportive agencies like non-governmental organisations (NGOs). While the irrigation department is a facilitator in various activities of the PIM, the NGO has a role in building up the skills and capacities of the farmers' organisation as shown below:

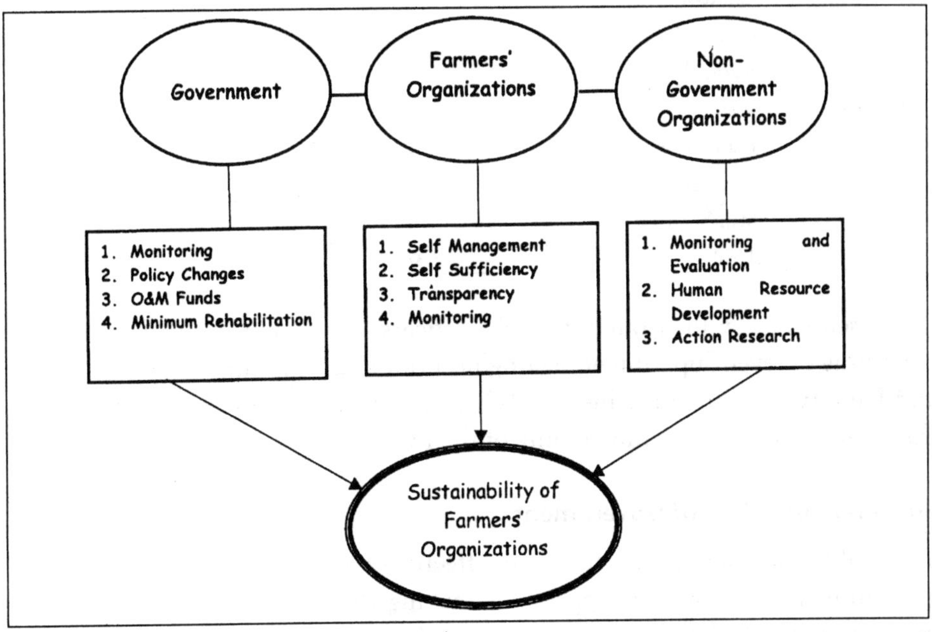

Methodology and Sample

To understand the functioning of the FOs in relation to their sustainability, a field survey spread out in 9 districts (out of 22 districts), covering around 37 WUAs under major and medium projects and 32 minor irrigation sources, was carried out. The sample covers all the three regions of the state (Table 1). In addition, information was collected from government sources.

Discussions were held with presidents, executive committee members and farmers in each of the sample WUA area.

For the effective functioning of the FOs, a constant and regular interaction with the irrigation department is essential, which provides the required technical assistance and management support to them. In APFMIS Act, as per clause 21, a field officer of the irrigation department who is called as "competent authority" is attached to each WUA/DC to provide this input. These officers have been assisting the WUAs in various aspects of their functioning.

Table 1
Sample Size

Region	*District*	*Sample WUAs*	
		Major/Medium Projects	*Minor Projects*
Coastal	Krishna	4	-
	West Godavari	5	-
Rayalaseema	Chittoor	1	5
	Cuddapah	2	5
	Kurnool	5	7
Telangana	Warangal	5	5
	Karimnagar	10	5
	Nizamabad	4	5
	Mahabubnagar	1	-
	Total	37	32

Discussions with around 30 field officers of the irrigation department, consisting of seven Superintending Engineers (SEs), 23 Executive Engineers (EEs) and Deputy Executive Engineers (DyEEs) across the state were held to gather their perceptions on the functioning of the FOs.

Sustainability: Role of Government

To enable the WUAs to function effectively on a sustainable basis, the government has provided support for making the system functional through maintenance and rehabilitation. Requisite policy changes through administrative orders were also made.

System Improvement

Due to various reasons, in all the irrigation projects the distribution system and drainage network have not been maintained and their operational efficiency has come down very much. To enable the water flow and its distribution to all the outlets in the system, a certain minimum maintenance is required.

The enhanced water rates in AP, operationalised from 1997-98 will be adequate to meet this expenditure. The enhanced water rates[5] are expected to flow to WUA from about 1999-2000 or so. However, in the meanwhile, for enabling the WUAs to function effectively funds were provided directly to WUA and DC for the years 1998-99 and 1999-2000. The O&M grants from government will cease beyond this period, as it will have to be a self-sustaining activity.

Earlier the little grant[6] under operation and maintenance (O&M) was spent in some isolated pockets as decided by the concerned irrigation authorities. With the ushering in of the WUAs, each organisation was provided funds for taking up maintenance activity directly as per their own priority and need felt at the ground level. The major advantage of this has been that the activity got spread out in the entire command area and large number of maintenance works were taken up in the past two years through the WUAs (Table 2).

Table 2
Maintenance Activities by WUA

Year	*Number of Maintenance Activities*		*Value (Rs. million)*	
	Planned	*Executed*	*Planned*	*Spent*
1998-1999	21,267	19,355	941	769.9
1999-2000 (Between April to June 1999)	11,722	7,041	457	304.4

These activities cover all the major and minor irrigation projects in the state[7] (Table 3).

Table 3
Maintenance Activities in Different Irrigation Systems

Year	*Irrigation Systems*	*WUA*		*DC*		*PC*		*Total*	
		A	*B*	*A*	*B*	*A*	*B*	*A*	*B*
1998-1999	Major	7943	7779	2120	2051	710	628	10773	10458
	Medium	868	653	-	-	208	161	1076	814
	Minor	9418	8083	-	-	-	-	9418	8083
	Total	18229	16515	2120	2051	918	789	21267	19358
1999-2000 (working season April/May/ June '99)	Major	2146	1003	1056	629	42	23	3244	1655
	Medium	211	112	-	-	20	8	231	120
	Minor	8297	5266	-	-	-	-	8297	5266
	Total	10654	6381	1056	629	62	31	11772	7071

Note: A = No. of works planned; B = No. of works completed.

It is estimated to have provided to an additional area of 4 lakh ha (Annexure 1).

A minimum rehabilitation programme is taken up to improve the performance of the system. In this the contribution of WUA is considered necessary and the activities should be transparent and approved by all farmers. This is mainly concentrated in major and medium irrigation projects and considerable work has been done in the past two years (Table 4).

The AP Government Order Ms. No.27 dated 6 March 1999 of the I&CADD

"It is decided that:

a. The WUA contributes 15% of the total cost of the works taken up under Minimum Rehabilitation in the area of their operation, in cash or kind, in terms of the stipulation in the Projects Report;
b. Works to be executed by the Water Users' Associations/Distributory Committees and not be sublet to any contractor;
c. Works taken up be restricted to the amounts allocated for each Farmers' Organization taken up works beyond the funds allocated to it;
d. Every work to be taken up by the WUA/DC be approved by the general body of the respective organization through a resolution to that effect."

Table 4
Minimum Rehabilitation Activities

Year	*Number of Works Under MR*		*Value (Rs. million)*	
	Planned	*Completed*	*Planned*	*Utilised*
1998-1999	2266	468.64	2051	394.84
1999-2000	5643	1708.70	4212	943.00

Policy Changes

The functioning of the WUAs in the past two years was periodically reviewed by the government. Based on this, changes were brought about to promote sustainability of WUAs.

Members of WUAs

For ensuing successful functioning of the organisation it is essential that all the users in the system are involved. This is possible when the members in the organisation are the actual users or beneficiaries. The conventional method is to prescribe a membership fee and forming a WUA only when more than half the group is enrolled. Enforcing any discipline among the non-members becomes difficult.

In AP, right from the beginning it has been considered essential that in a WUA all the landholders and tenants in its operational area, whose names are recorded in the official records, should be the members with voting rights.[8]

However, it was observed that the names of individuals who are actually cultivating the land under a water source are not recorded due to one reason or other. Hence, an amendment was made to remove this lacuna, so that all farmers who are using water under the system will be members of a WUA.

> *Amendment through A.P. Legislative Bill No. 32 dated 24 November 1998 to Section 3(4) (i)*
>
> It provides:
>
> "any person who is in lawful possession and enjoyment of the land under a water source, on proof of such possession and enjoyment in a crop year may claim membership notwithstanding, whether he is a recorded landholder or not".

Filling Up Vacancies of Elected Members in WUA/DC

For filling up vacancy of a member, caused due to resignation/ recall or any other reason, it was originally indicated in the Act that the next higher body would take action to fill it up. In case of minor irrigation projects there is no next higher body and problems were faced in the immediate action. To remove this bottleneck, an amendment was made empowering the members of the managing committee to take action.[9]

> *AP Government Order Ms. No.362 dated 10 October 1998 of I&CADD*
>
> "In rule 9 after clause (c), the following shall be added, namely:
>
> d(i) If any vacancy in the office of the President of the WUA of *Minor Irrigation* arises, the members of the Managing Committee shall choose one of the Managing Committee members to officiate as President during the absence of the elected President.
>
> (ii) If a vacancy in the office of a President of WUA of *Major Irrigation* arises, the President of the Distributory Committee may nominate one of the existing Managing Committee members of the WUA to officiate as the President during the absence of the elected President.
>
> (iii) If a vacancy in the office of a President of WUA of *Medium Irrigation* arises, the chairman of Project Committee shall nominate one of the existing Managing Committee officiate as President."

Administrative Expenses

All WUAs are in the early stage of their growth and need some financial support to establish themselves. It was thought desirable that some initial support is

> *The AP Government Order Ms.No.74 dated 17 May 1999 of the I&CADD*
>
> "In order to meet the expenditure towards the administrative expenditure of Farmers Organizations and as represented by them, government approved the following for 1999-2000:

i. For Distributory Committees 174 Nos. @ Rs.25,000	Rs. 43.50 lakhs
ii. For WUAs (Major Irrigation) 1729 Nos. @ Rs.10,000	Rs.172.90 lakhs
iii. For WUAs (Medium Irrigation) 435 Nos. @ Rs.5,000	Rs. 21.75 lakhs
iv. For WUAs (Minor Irrigation)8128 Nos. @ Rs.2,500	Rs. 203.20 lakhs
Total	Rs.441.35 lakhs

provided. This is made as a grant by the AP government to meet the administrative expenditure.

Planning Irrigation and Drainage at District Level

So far, at the district level (administrative unit with about 1,000 to 1,200 villages) an official body called District Irrigation Development Board has been functioning. To provide an effective role to WUA this is reconstituted as Advisory Board.

The AP Government Order MS. No.18 dated 28 November 1999 of the I&CADD

Functions: The functions of the Board are to advise the Government on the following matters pertaining to district:

1. Irrigation and drainage
2. Canal closure and opening -- crop calendar for project
3. Planning of the irrigation and drainage schemes and works
4. Environment issues relating to irrigation and drainage

"Government reconstituted District Irrigation Development Boards which is hereby renamed as 'Irrigation Advisory Boards' in modification of the orders issued in the Government Order read above as under:

The Board consists of the following members:

i.	All Members of Parliament of the District	Members
ii.	All Member of the Legislative Assembly of the District	Members
iii.	Superintending Engineer of the District	Member Convener
iv.	Collector of the District	Chairman
v.	Chairman, Project Committee	Member
vi.	All Distributory Committee Presidents within a District	Members
vii	Two Water Users Association Presidents (Minor Irrigation) from each of the Revenue Division in the District	Members
viii	Joint Director of the Agriculture Department	Member
ix	Assistant Director of Horticulture	Member
x	Revenue Divisional officer	Member
xi	Chief Planning officer	Member

Sustainability: Role of Farmers' Organisations

Self Management

The sustainability of the FOs essentially depends on their capacity to operate and manage the system on their own and equitably serve all the farmers in the command.

One of the important factors for the sustainability is to convey the water to various points within the command. For this, the distributory network needs to be made functional by way of removal of silt, vegetation, repairs of structures and sluices, strengthening the bunds, etc.

All the WUAs have taken up the maintenance activity in the last two working seasons with the financial support provided by the government (Table 5).

Table 5
Maintenance Activities

Activity	*Irrigation Project*	
	Major/Medium (%)	*Minor (%)*
Silt removal	96	84
Vegetation clearing	95	62
Repairs to structures	73	81
Repairs to sluices	81	43
Strengthening bank	95	81
Breaches filling	70	25
Improving drainage	65	19

Water Management

Due to the physical improvements in the irrigation system, the water availability in all the schemes has improved (Annexure 1). However, detailed operational plans are still missing. This needs to be pursued further.

Self Sufficiency

Self-sufficiency in finance plays an important role in the sustainability of an organisation. At present, for maintenance the support is largely from the government. However, with the transfer of water cess, which is approved by government, they will be self-sufficient. However, as of now, there is little effort to generate their own resources.

Transparency and Accountability

The general body meeting is the main forum for achieving transparency and accountability in the functioning of a WUA. The managing committee meetings, which are held more frequently, enable them to firm up working details.

In the past two years of their working, each WUA is expected to have organised four general body meetings (Table 6a).

There is need to encourage the WUA to conduct more number of general body meetings so that the transparency is increased.

Table 6a
Meetings

Number of General Body Meetings	*Irrigation Project*	
	Major/Medium (%)	*Minor (%)*
Less than 2	24	19
2 to 4	54	59
Above 4	22	22

Similarly the frequency of managing committee meetings can be an indication of the democratic pattern of working of the WUA. Even though the numbers vary in each WUA, its overall appreciation is satisfactory (Table 6b).

Table 6b
Meetings

Number of Managing Committee Meetings	*Irrigation Project*	
	Major / Medium (%)	*Minor (%)*
Above 15	22	3
10 to 15	32	47
Below 10	46	50

Sustainability: Role of Non-Governmental Organisations

Non-governmental institutions like IRDAS are giving support in the Training of Trainers (TOT) programmes for the promotion of PIM. There is, however, a need to involve local NGOs at the field level to give an intensive support to the FOs in skills and capacity building. They could also help in the monitoring and evaluation of the FOs so that corrective measures can be taken to promote sustainability.

Impact of Working of WUAs

The sustainability ultimately depends on the benefits derived by the farmers in the command. These benefits are perceived in different ways (Table 7).

Table 7
Benefits Derived

Benefits	*Irrigation Project*	
	Major/Medium (%)	*Minor (%)*
Improvement in the irrigation system	81	100
Increase in the area irrigated	54	31
Availability of irrigation to tail-end areas	46	37
Increase in crop yields	48	37

Analysis and Conclusions

The study throws light on the fact that there is some positive impact of the functioning of the WUAs. While there have been factors which have contributed to this, the study also highlights the weaknesses which need to be tackled for achieving sustainability of the FOs.

The contributing factors are essentially the legal status of the WUA and freedom to decide its working. The weaknesses, which need to be taken care of, are political interference, presidents turning out to be contractors, lack of operational plans, etc.

Contributing Factors for Sustainability

1. Legal status of the organization.
2. Social status of the elected members. Recognition in the society as people's representatives.
3. Platform for expressions of the members.
4. Freedom in deciding about maintenance activities and better quality of works.
5. Strong will of the government to provide the necessary support.
6. Improvement in the conveyance system resulting in increase in irrigated area.
7. Recall provision -- enabling replacement of errant members.
8. Sense of ownership of the system.

Weaknesses

1. President working as contractor.
2. Members not attending general body meetings.
3. Political interference.
4. Non-availability of maps of the command.
5. Lack of operational plans and wastage of water.
6. Full dependency on competent authority.
7. Misuse of power and money (sometimes).
8. Lack of training to farmers.
9. Lack of regular monitoring mechanism.
10. Not tapping the resources of the local NGOs.

Future Perspectives

For the sustainability of the WUAs, even though adequate groundwork and the required environment is created, there is a need to look into the weaknesses and remedy them. More importantly, a regular monitoring mechanism has to be put into operation to initiate corrective measures as and when needed. Training has to be a continuous process to build up the skills and capacities of the WUA to meet the day-to-day challenges.

Annexure-1

Andhra Pradesh Economic Restructuring Project (Irrigation Component)
Bridging of Gap Ayacut

S. No	*Name of Project*	*Gap Ayacut Bridged in 1998-99 (in acres)*	*Gap Ayacut Anticipated During 1999-2000 (in acres)*	*Total (in acres)*
I	*Major Irrigation*			
	Vamsadhra Project	17,000	874	17,874
	N.S.Left Canal	50,000	40,000	90,000
	N.S Right Canal	53,000	30,000	83,000
	Tungabhadra Project (H.L.C)	10,000	13,000	23,000
	Tungabhadra Project (L.L.C)	11,000	5,000	16,000
	Rajolibanda Diversion Scheme (RDS)	2,000	2,500	4,500
	Nizamsagar Project	74,000	17,000	91,000
	Kadam Project	23,000	12,000	35,000
	Sriramsagar Project (SRSP)	2,51,000	75,000	3,26,000
	K.C.Canal	21,000	10,000	31,000
	Sub-Total – I	5,21,000	2,05,374	7,17,374
II	*Medium Irrigation*			
	Coastal Andhra Region	-	30,640	30,640
	Rayalaseema Region	-	8,649	8,649
	Telangana Region	-	27,428	27,428
	Sub-Total – II	-	66,717	66,717
III	*Minor Irrigation*			
	Coastal Andhra Region	-	13,370	13,370
	Rayalaseema Region	-	42,000	42,200
	Telangana Region	-	1,68,100	1,68,100
	Sub-Total - III	-	2,23,670	2,23,670
	Grand Total	5,12,000	4,95,761	10,07,761

Note: Status Note of the Chief Engineer, dated 25 October 1999.

NOTES

1. Raymond Peter J. (1998) "Management of Irrigation Systems by Farmers in AP, India -- the Process", Fourth International Seminar, Bali, Indonesia.
2. The APFMIS Act No.11 of 1997 was published in the Andhra Pradesh Gazette on 9th April, 1997.
3. Raymond Peter J. (1998).
4. Gautham Pingle (1997), a senior faculty in the area of Public Policy and System, Administrative Staff College of India, elaborates this in his article on "Tiding Over Water Crisis", published in the daily *Newstime* of 11th November,1997.
5. The water rates have been raised from Rs.150 per ha, per crop season to around Rs. 500 per ha, per crop as per A.P. Government Act No.13 of 1997.
6. The earlier O&M grant was Rs. 100 per ha, and around 70 to 80 percent was used to meet the salaries of the field staff.
7. The information is gathered from the Status Notes on AP Economic Restructuring Project prepared in October, 1999.
8. Under Section 3(4) (I) of the APFMIS Act, 1997, this is specified.
9. An amendment was made through Government Order No.362 in November, 1998.

6

Performance of Water Users Associations: A Case Study in Krishna District

M. JAYA RAJ

The demand for water from non-agricultural sectors is on the rise as it is essential for improving the quality of life. Even though it reduces the availability of water for irrigation purposes, this demand cannot be overlooked. This certainly warrants improvement in storing, handling and delivering water as per the requirements of crops for increasing production per unit of water (FAO, ICD Paper 24, 1984). Huge investments have been made for creating irrigation potential in the state but the experience revealed by eminent irrigation engineers shows that the gap between potential created and utilised, under major and medium irrigation projects, was increasing. This gap was said to be around 35 to 55 percent under major and medium irrigation projects (Rao, 1988).

The Commission on Irrigation Utilisation and Command Area Development (1982) has stated the following reasons for the gap between potential created and utilised:

- Inadequate allocation of funds for the maintenance of irrigation systems.
- Defective localisation of area in the command into irrigated dry (ID) and wet cultivation.
- Conversion of ID cops into wet crops in the commands which requires more quantity of water.
- Head-reach farmers on the canals drawing more water and tail-end farmers deprived of their share.
- Performance efficiency of the irrigation systems was at a low level due to social and technical problems in management of irrigation water.
- Due to the diversified crops and cropping systems in the command, the irrigation scheduling became a major problem.
- Unauthorised irrigation by farmers lead to the inequitable disbursement of irrigation services.

There may be many other problems which lead to the gap between potential created and utilised under major and medium irrigation projects in the state.

There was a great scope for improving the performance in terms of productivity and earning through improved management of services to farmers in the field of water distribution in the commands of various irrigation projects (Bottral, 1981). Narayana (1986) stressed the need for people's participation in operation and maintenance of the irrigation systems. The lack of involvement of farmers paved the way for increasing the gap between potential created and utilised. The discontent in the minds of the farmers over the management of irrigation systems and inequitable distribution of water to the crops increased the desire to manage the systems on their own.

Abel (1997) identified four kinds of inputs for good water management:

- Recognition of water as a scarce factor for productivity.
- Centralised planning for irrigation investments and decentralised management of the irrigation systems.
- Exchange of agronomic and engineering information between users of water and managers of the systems.
- Incentives for both managers and users of water.

To overcome many of the problems of management of irrigation systems, the Andhra Pradesh government constituted 10,292 WUAs under the provisions of APFMIS Act, 1997. The WUAs were empowered with administrative, legal and technical powers to mange the irrigation systems in the state.

Since the inception of WUAs, farmers were involved in management of irrigation systems with the financial support of Andhra Pradesh government. Many minimum rehabilitation works were undertaken by the WUAs.

The success or failure of management of irrigation systems in the state depends upon the performance abilities of WUAs. This, in turn, depends on sociological, physiological and economic factors of the irrigation systems, which make them a viable social unit at the command area level. To study the extent of benefits accrued by involvement of farmers in management of irrigation systems, an attempt has been made with the following objectives:

(1) To assess the benefits accrued to WUAs by involving the farmers in management of irrigation systems.
(2) To critically examine the farmers participation in management of irrigation systems.

The Study Area, Methods and Tools

The Krishna district was selected as it is endowed with many resources and the farmers here are progressive in nature. It was purposively chosen as there was a significant appreciation of performance of WUAs in this region during the period of floods in 1997.

Multi-stage random sampling technique was adopted to elicit primary data from the selected sample of farmers in the first stage. In the second stage, two among 465 WUAs were selected at random for the purpose of the study. From the stratification of the territorial constituencies of WUAs, 80 farmers were selected randomly (Table 1).

Table 1

Selected Water Users' Associations and Number of Territorial Constituencies and Farmers

Distributory Committee	*No. of WUA*	*Name of TC*	*No. of Farmers Selected*
(1) DC-17	(1) WUA-192	(1) Hanumallanka	10
		(2) Kanumoor	10
		(3) Kondur	10
		(4) Kottapalli	10
(2) DC-10	(2) WUA-59	(1) Gundupalem	10
		(2) Mallavolu	10
		(3) Rudravaram	10
		(4) Dodalapalem	10
	Total		80

The approach adopted for the case study was the comparison of pre- and post-WUAs period in the state. The data for the years 1996-97 and 1997-98 (rabi) and 1997-98 and 1998-99 (kharif) was collected to assess the impact of WUAs.

The data thus collected from the selected sample through interview schedules (which were specially designed for the study) were coded and analysed with appropriate measuring tools. The simple averages and percentage were calculated to study the general characteristics. The productivity levels were calculated with the following formulae:

1. $\text{Productivity} = \dfrac{\text{Yield in 'q'}}{\text{Area in Acs.}}$
2. Average difference in productivity (Dp) = Average production before formation of WUAs – Average productivity after formation of WUAs
3. Calculation of additional income:
   Income before formation of WUAs (Ib) : production in q× price/q,
   Income after formation of WUAs (Ia): production in q× price/q
   Therefore the Difference in Income (Id) = Ia –Ib
4. Income from land brought under cultivation after formation of WUAs = Li = land in acres × productivity × price.
   Additional income (Ai) = Id + Li

Results of the Study

Socio-economic Profile of Farmers

The socio-economic profiles of the members (farmers) of WUAs were some of the most important parameters in the study area. WUAs were constituted with a primary aim to increase the socio-economic conditions of the farmers by an effective management of water for increasing agricultural production. The analysis of the data collected for the study area has been discussed and results were drawn.

Status

The majority of farmers (57.5%) of the WUAs were small farmers (2.5-5 ha), 55.5 percent were big farmers (more than 5 ha), and 20 percent belonged to the marginal category (Table 2).

Table 2
Category of Farmers in the Study Area

Category	*WUA-192*		*WUA-59*		*Total*	
	Frequency	*%*	*Frequency*	*%*	*Frequency*	*%*
Marginal farmers (1-2.5 ha)	11	27.5	5	12.5	16	20.0
Small farmers (2.5-5 ha)	21	52.5	25	62.5	46	57.5
Big farmers (more than 5 ha)	8	20.0	10	25.0	18	100.0
Total	40	100.0	40	100.0	80	100.0

According to Niranjan Pant (1981), the majority of farmers in the WUAs belong to small and marginal farmers category (77.5%). From this, it is evident that the small and marginal farmers with small holdings in WUAs are playing an important role in management of irrigation water. About 22.5 percent of the farmers (economically well placed) having more than 5 ha of land are also involving themselves in the management of the irrigation system.

Education

From Table 3 it can be seen that about 86.5 percent of the farmers in WUAs were having secondary school certificate education and only 13.75 percent of them were having no education at all. From the above, it can be concluded that the farmers having education are participating in large numbers in management of the irrigation systems as they understand the government policies immediately and involve themselves in productive activities, whereas the farmers who are illiterate take much time in following the development policies of the government.

Table 3
Literacy Levels of Farmers in WUAs

Category	WUA-192		WUA-59		Total	
	Frequency	%	Frequency	%	Frequency	%
Literates	33	82.5	36	90.0	69	87.25
Illiterates	7	17.5	4	10.0	11	13.75
Total	40	100.0	40	100.0	80	100.00

Area Under Different Crops

From Table 4 it can be seen that the total study area is 720 acres, out of which 248 acres of land was under rice, cultivation, 172 acres under cotton, 130 acres under orchards (mango), 62 acres under groundnut and 53 acres was under aquaculture.

Table 4
Particulars of Land Holdings of Respondents and Area Under Major Crops in the Year 1998-99

(land in acres)

Category	WUA-192	WUA-59	Total
Rice	120	128	248
Cotton	172	-	172
Groundnut	62	-	62
Mango	130	-	130
Fish	-	53	53
Total	539	181	720

From the above it can be inferred that about 35.5 percent of farmers of WUAs were cultivating rice as their main crop and 65.5 percent of the farmers were involved in cultivation of commercial crops. The farmers of WUAs were progressive in nature and were cultivating different combinations of crops for increasing their economic conditions by adopting effective management of water.

Empowerment with Voting Right

Table 5 reveals that under the provisions of APFMIS Act, 1997, the farmers in the area of operation were empowered to elect their leaders to the office of the WUAs. Majority (73.75%) of the water users (farmers) were having voting right and remaining 26.25 percent were the only users of water without voting right.

It can be inferred that majority of the farmers in the area of operation of WUAs were having land on their name under the revenue rights. Whereas, only a small percent of the farmers were having land but were not registered due to which they were not given the power of electing their leaders to WUAs.

Table 5
Farmers with Voting Rights in WUAs

Category	WUA-192		WUA-59		Total	
	Frequency	%	Frequency	%	Frequency	%
Farmers with voting right	29	72.5	30	75.0	59	73.75
Farmers without voting right	11	27.5	10	25.0	21	26.25
Total	40	100.0	40	100.0	80	100.00

Financial Support to the WUAs for Rehabilitation of the System

After constitution of the WUAs, the Andhra Pradesh government financially supported them for undertaking the minimum rehabilitation of the irrigation system in their area of operation.

Table 6
Cost of Various Works Undertaken by WUAs in the Financial Year 1998-99

Type of Work	Cost Incurred (in Rs.)	Type of Work	Cost Incurred (In Rs.)
Providing shuttering arrangement on various channels in operational area	2,90,000	1. Maintenance works, i.e., desilting	3,40,000
MR to Mallavalu main channel	4,95,000	2. Embankment repairs	2,00,000
MR to Kolimakodu channel	2,16,000	3. Repairs to shutter sluices	1,50,000
MR to Dressercoda channel	84,000		
MR to good water channel	50,000		
Total	11,35,000		7,00,000

As seen form the Table 6, about Rs. 18,35,000 was allotted to the WUAs in the study area and the entire amount was spent on the following works of the irrigation system:

1. providing shutters and repairs to the shutters in the canals,
2. strengthening of the embankments of the canals,
3. desilting of the canals, and
4. repairs to the canals.

All the above works were undertaken in order to provide water equitably to all farmers in the area of operation.

Advantages and Benefits of Constituting the WUAs

The constitution of WUAs in the state paved the way for:

1. increase in the performance of the irrigation systems,
2. efficient use of irrigation water, and
3. improving the socio-economic conditions of the farmers by increasing the production.

The benefits accrued in involving the farmers in management of the irrigation systems have been discussed below.

Increase in Land Value

It is evident from Table 7 that before formation of the WUAs, the cost per acre of land in irrigated area was Rs. 77,584 and in irrigated dry land was Rs. 52,596. Later, after formation of the WUAs, during the year 1997-98 the cost per acre of irrigated land was Rs. 86,472 and irrigated dry land was Rs. 76,251. On an average, the increase in the land value of irrigated and irrigated dry land was Rs. 8,800 and 23,745 respectively.

Table 7

Average Land Value Before and After Formation of WUAs

(in Rs.)

WUA No.	*Irrigated*			*Irrigated Dry*		
	Before	*After*	*Increased*	*Before*	*After*	*Increased*
WUA-192	75002	85203	10201	52506	76251	23745
WUA-59	80166	87741	7575	-	-	-
Total (average)	77584	66472	8888	52506	76251	23745

Increase in the land value especially in irrigated dry area may be due to the fact that after formation of WUAs the land which was not receiving water was assured of irrigation water. The irrigated land value was marginally more in irrigated area due to the field drainage provided by the WUAs which also arrested the waterlogged conditions in the area of operation of WUAs.

Increase in the Productivity of Crops

After formation of the WUAs, the Andhra Pradesh government provided financial support for operation and maintenance of the irrigation systems in the area of operation of WUAs. The productivity in rice increased by three bags (i.e., 210 kgs) per acre, in cotton it was 200 kgs per acre, in chillies 100 kgs per acre, in groundnut 200 kgs per acre and in orchards (i.e., mango) the productivity levels increased by 1 ton per acre. Table 8 below gives the details:

Table 8

Productivity Levels in Major Crops Grown, Before and After Formation of WUAs

Time	*Rice (bags per acre)*	*Cotton (kgs per acre)*	*Challis (kgs per acre)*	*Groundnut (kgs per acre)*	*Mango (tons per acre)*
Before WUAs	320	10	120	9	3
After WUAs	35	12	110	11	4
Increase	3	2	10	2	1

The comparative productive levels in different crops reveal that before formation of WUAs the productive levels were below the present productive levels. The reasons for this may be:

- assured irrigation water to crops.
- farmers were confident of getting water supply at the different stages of crops.
- economic use of water for increasing the production.

Hence, the increase in productivity levels of different crops in the area of operation of WUAs may be attributed to the effective involvement of farmers in optimum utilisation of water.

Increase in the Income Levels

Table 9 reveals that the income levels in different crops grown in the area of operation of WUAs were more when compared with income levels in crops grown prior to the formation of WUAs. An additional income of Rs. 3.86 lakh was attained by the farmers after formation of WUAs. The highest level of income (9.58 lakh) was observed in paddy and lowest level (0.65 lakh) in groundnut. Even though the production level in chillies was more, the farmers could not get additional income due to fluctuation of prices. Mango was grown under rainfed conditions. The income was less in mango due to weather fluctuations during the flowering stage due to which less production was recorded.

It can be concluded that there was an additional income of Rs. 3 lakh in the study area of 720 acres. Further, the area of operation of the WUAs was under rice-growing belt, and according to the farmers the cropping pattern may conveniently be shifted to some extent under commercial crops like chillies, cotton and groundnut for increasing their economic viability.

Table 9

Income of the Respondents Before and After Formation of WUAs

Crop	*WUA-59*		*WUA-192*		*Total*		*Difference*
	After	*Before*	*After*	*Before*	*After*	*Before*	
Rice	18.77	13.80	35.02	30.41	53.79	44.21	9.58
Cotton	-	-	38.30	29.34	38.30	29.34	8.56
Chilli	-	-	16.94	22.77	16.94	22.77	-5.83
Groundnut	-	-	7.60	6.95	7.60	6.95	0.65
Mango	-	-	17.80	27.80	17.80	27.30	-9.50
Total	18.77	13.80	115.66	116.77	134.43	130.57	3.86

Equity in Distribution of Irrigation Water

The analysis of the responses of farmers in the study area, as shown in Table 10, reveals that about 86.25 percent of them farmers were of the opinion that the water was distributed equally in the area of their operation and only a negligible percent (13.75%) believed that there was no equity in distribution of water.

Table 10

Responses of Farmers Towards the Equity in Distribution of Water by WUAs

(in Rs.)

WUAs No.	*Irrigated*			*Irrigated Dry*		
	Before	*After*	*Irrigated*	*Before*	*After*	*Irrigated*
Equity in distribution	33	82.3	36	90	69	86.25
No equity in distribution	7	17.5	4	10	11	13.75
Total	40	100.0	40	100	80	100.00

It can be inferred that the WUAs in the study area experienced equitable distribution of water based on the land holdings in the area of their operation.

The findings of various researchers indicated that the bigger and economically viable farmers have a greater share of water in the government-managed canal systems but, contrary to the above, the results show that in WUA-managed canal irrigation systems the water was distributed equally to all the farmers in the area of operation of the WUAs, though there was an unscrupulous distribution here and there.

In general, it can be concluded that the farmer-managed irrigation systems definitely provide an assured and equitable distribution of water.

Conflict Management after Formation of WUAs

One of the most important objectives of the formation of water users' associations was to minimise the disputes and conflicts in distribution of water to farmers. Table 11 reveals that 91.25 percent of the farmers in the area of operation confirmed that the disputes or conflicts over distribution of water were kept at a minimum level in the area of operation of WUAs, and only 8.75 percent of the farmers were of the opinion that the disputes were still persisting.

Table 11

Responses of the Farmers Towards the Disputes After Formation of WUAs

Disputes among Farmers	*WUA-192*		*WUA-59*		*Total*	
	Frequency	*%*	*Frequency*	*%*	*Frequency*	*%*
Increased	-	-	-	-	-	-
Decreased	38	95	35	87.5	73	91.25
Remained as it was in past	2	5	5	12.5	7	8.75
Total	40	100	40	100.0	80	100.00

The result of the study is in accordance with the statement of Albert, that if the farmers manage the irrigation systems on their own, the water allocation and distribution will be equal to all the farmers and the disputes and conflicts can be prevented or controlled.

Transparency of the Activities of WUAs

Transparency is one of the key principles of Participatory Irrigation Management. In order to strengthen confidence over the functioning of WUAs, the activities undertaken by them have to be informed to all the water users in the area of their operation. Some of the results obtained in the study area over the transparency of WUAs are discussed below:

Dissemination of Information to the Water Users

Table 12 reveals that the majority (93.75%) of farmers in the area of operation of WUAs have said that they have been informed in advance by the WUAs about the various activities undertaken, whereas only 6.25 percent felt that they were not informed about the various works undertaken by the WUAs.

Therefore, it can be concluded that the managing committee of WUAs were disseminating information on various activities undertaken for efficient management of the irrigation systems in the area of their operation. From this, it can be inferred that transparency is existing in management of the irrigation systems. The dissemination of information was through distribution of brochures and regular general body meetings being held.

Table 12

Dissemination of Information on Works Undertaken by WUAs

Information	*WUA-192*		*WUA-59*		*Total*	
	Frequency	*%*	*Frequency*	*%*	*Frequency*	*%*
Disseminated to farmers	37	92.5	38	95	75	93.75
Not disseminated to farmers	3	7.5	2	5	5	6.25
Total	40	100.0	40	100	80	100.00

Prioritisation of Works Undertaken by the WUAs

For operation and maintenance of the irrigation systems, the Andhra Pradesh government has laid down procedures for undertaking the repair works. Before undertaking of the works by the WUAs, it was made compulsory to conduct the walkthrough survey of the system in their area of operation. The analysis of the data revealed that majority (77.5%) of the farmers were of the opinion that walkthrough survey was essential to prioritise the repair works on the system so that the funds made available could be utilised precisely.

Table 13
Responses Towards the Quality of Works Undertaken by WUAs

Quality of Works	*WUA-192*		*WUA-59*		*Total*	
	Frequency	*%*	*Frequency*	*%*	*Frequency*	*%*
High	30	75.0	35	87.5	65	81.25
Medium	5	12.5	5	12.5	10	12.50
Low	5	12.5	-	-	5	6.25
Total	40	100.0	40	100.0	80	100.00

The farmers were the best judges at the field level in undertaking the repair works of the system. Therefore, it can be observed from the majority opinion that the prioritisation of work is essential keeping in view the limited funds.

Quality of Works Undertaken by WUAs

Prior to the formation of the WUAs, all the repair works were undertaken by the contractors and the quality of the repair works was low as revealed by many of the farmers. Later, after formation of the WUA, the repair works were undertaken by the farmers under the provisions of APFMIS Act. In the study area, majority (81.25%) of the farmers revealed that the repair works undertaken by the WUAs were of a high quality. Only 12.5 percent informed that the works were of medium quality and about 6.25 percent were of the opinion that the works undertaken were of poor quality (Table 13).

The conveyance losses in the canals as well as in the field channels were high (30-60%) due to various reasons (Jain et al). If the canals and field channels are repaired and strengthened the losses in the conveyance will be minimised. This was achieved by WUAs by undertaking the works with responsibility and qualitatively. This lead to the distribution of water up to the tail end area. The response of the farmers revealed that quality of the works undertaken by WUAs were better than the works undertaken by the contractors through irrigation agencies.

Precision in Utilisation of Funds Allotted for Undertaking Works

Under minimum rehabilitation of the irrigation systems, prioritisation of the works was given much importance by the WUAs. The precision with which the funds were utilised was tested in the study area. According to Table 14, the majority (97.5%) of farmers revealed that the WUAs have utilised the funds in precise manner by prioritising the works through walkthrough survey, and 25 percent informed that the funds were not utilised adequately.

Table 14

Responses of Farmers Towards the Utilisation of Funds

(N = 80)

Funds Utilisation	*WUA-192*		*WUA-59*		*Total*	
	Frequency	*%*	*Frequency*	*%*	*Frequency*	*%*
In precise manner	38	95.0	40	100.0	78	97.5
Not in precise manner	2	5.0	-	-	2	2.5
Total	40	100.0	40	100.0	80	100.0

From the above it can be concluded that the procedure laid down by the government in utilisation of funds was strictly followed by the WUAs and also in undertaking the minimum rehabilitation of the systems.

Farmers' Participation in Management of Irrigation Systems

Participatory Irrigation Management (PIM) revolves round the sharing of the responsibility in management of the irrigation systems. The mode of participation of the farmers in management of irrigation systems was analysed in the study area.

Participation of Farmers in Election of the Office-bearers of WUAs

The president was elected directly from the area of operation of the WUAs and members of the managing committee were elected from the territorial constituencies of the area of operation through secret ballot. About 76.27 percent of the farmers have exercised their voting right and only 23.73 percent have not used their voting right (Table 15).

Table 15

Farmers who Used their Voting Power in Electing WUAs

Category	*WUA-192*		*WUA-59*		*Total*	
	Frequency	*%*	*Frequency*	*%*	*Frequency*	*%*
Farmers who used their voting power	20	68.9	25	83.4	45	76.27
Farmers who did not use their voting power	9	31.10	5	16.6	14	23.73
Total	29	100.0	30	100.0	59	100.0

The participation of farmers in large numbers in electing the office bearers of WUAs may be attributed to the discontent in their minds over the management of irrigation systems by the government agency and shows their desire to manage the systems on their own.

Participation of Farmers in Social Audit

Accountability is one of the most important parameters of PIM. The WUAs are accountable to the farmers in their area of operation for providing irrigation services. At the end of the irrigation season the WUAs have to conduct the social audit. In the study area, the opinion of the farmers was gathered in order to know whether the social audit is essential or not. The majority (95%) of members of WUAs were of the opinion that social audit was essential along with the financial audit, and 5 percent were of the opinion that the social audit or audit of funds was not required (Table 16).

Table 16

Response of Farmers Towards the Auditing of Funds Utilised by WUAs

Response Category	*WUA-192*		*WUA-59*		*Total*	
	Frequency	*%*	*Frequency*	*%*	*Frequency*	*%*
Audit is essential	36	90.0	40	100.0	76	75.0
Audit is not essential	4	10.0	-	-	4	5.0
Total	29	100.0	40	100.0	69	100.0

From the opinion of the farmers it can be concluded that whatever the financial resources are received should be audited at the end of the working season by an agency appointed by the government. The auditing and social auditing as per the majority will help in creation of confidence in WUAs and accountability can be ensured in spending the financial resources by them.

Farmers' Participation in Collection of Water Charges

One of the main objectives of implementation of PIM was to increase the collection of water charges. At present, the collection of water charges from the farmers has been based on the area of crops grown by them in the area of operation of WUAs. In the study area, to what extent the farmers are involved in the collection of water cess was studied. From Table 17 it can be seen that 72.5 percent of the farmers voluntarily come forward to pay the water charges whenever the revenue officials came for the payment and 27. 5 percent of them paid the water charges only when they were repeatedly contacted for payment.

Table 17

Opinion of Farmers in Collection of Water Charges

Water charges	*WUA-192*		*WUA-59*		*Total*	
	Frequency	*%*	*Frequency*	*%*	*Frequency*	*%*
Pay voluntarily	30	37.5	28	35.0	58	72.5
Pay on request	10	12.5	12	15.0	22	27.5
Total	29	100.0	30	100.0	59	100.0

From the above, it can be concluded that if the responsibility of collection of water charges is bestowed upon WUAs they would collect it in full, if assisted by the revenue department. Further, it was also revealed that if the water was assured and reliable, cent percent collection of water charges can be ensured.

Forms of Participation of Farmers in Execution of the Works

Participatory irrigation management was introduced in the state to involve the farmers at every stage of management of irrigation systems in the area of operation of WUAs. The analysis of the data of Table 18, on farmers participation in execution of works, revealed that 72.5 percent of the farmers provided labour in the form of cash (i.e., 15% of the estimated cost) and 27.5 percent worked physically in attending to the execution of works for maintenance of the system.

From the above it can be inferred that most of the repair works in the area of operation of the WUAs were executed with the active involvement of farmers. However, it was also observed that the president of the WUAs were taking keen interest in execution of the repair works rather than the territorial constituency members and members of WUAs. Political interference and groupism may be the separating force which may in future hinder the unity and participation of members in sharing their responsibilities in management of the irrigation system.

Table 18
Forms of Participation of Farmers

Participation	*WUA-192*		*WUA-59*		*Total*	
	Frequency	*%*	*Frequency*	*%*	*Frequency*	*%*
Contribution of 15%	30	37.5	28	35	58	72.5
Contribution of labour	10	12.5	12	15	22	27.5
Total	40	50.0	40	50	80	100.0

Summary and Conclusion

A dynamic management of the irrigation systems requires the active involvement of the farmers at every stage of planning, designing, execution and distribution for increasing agricultural production. The farmers participation moved from peripheral stage of irrigation management to the central state of management of the irrigation system. The conventional wisdom that dominated the management of the irrigation system with a centralised control was decentralised to the local farmers' organisations by implementing the PIM program in the state. The enactment of the APFMIS Act, 1997, provided the farmers' organisations the administrative, legal and technical autonomous status to take the decisions independently and the organisational reforms made the engineers to become facilitators in management of the irrigation system. The empowerment of the farmers ensured the effective participation in rehabilitating the irrigation system to provide water to the unprivileged farmers in the tail end of the command.

The present study, therefore, concentrated more on the improvement of the economic viability of farmers in sharing the responsibility in management of the irrigation systems in the state of Andhra Pradesh, with the following objectives:

- To assess the benefits accrued to the WUA by involving the farmers in management of the irrigation systems.
- To critically examine the farmers participation in management of the irrigation systems.

Keeping in view the above objectives, the literature was reviewed and a multirandom sampling technique was adopted for selection of the WUAs and respondents (farmers) for the study. The selected WUAs were two and sampling size was 80 (i.e., 10 respondents from each territorial constituency of the two Water Users' Associations).

The data was collected with a specially developed interview schedule for this purpose. The data thus collected was coded, analysed and tabulated for arriving at results. The results that emerged are discussed below:

(1) The Government of Andhra Pradesh empowered the farmers in management of the irrigation systems under the various provisions of APFMIS Act, 1997, and enabled them as decision makers and implementors of effective management of irrigation water for increasing the production in the commands of various irrigation systems.

(2) Small and marginal farmers constitute major part of the members of WUAs and have elected their leaders to become the office-bearers of WUAs.

(3) The majority of farmers (members) of WUAs were aware of the rights provided to them under APFMIS Act, 1997, in management of the irrigation systems.

(4) Majority of the farmers were cultivating rice and tend to change the cropping pattern based on the availability of water.

(5) Majority of the farmers were aware of the financial support made available by the government and involved themselves in execution of works so that the water reached to the tail end area of the command.

The author was not able to quantify the benefits accrued in terms of all valid parameters, but some key parameters were considered for examining the economic benefits before and after formation of WUAs.

(6) Quantified benefits:
   (a) The irrigated and irrigated dry land value was increased to a greater extent due to the assured irrigation water by WUAs.
   (b) The productivity levels of various crops like paddy (210 kgs per acre), chillies (100 kgs per acre), cotton (2 quintals per acre) and groundnut (two quintals per acre) were increased, due to which the study area could get Rs. 3.86 lakh additional income from marketing of the produce.

(7) Non-quantified benefits:
   (a) Majority of the farmers in the study area were highly satisfied with the distribution of water and the disputes for the same were minimised.
   (b) Majority of the farmers were highly satisfied with the utilisation of funds in rehabilitating the irrigation systems for pushing the water up to the tail end of the command.
   (c) In general, majority of the farmers are satisfied with the transparency and accountability existing among the office-bearers of WUAs and satisfied with the quality of works undertaken by the WUAs.

(8) The participation of the farmers in large numbers was observed in:
   - electing their leaders
   - attending to the general body meetings to take decisions
   - walkthrough survey for prioritising the works
   - execution of the works
   - payment of water cess, etc.

Users' participation is an essential feature of management of the irrigation systems through formal or informal channels, but it is not the only essential feature. Much of the impact of the participation, particularly in management transfer programme, ultimately depends on the ability of the state to provide the space and provide greater responsibility to the WUAs in management of the system, water control and distribution, so that the objective of implementation of PIM will be achieved with a greater success.

REFERENCES

Abel, Martin, E. (1977), Irrigation System in Taiwan: Management of a Decentralized Public Enterprise in Water Resource Problem in Developing Countries, Economic Research Center, University of Minnesota.

Andhra Pradesh Farmers Management of Irrigation Systems, 1997 (1998) Published by the Commissioner, CAD, Government of Andhra Pradesh.

Bottrol, Anthony, F. (1981), Comparative Study of Management of Organisation of Irrigation, *WAMANA*, 1 (1): 5.

FABO (1982), Organisation, Operation and Maintenance of Irrigation Schemes, USDA, CWC, New Delhi.

IU&CAD Commission (1982), Report on Irrigation Utilisation and Command Area Development in Andhra Pradesh (submitted to Government of Andhra Pradesh).

Niranjan Pant, G. and Siva Mohan M.V.K. (1986), Farmers Involvement in Irrigation Development: A Study of WARABANDI System in a Village Setting, Issues in Irrigated Agriculture and CAD. BIDs 145-190.

Rao, T.H. (1988), Objectives of NWMP, a paper presented during the workshop on objectives of NWMP from 9-6, July 1988.

III

The Political Dimension in Andhra's Irrigation Reforms

7

Politics of Irrigation Management Reforms in Andhra Pradesh

DAVULURI VENKATESWARLU

The recent initiatives on participatory irrigation management (PIM) in Andhra Pradesh (AP) have received considerable attention from policy makers, activists and researchers within and outside the country. The PIM programme initiated and promoted by the government of AP has been carried out at great magnitude covering all irrigation systems in the state. This is considered to be the recent major evidence of reforms in the arena of irrigation management systems.[1] Since 1996, the Telugu Desam government in AP has introduced drastic policy changes in irrigation sector and has been trying to implement them vigorously. The reforms in irrigation sector include enacting a new legislation on participatory irrigation management (AP Farmers' Management of Irrigation Systems Act, 1997) which provides the creation of and functioning of WUAs (Water Users' Associations) as independent legal entities with responsibilities of maintaining and managing public irrigation systems, constitution of a Water Charges Review Committee (WCRC) and adoption of new irrigation policy for the state (reforming the irrigation sector for sustainable development). Over 10,000 WUAs were constituted in 1997, covering the entire surface irrigation area of the state. In this connection, state-wide elections were held for the office-bearers of WUAs. A minimum rehabilitation and maintenance activity of irrigation at large-scale, involving WUAs, has been undertaken as a part of AP Economic Restructuring Project supported by the World Bank.

The paper is divided into four sections. Section I presents the background and broad context in which the AP government introduced management reforms in irrigation sector. The main focus would be on the process of policy formulation and various factors influencing it. The main objectives and features of the Andhra Pradesh Farmers' Management of Irrigation Systems (APFMIS) Act[2] 1997 and other policy interventions are discussed in section II. It also discusses various strategies and measures initiated by the government to implement the policy changes and gain support for the programme.[3] The responses and reactions of the local bureaucracy and water users are analysed in section III. Some observations on the present status of the programme and its implementation at grassroot level by focusing on how the newly created WUAs are actually functioning, and factors influencing their functioning are also presented in section III. Summary and concluding remarks are presented in section IV.

I
Background to the Irrigation Management Reforms

Since its formation in 1956, the expansion of irrigation has been one of the highest development priorities of Andhra Pradesh. Substantial budgetary allocations have been made to irrigation sector, much of it to public canal irrigation. During the Eighth Plan (1992-93 to 1996-97), expenditures on irrigation amounted to 24 percent of total plan expenditure (second only to power sector). The gross irrigated area has increased from 3.1 mha in 1955-56 to 5.5 mha in 1989-90. However, there is a large gap between irrigation potential created and actual area being irrigated under surface irrigation systems in the state. According to World Bank Report (1997) on AP, the gap between the potential created and utilised under surface irrigation is 20-25 percent, which is quite substantial and indicates significant inefficiency in the surface irrigation systems.

A number of factors have been attributed as responsible for the inefficiency of public surface irrigation systems in AP. Critical among them are: (a) lack of sufficient fund allocations for O&M works;[4] (b) poor water management and unequal distribution of water between head and tail end farmers; (c) lack of user (farmers') participation in maintenance and management of irrigation systems; (d) inadequate cost recovery; and (e) low priority to minor irrigation and the disappearance of local institutions and resources for the maintenance of thousands of irrigation tanks. This is the context in which the current reforms in irrigation sector have been justified.

The problem of low efficiency in surface irrigation systems is not a sudden development. The problem as well as factors contributing to it have been well recognised by the people who are concerned with irrigation policy matters in the state prior to 1996. Since early 1980s, a number of committees and commissions have been suggesting drastic reforms in surface irrigation systems that would involve raising of water charges and participation of water users (farmers) in the operation and maintenance of irrigation systems. There has been a strong pressure for management reforms from agencies outside the state, particularly from the Government of India and World Bank. All the state governments have accepted the National Water Policy formulated in 1987 which has advocated participatory forms of irrigation management. While extending support to AP Irrigation Project III, the World Bank also suggested the AP government to introduce reforms in irrigation management by involving user participation in O&M activities (World Bank Report, 1998).

In addition to the external influences, the local NGOs like IRDAS and Sonar[5] also played a significant role in propagating the concept of PIM through their activities and tried to influence people who are in policy making bodies. In 1991, IRDAS introduced Pipeline Committee experiment in selected areas of Sriram Sagar Project. The farmers falling under selected command areas were organised into a committee and it was envisaged that this committee would supervise the distribution operations and ensure equity. The experiment was successful and the utilisation capacity in the selected outlets went up to 80 percent without any substantial investment (Singh et al., 1994).

Despite all these external and internal influences, except a small initiative in Sriram Sagar project on a pilot basis[6], the PIM programme did not really become an issue of serious debate in the state till the government came up with a White Paper on irrigation sector in June 1996. In the same year, the TDP government come up with series of White Papers on different sectors, i.e., power, irrigation and social welfare, in order to explain the critical problems faced by these sectors. The decision to publish White Papers and initiate a public debate was the outcome of a major shift in policy priorities of the ruling party, and its intention to go for drastic economic reforms in the state.

Change of Leadership and Shift in Policy Priorities of the Ruling Party

The TDP had witnessed a major split in 1995. The TDP president and chief minister N.T. Ramarao, who was the founder and undisputed leader of the party since its inception in 1982, was replaced from party presidentship and chief ministership by a majority group in the party led by Chandrababu Naidu in 1995. With the change of leadership the policy priorities of the government and party also witnessed a significant change.

The TDP government under the leadership of N.T. Ramarao (from 1983 to 1989, and again from January 1994 to October 1995 except for a brief spell of six month rule by Nadendla Bhaskara Rao in 1984) gave the highest priority to social welfare programmes with little attention to power and irrigation sectors (Venkateswarlu, 1988). The welfare programmes like the Rs. 2 per kg rice scheme, prohibition, subsidised power tariff to irrigation pumpsets, have brought distinct identity to the party in general and N.T. Ramarao in particular. The subsidised rice scheme and prohibition policy did play a significant role in the landslide victory of the party in 1994 Assembly elections (Srinivasulu and Sarangi, 1999). The subsidy rice scheme had become a pet scheme of N.T. Ramarao and he never yielded to any pressure both inside and outside the party to modify this scheme (Venkateswarlu, 1988). The total cost of the subsidy rice scheme was about Rs. 12.7 billion in 1995-96, 1.9 percent of the state's GDP and about half the total budget allocations for education and health. Loss of revenue to the state due to total prohibition was estimated at Rs. 14 billion, about one quarter of its tax revenue in 1994-95. While welfare programmes were consuming a substantial proportion of state expenditures, the development in other sectors particularly irrigation and power suffered due to lack of sufficient funds (World Bank Report, 1997).

Though the main reason for the change of leadership were obviously not the policy priorities of N.T. Ramarao, but they seem to have caused some discontent (though never expressed openly) among his party men and finally led to change in leadership.[7] This can be seen through the overwhelming support Chandrababu Naidu had received from his party men when he announced his intention to bring changes in some of the important policies (subsidy rice scheme and prohibition) initiated by N.T. Ramarao.

Political and Economic Compulsions

Chandrababu Naidu had strong political and economic compulsions to go for major economic reforms. The economic compulsion include the deteriorating financial situation of the state and the need to raise sufficient resources for developmental activities.[8] The political compulsions include the need to consolidate his power and strengthen the social base of the party following the spilt in the party in 1995. Till the split, the TDP had the advantage of having a charismatic leader, N.T. Ramarao, who had tremendous mass appeal in leading the party. The new leadership realised this fact and worked out a strategy to fill this gap by (a) building a strong cadre to strengthen the party from grassroots level, and (b) actively working for the development of the state which would in turn help the party to gain popular support. The political situation in the state in 1996 was also very conducive to the ruling party to go for major revision of previous government decisions. N.T. Ramarao who would have severely opposed any decision to modify the subsidy rice scheme, prohibition policy and irrigation charges passed away in January 1996. The elections to Lok Sabha were also over by March 1996. Contrary to the expectations, the ruling party had done well in this elections. The other factors like having two-thirds majority in the State Assembly and no need of facing assembly elections in the near future were also favourable for the ruling party. The major boost to the economic reforms in general and initiatives of irrigation management reforms in particular, in the state in 1996, needs to be understood in this context.

The strategy adopted by the ruling party to gain popular support for its irrigation reform programme is noteworthy. In order to avoid or minimise the resistance from opposition parties, it first issued a White Paper on irrigation sector explaining the critical problems faced by this sector and initiated a public debate. It organised a number of public meetings inviting suggestions from intellectuals, academicians, NGO activists, etc. The ministers including chief minister and officials from irrigation department actively participated in the meetings. The NGOs and the top officials in irrigation department who had been advocating PIM and were involved in pilot project experiment also actively participated in the meetings and played an important role in trying to get support for their initiatives. The White Paper was also put forward for discussion in the State Assembly. An expert committee was appointed for studying the existing laws and provisions. A new draft legislation was prepared and was widely circulated for discussion throughout the state. Consultants were appointed for different regions to guide the consultation process and provide the feedback. During the consultation process, besides the discussion on proposed Act and land delineation, the issue of raising the water charges was also discussed and negotiated. On the basis of initial feedback, the government first proposed to raise the water charges from Rs. 60 to Rs. 350 per acre in the case of paddy and indexed to the value of one bag of paddy (70 kgs). The proposal met with stiff opposition from various quarters – opposition parties, farmers organisations including farmers lobby within the ruling party. Responding to the reaction the government agreed to reduce the rates to Rs. 250 and to withdraw indexation of one bag of paddy. The issue was further negotiated during the Assembly discussion in 1997 and proposed water rates were further modified to Rs. 200 per acre in the case of paddy and received approval from the Assembly. The consultation process took about 10 months and the final legislation (the APFMIS Act) was passed by the State Assembly on March 27, 1997.

World Bank Assistance and Conditionalities

While this process was going on, realising the need to have large funds to implement the PIM programme, the government simultaneously initiated a dialogue with World Bank for securing financial assistance for its reform programme. The chief minister personally met the World Bank officials in October 1996 and explained his government's proposed plans for major reforms and sought the Bank's assistance for the same. The subsequent discussions between Bank and AP government resulted in the former agreeing for conducting a detailed study on the economic situation of the state and making suggestions to carry out economic reforms. The World Bank report 'Andhra Pradesh: Agenda for Economic Reforms', prepared in January 1997 was an outcome of this. In this report the Bank has made a number of suggestions for the overall development of the state in general and improvement of irrigation sector in particular. The important suggestions for improving the irrigation sector include (a) enactment of new legislation enabling the participation of water users in the management of irrigation systems, and (b) increasing the water charges to meet the full O&M costs.

Impressed by the initiatives already taken by the government since early 1996 and its commitment and assurance to carry the reforms further (which are in tune with the World Bank's objectives), the Bank has agreed to extend financial assistance of 543.2 million US dollars to AP government in 1998 for its project named Andhra Pradesh Economic Restructuring Project (APERP). A substantial portion (142 million dollars) of this assistance was meant for reforming the irrigation sector. Before finalising the APERP the Bank and government of AP had several rounds of discussions. The Bank assistance was conditional. The important conditions of the Bank with regard to irrigation sector include: (a) the government should continue with its reform programme, (b) review the water charges annually and implement the decisions of Water Charges Review Committee, (c) increase the collection rate of water charges at least by 90 percent by March 2001, and (d) implement a system for collection of water charges on a volumetric basis by WUAs on a pilot basis by March, 2001. When project preparations were going on, even before finalising the assistance of APER project the Bank tried to exert its influence on the government to speed up the reforms in irrigation sector. Constitution of WCRC and approving the new policy statement on irrigation sector have all been done with great speed so as to meet the requirements of the Bank, which made completing these measures conditional for finalisation of project appraisal report.[9]

The process of policy formulation for irrigation management reforms in AP described above indicates the following. Though a number of external and internal agencies tried to influence the government they were not quite successful till the political commitment for the reforms was generated and people who were interested in reforms occupied right political positions. The policy formulation process seems to be top down in its approach, which was manifested in the way the government has legislated an Act that has made the formation of WUAs compulsory to maintain the irrigation systems in the entire state. However, the way the AP government held extensive consultations by involving political parties, bureaucracy and water users to muster political will, bureaucratic commitment and people's support and involvement indicate that though the process seem to be initially prescriptive, the government had made a sincere attempt in taking the people's view in cognisance in obtaining their support by initiating extensive debate.

II

Policy Interventions and Implementing Strategies

The present section discusses various policy interventions by the government in reforming the irrigation sector and steps initiated to implement them.

The Andhra Pradesh Farmers' Management of Irrigation Systems Act, 1997

This Act was passed in the State Assembly in April 1997. The main objective of this Act is to provide a legal basis for the participation of water users in the operation and maintenance of irrigation systems. It was envisaged that the participation of water users in the management would lead to considerable improvement in the efficiency of irrigation systems. The main features of this Act are as follows:

- It enables the formation of Water User Associations (WUAs) in all surface irrigation projects of the state, i.e., major, medium and minor irrigation projects except those under Panchayat Raj Institutions (PRIs) and all minor bodies in the scheduled areas of AP.[10]
- Every irrigation project would be divided into convenient units of operation of the project, Distributory Committees (DCs), and WUAs at primary level. The jurisdiction of the WUA would be delineated on hydraulic basis and typically cover a minor or a group of minors or a small distributory, or in the case of minor irrigation, the entire system. Depending on the type of the irrigation system there may be single or two or three tier associations/ committees. The command area under a WUA would be divided into territorial constituencies depending upon hydraulic units. Each constituency shall elect a member of the managing committee of WUA and shall directly elect the president of WUA as well.
- All farmers within the command area including title holders and tenants are compulsorily made members of the WUA with voting rights. All other water users either from non-command areas or users for non-agricultural purposes will also be coopted as members without voting rights.
- The president and other office-bearers of WUA are to be elected directly by the members. All the presidents of WUAs will be members of the Distributory Committee (DC) and they will directly elect the president and managing committee members of DC. All the presidents of DC will be the members of Project Committee (PC) and they will in turn elect the president and managing committee members of PC.
- The main responsibilities of the WUAs include: (a) maintaining of irrigation system within its command area, (b) preparation of operational plans at the beginning of the crop season, (c) ensuring equitable distribution of water and resolving conflicts among members, (d) maintaining the records, accounts and list of members, and (e) conducting social audit, mobilising the resources, etc.
- WUAs are given powers to raise funds locally for its activities. Apart from grants they receive from the government (share in the water charges and developmental grants), they can also raise loans from the banks, impose levy or special taxes on members for taking up specific works with the approval of

general body. They are also entitled to get income from sources (trees, government buildings, etc.) attached to irrigation system.

- Under this Act the irrigation department officials will have a completely new role to perform. It makes irrigation department accountable to the WUAs as the competent authority, requiring its staff to implement the decisions of WUAs.
- The most significant feature of this Act is that it has a provision to recall the elected members of the WUA. No other elected bodies, i.e., PRIs, state legislatures and parliament in the state as well as in the country have this kind of provision to recall the elected members.

Three-fold Increase in the Water Charges

Inadequate cost recovery has been attributed as the main cause for under-funding of O&M works. During 1988-1996 there had not been any change in the water charges for irrigation in the state. In August 1996, as part of a reform programme, the government first proposed to increase the water charges highest upto Rs. 350 per acre and indexed to the value of one bag of paddy. This was almost six times higher than the previous rates (from Rs. 60 to Rs. 350 per acre). However, due to heavy pressure from the interest groups, the charges were later reduced to highest Rs. 200 per acre and indexation was withdrawn.[11] The new water rates were ratified in April 1997. The new water rates would cover about three-quarters of the required O&M needs.

Table 1

Revised and Pre-revised Water Charges for Irrigation

(Rs. per acre)

Nature of Crop	*Category I*		*Category II*	
	Pre-revised	*Revised*	*Pre-revised*	*Revised*
First or single wet crop	60	200	40	100
Second and third wet crop	60	150	40	100
First crop irrigated dry	40	100	20	60
Second and third crop irrigated dry	40	100	20	60
Dufassal crop in falsi year	120	350	80	350
Aquaculture per acre	—	500	—	500

Note: Pre revised = with effect from 1-7-1986, Revised = with effect from 1-7-1996
Category I = Major and Medium irrigation systems, Category II = Minor irrigation systems (tanks)

Constitution of Water Charges Review Committee

A permanent body of Water Charges Review Committee (WCRC) was constituted in December 1997. The main function of this body is to review and make recommendations on all aspects of water charges and maintenance funding on an annual basis. The WCRC is responsible for overseeing the transition to full cost recovery, with specific actions including: (a) undertaking annual reviews of O&M needs, expenditures, collection rates and revenues, (b) guiding the planned transition from the current assessment of water charges on per acre basis to volumetric charges, and (c) monitoring the process of shifting responsibility of collection of water charges from revenue department to WUAs. The WCRC has not come out with any recommendations so far.

Approving Irrigation Sector Policy

A policy statement for irrigation sector, 'Reforming the Irrigation Sector for Sustainable Management and Development', was approved in May 1998. This policy statement outlines the need for reforming the irrigation sector, steps already initiated by the government, future plans and strategy of the government to carry the reforms further. The long-term goal of the state, the policy states, "is to create financially autonomous irrigation systems fully managed by WUAs". The immediate priority of the state is to strengthen the reforms already initiated by making the newly created WUAs functional through extensive training and capacity building and releasing adequate funds for them to take up rehabilitation and maintenance activities.

Table 2

Details of Water User Associations Notified and Formed (6-7-99)

Type of Irrigation System	*Total No. of WUAs Notified*	*Total No. of WUAs Formed*	*Elections Withheld*
Major	1700	1645	55
Medium	411	401	10
Minor	8181	7754	427
Total	10292	9800	492

Note: The reason for not holding elections for 492 WUAs are as follows: 86 are due to stay orders by the court, 141 are due to stay by the government and remaining 265 are due to other reasons.

Source: Irrigation and Command Area Development Department, Government of AP.

Formation of Water User Associations Across the State

Soon after the enactment of legislation the state government has undertaken a massive publicity and extension campaign to create awareness about the new Act. In July 1997 state-wide elections were held to the president and territorial constituency members of WUAs for all the surface irrigation systems in the state. Of the total identified 10,271 WUAs, elections were held to 9,795 WUAs. The elections to remaining 476 WUAs have been postponed due to various problems like court litigation. Out of the 10,271 WUAs a large proportion (8159) come under minor irrigation alone. In November 1997, elections were also held to the president and managing committee members of the distributory committees. The constitution of project-level committees is yet to be completed.

Massive Propaganda and Training Programmes

As PIM concept was new to the irrigation department officials and farmers, extensive training and propaganda was recognised as a must to ground the PIM programme. The creation and conducting elections to water user bodies was followed by massive propaganda through media, public meetings, posters, etc., and extensive training programmes for irrigation department officials and elected members to create awareness about the new Act and its implications. In August 1995 an initial training was given to officials of irrigation department. During September and October 1997, all the elected presidents of WUAs were given initial orientation training through two-day workshops. In addition to this, state, regional and district level conferences with all the presidents of WUAs were conducted. The main purpose of these conferences was to create further awareness and discuss the implementation of the PIM programme. During July and October 1999, further detailed training programmes for all the presidents and other office-bearers of WUAs were held.

More than training and usual propaganda techniques, it is the Janmabhoomi (motherland) programme of state government launched in 1997 which has helped to create mass awareness and generate large-scale grassroots support to the PIM programme.[12] Since its launching, the state government has conducted ten rounds of this programme covering all the regions of the state. In each round one specific theme was chosen and week-long activities were undertaken. During this programme all the elected representatives and government officials were to meet the people and participate in the activities undertaken by them. The occasion of Janmabhoomi programme gave good opportunity for the officials and politicians to popularise the concept of PIM among the people.

Irrigation Rehabilitation and Maintenance Activities

One of the important factors which has been attributed for the low efficiency of irrigation systems is lack of sufficient fund allocations for O&M activities. Since 1998 the government has undertaken irrigation rehabilitation and maintenance activities on a large scale, involving the water user bodies throughout the state. The government has received funds from the World Bank for this purpose under APER project. The funds for O&M activities have been significantly raised from Rs. 40 per acre to Rs. 100 per acre in 1998. The earlier allocation of Rs. 40 per acre was inclusive of salaries paid to irrigation department employees. It was estimated that out of Rs. 40, Rs. 35 were spent on salaries and only Rs. 5 per acre were left for repairs and maintenance of structures. A total of Rs. 106 crores was released in 1998 to WUAs and DCs for taking up repairs and maintenance activities. The identification of civil works to be undertaken in different irrigation systems was done through joint walks to the fields by the officials and representatives of farmer organisations. To speed up the whole process of execution of these works, the government has released 40 percent of the estimated cost as advance money to the farmers' organisations. The various activities undertaken included desilting, construction of lining, drops, pipe out lets, clearing of weeds and strengthening of embankments of channels. For the first time in the state the works are being undertaken simultaneously in all the irrigation systems and this has considerably improved the work output. Within a short span of time much higher work output was achieved. The most significant feature is that contractors have been eliminated from minor works.

All these policy interventions and steps initiated to implement them do indicate the strong commitment on the part of the government especially from the policy makers and top government officials in the I&CAD department which is the implementing agency of the PIM programme. The main political support for this programme has come from the chief minister of the state who has shown active interest in it.

The strong political commitment and top level official support though very important but by itself is not sufficient to make the PIM programme a success. Its success largely depends upon the active involvement of the key stakeholders at grass roots level, i.e., farmers and local irrigation bureaucracy. An attempt is made in the following section to study the impact of the programme at grass roots level by looking at the functioning of newly created WUAs.

III
Functioning of Water User Associations

Considering the fact that the implementation of the PIM programme is in the initial stage, in this section an attempt has been made to capture the initial trends in the formation and functioning of the WUAs. The observations reported in this section are based on the study of 14 WUAs selected from divergent regions/irrigation systems. Of the 14 WUAs, five are under major irrigation systems, two under medium and remaining five under minor irrigation systems.[13]

Elections to Office-bearers of WUAs

Of the 14 WUAs in the study area, elections to office-bearers of 13 were completed in July 1997. Elections did not take place for one due to outbreak of violence during election between local factional groups who competed for the positions in WUA. Of the 13 WUAs, elections for five WUAs were held unanimously and in the rest elections were held through voting.[14] There was an active competition for the post of president in two associations and contestants seem to have spent lot of money in the elections. In four cases though, there was a competition but it was less intensive.

On the whole, the elections to WUAs did not seem to generate the kind of enthusiasm, interest and intense competition which is generally associated with any local body elections in the villages. The timing of elections to WUAs had a significant impact on the nature of people's participation and competition among contestants. Two months after the enactment of the law the elections for WUAs were announced. Neither people had enough time nor the government could create enough awareness about the details of new policy. Though there was a little awareness in some areas, the implications of the policy were not realised by majority of the farmers. In many places it helped the ruling party cadre, local civil contractors and people who are connected to irrigation department, who had greater access to information, to take advantage and secure positions in WUAs. This can be clearly seen from the socio-economic and political background of the people who were elected as presidents and technical constituency members.

The financial assistance offered by the government for unanimous elections for the president of WUA also had its impact on the elections. In order to avoid possible local conflicts in the elections and encourage consensus among farmers, the government has announced incentives to the WUAs for electing their members unanimously. Each WUA was offered Rs. 20,000 as an incentive if they elected their president unanimously. In four out of the five WUAs where candidates were chosen unanimously; the main reason for unanimous selection of candidates seems to be the financial incentive factor.

Though unanimous election process has yielded some financial incentives as well as avoided any kind of conflict, in such cases the farmers are found to be less aware of the details of the programme, whereas among the farmers of WUAs where elections were held there is greater awareness. It is due to campaign as well as constant persuasion by the contestants, who had created considerable awareness among the members. Interestingly, in some areas where elections were not contested, some of the Ayacatdars (members) do not even know that they are the members of the WUA in their area. The financial incentive offered by the government for electing a consensus candidate for the post of president in WUAs seems to have worked negatively in some ways by avoiding the election process which would have really helped the members to become aware of their membership and also the programme.

Due to the lack of information and awareness on the strengths of WUA positions, the participation of people and contestants in the election was not as it was expected to be. But, at the time of investigation, it is time and again informed that they have realised later that these positions involve considerable power and financial dealings. It is to be noted that elections for WUAs in the following years are anticipated to be with more active contestants and people's participation.

Socio-economic and Political Background of the Office-bearers of WUAs

Tables 3, 4 and 5 present the socio-economic and political background of the office-bearers of the WUAs. The background of these people reveals that a majority of them have come from rich, upper caste and are affiliated to ruling Telugu Desam Party (TDP). Of the 13 WUAs that are studied, seven presidents belong to TDP. Three belong to Congress party and remaining three do not have any party affiliation. Of the total 78 TC members, 36 belong to TDP and 20 to Congress. The caste background of these members indicates that the upper castes like Reddy, Kamma, Velama, Vaisya, etc., have greater representation than backward and Scheduled Castes. Nine out of 13 presidents and 39 out of 78 TC members belong to upper castes. The economic background of these members

Table 3

Party Affiliations of the Office-bearers of WUAs

Name of the Party	*President*	*TC members*
TDP	7	36
Congress	3	20
Other	0	4
Independents	3	18
Total	13	78

Other = Bharatiya Janata Party, Communist Party of India and Communist Party of India (Marxist)
TC = Technical Constituency

Note: Though the elected members have respective party identities and affiliations but elections were not held on party basis and party symbols were also not used.

Table 4

Economic Background of the Office-bearers of WUAs

Category	*President*	*TC members*
Rich	9	32
Middle	3	25
Poor	1	21
Total	13	78

Note: The above classification shown in the table indicates broad categorisation. This categorisation is adapted to present a more accurate economic status of the families of the office bearers of WUAs. In this context it is felt that the size of the land holding is inadequate for classification of economic status of the members. Such that, the present classification is based on the subjective perceptions provided by the informants. It seems to be the broad categorisation which presents the subjective perceptions of the villagers. The villagers use the terms 'ounnavallu/dhanavanthulu, okamosturu/madhyarakam, and lenollu/pedollu' which denote broad economic grouping of households. The perception of economic grouping based on not only the size of the land holding but also non-land assets and income from both agriculture and non-agricultural activities. This local categorisation is translated into rich, middle and poor categories in the present study. This information is collected through group discussions and subsequently cross checked.

indicates that almost two-thirds of the presidents belong to rich families. Among TC members also, majority are from rich family background. Of the nine presidents from rich family background, two are ex-civil contractors of irrigation department. An important aspect to be noted here is that women representation is almost negligible. Except four TC members all presidents and TC members are males.

Table 5
Caste Background of the Office-bearers of WUAs

Caste	*President*	*TC members*
Upper/Forward Castes	9	39
Backward Castes	4	21
Scheduled Castes and Scheduled Tribes	0	8
Total	13	78

There are clear indications that the party, caste and class background of the office-bearers are coming in the way of functioning of these institutions. In several WUAs we have noticed that there is a clear lack of coordination and cooperation among the members belonging to different backgrounds. For example, in Giddalur WUAs the president and majority TC members belong to TDP and they dominate in all the activities of the association. The other members who belong to Congress are completely sidelined and their involvement in the decision-making and activities of the association is negligible. But this kind of situation is not a general phenomena. In some places it is also noticed that the members are working together irrespective of their party affiliations.

An important aspect is that though it is the TDP which is responsible for bringing the reforms and creating WUAs, the programme as such is not seen as the programme of the TDP only. It is evident that the local Congress party leadership also extended its support and actively participating in the programme. The three WUA presidents from Congress party in the study area are as much active as the other TDP presidents in implementing the programme.

As a whole, though the newly formed WUAs and their representation are somewhat the reproduction of existing local power structure, the impact of the programme seems to have devolved the existing power structure and influenced the local dynamics to some extent. As the programme is very much in the beginning stage, these dynamics are not tangible.

Maintenance and Rehabilitation Activities

On the whole, compared to several other government programmes, PIM seems to have generated much greater awareness and interest among the farmers.[15] The most important element, which helped to generate support and interest and also improved the credibility of the PIM programme among the farmers is the financial resources allocated to WUAs for maintenance, rehabilitation works and also towards administrative expenses. Each WUA is given Rs. 30,000, and in case of unanimous election Rs. 50,000, towards administrative and other expenses. In addition to this, since 1998 all the WUAs have been receiving funds at the rate of Rs. 100 per acre for O&M activities.

With the money received from the government the WUAs have taken up a number of maintenance and rehabilitation works to improve the irrigation systems. The works taken up by the WUAs include (a) desilting, (b) weed removal, (c) repairs to embankments, shutters, drop regulators, and (d) repairs and construction of sluices. The WUAs have taken active interest to complete these works within time. As a result of this, in many places there is an improvement of physical structures which are necessary for improving the water supply.

In some places there are allegations against the presidents and some TC members of the WUAs that they are misappropriating the funds and acting irresponsibly. There seems to be some element of truth in these allegations.[16] Though contractors have been eliminated for minor works, in some places they are being replaced by the presidents and their nominees. In four WUAs it is noticed that presidents are taking the entire responsibility of carrying the maintenance works without involving the other members.

For all the repair works undertaken by the WUAs, in addition to the government funds the farmers need to contribute 15 percent either in the form of cash or kind. Except one, none of the WUAs in the study area made any cash contribution for the works. The contribution in the form of free labour services by the members has taken place in three places only. In the cases where farmers contribution in cash or kind is absent, no efforts have been initiated by the president and TC members to extract their involvement. They also did not make any serious attempt to discuss this issue in their meetings and collect the contributions from the members.

Despite all these problems of misappropriation of funds and lack of participation of members, a number of positive things about the maintenance and rehabilitation works taken up by the WUAs have been acknowledged. The works have been completed within time and quality of the works also seems be to quite impressive. One Assistant Engineer who is otherwise very critical of WUAs states that it is true that there is some misappropriation of funds by the WUA presidents, but it is not affecting the quality of works in any significant way. In many places WUAs have completed works within specified time. The quality of works done by the WUAs is quite impressive and much better than the quality of works done by civil contractors.

The programme was not able to create as much interest among the tank farmers as it did with canal farmers. The main reason seems to be the inadequate fund allocations for taking up minimum required maintenance and rehabilitation activities in tank irrigated areas. In several places the members complained that the funds they received from the government were quite insufficient to carry the minimum required repairs resulting in no significant improvement in water supply.

As a result of maintenance and rehabilitation activities, in many places there is a considerable physical improvement of irrigation systems which are necessary, although not sufficient, for improving the water supply and distribution. The water supply position seems to have improved somewhat better in canal than tank irrigated areas. Due to better water supply in many areas, the area under irrigation has slightly increased during the last two years. Regarding the extent of additional land brought under irrigation after reforms, there is a large gap between the data provided by the revenue department and data gathered for the present study through field interviews. The revenue department data shows that there is a significant increase in the area under irrigation during the post-reform period. The revenue department data for two WUAs show that there is an increase of about 28 percent in the area under irrigation after 1998. But, in reality the actual increase is only about 5 percent as reported by the farmers. As pointed out by Jairath, the significant increase in the area under irrigation according to the revenue department figures is mainly due to the problems in reporting the data and not necessarily due to actual increase in the area (Jairath, 1999).

Lack of Interest in Water Regulation

According to the Act, WUAs have two key functions to perform, i.e., maintenance of irrigation systems and distribution of equitable water supply (per acre basis) among all the users. As of now, WUAs seems to be much more interested in taking up maintenance works than water distribution activities. Water regulation is a crucial function which requires lot of motivation and self-discipline among the water users. Of the 13 WUAs, except two none seems to have made serious attempt to motivate members about judicious use of water and control over unauthorised water users. Of the two WUA areas where water regulation systems is working somewhat better, one is in tank-irrigated area and another in tail-end area of major irrigation system. Even before the WUAs formed in these places, there existed a system of 'common irrigators' which is locally termed as 'neerukuttu' and is appointed by the village council to look after the distribution of irrigation water to all the fields in the village. Robert Wade's (1987) has carried out a detailed study in one of these two villages. The system of 'common irrigators' described by Wade is still operating in these villages and close interaction and cooperation between the newly created WUA and the old village council which appoints the common irrigators is present.

Given the disparities and complexities involved in composition of WUAs, water use practices among the farmers (head-end farmers consuming more water than they require depriving the tail-end farmers of their share) and lack of any systematic method for equal sharing, the question of equal distribution among all members is a crucial one and needs to be addressed.

WUAs and Local Bureaucracy

Though the programme seems to have received strong support from the higher bureaucratic circles, the response of lower level bureaucracy is not that encouraging.

WUAs come in contact with three government departments, i.e., irrigation, revenue and agriculture departments, which are connected with irrigation management, collection of water charges and irrigated agriculture. Before enactment of the PIM Act, the irrigation department and revenue department had complete control over irrigation management and collection of irrigation charges. With the introduction of the Act, the role of these departments has changed radically. The Act makes Irrigation Department accountable to the WUAs as a 'competent authority', requiring its staff to help the WUAs in their activities and implement their decisions. Though the responsibility of collection of water charges is still with the revenue department, now they require to share the information with WUAs and involve them in collection of water charges.

Except a small group of enthusiastic and progressive officials, the lower level staff in irrigation and revenue departments in general does not seem to be very happy with the PIM programme. Though majority of the people interviewed did not openly admit the main reason for their unhappiness, it seems to be the perceived threat of loosing their control and power. Though they are not openly opposing the reforms, their unhappiness is expressed through several ways like making delays in implementing the measures, non-cooperative attitude with newly elected members of WUAs, etc.

In several places the newly elected members of WUAs complained that they were not receiving proper support and cooperation from irrigation and revenue departments. In addition to help the WUAs in their multiple activities, the irrigation and revenue departments are supposed to provide necessary information required by the WUAs. According to the Act, the WUAs are required to maintain a number of registers and records for which they need the help of irrigation and revenue department staff. As of now, several WUAs are not maintaining these records. Part of the reason lies in the lack of interest among WUA members in maintaining these records, but in many places it is the general attitude and lack of interest of local staff in extending cooperation. One WUA president reported that he had to make six visits and wait for about three months to get the information about area actually irrigated and amount of revenue collected from the local revenue staff.

Several of the irrigation department staff interviewed also expressed their strong reservations and doubts about the capability of WUAs to perform the tasks entrusted to them and viability of the programme in the long run. In their opinion, the WUAs are more interested in receiving funds from the government and maintenance activities than water regulation and equitable distribution among all the members.

IV
Summary and Concluding Remarks

In the preceding sections we have made an attempt to understand the politics of irrigation management reforms in Andhra Pradesh. The background to reforms focusing on the process of policy formulation was explained in section one. Various policy interventions and strategies adopted to implement these policies were discussed in section two. Some observations on the impact of the programme at grassroots level, by focusing on how the newly created WUAs are working and factors influencing their functioning, were presented in section three. The following are some of the important observations of the study.

In AP, the reforms in irrigation sector are long overdue. The problem of low efficiency in irrigation systems and the need to bring management reforms to correct this was well-recognised by the people who were in policy making bodies, even before 1996. A number of agencies, both external (World Bank and Government of India) and internal (few NGOs and a section of top bureaucracy in I&CAD department), tried to influence the state government to go for reforms in irrigation sector. Despite these external and internal pressures, except a small initiative in Sriram Sagar project on pilot basis, the previous ruling parties including the present one, which enjoyed clear electoral support, did not make any serious attempt to introduce management reforms before 1996. What made the present government go for the reforms in 1996 is the strong political commitment of the new leadership of the ruling TDP which came to power in 1995. Since then, there has been a major shift in the policy priorities of the ruling party. The new leadership had strong economic and political compulsions to go for major economic reforms in the state. The economic compulsions included the deteriorating financial situation of the state and the need to raise sufficient resources for developmental activities. The political compulsions included the need for new leadership to consolidate its power and strengthen the social base of the party. The major boost to the economic reforms in general and initiatives of irrigation management reforms in particular, in 1996, needs to be understood in this context.

The process of policy formulation for irrigation management reforms in AP indicates the following: (a) the major reforms were initiated only when people with strong political commitment occupied right political positions; and (b) though the process seems to be prescriptive in nature and top-down in its approach, the government made a sincere attempt in involving farmers who are major stakeholders in the process

The various strategies adopted by the government are noteworthy. Publishing White Paper on irrigation sector, initiating public debate on reforms and consultations with water users before finalising the policy, have all played an important part at policy formulation level to get support from the people as well as from opposition parties. This strategy seems to have worked out and the PIM Act was passed in the Legislative Assembly without any opposition.

The enactment of PIM law was followed by creation of farmers' organisations across the state and completion of elections to these bodies. A massive outreach programme through media, public meetings and extensive training programmes to irrigation department officials and elected members was undertaken to create awareness about the new Act and its implications. In addition to this, the PIM programme has been included as one of the main aspects of government's new populist programme 'Janmabhoomi' which is aimed at mobilising people's support and their active involvement in developmental programmes. In order to make the newly created WUAs functional, the government provided them with substantial amount of funds for taking up maintenance and rehabilitation works. All these initiatives have generated a lot of excitement, interest and activities at the grassroots level. The strong political commitment for the programme and large-scale undertaking of maintenance and rehabilitation works to improve the irrigation systems have increased the credibility of the PIM programme among the farmers who are the key stake holders of the programme.

The response and support from the farmers to the programme vary from one type of irrigation system to another. The farmers' awareness and participation levels are quite low in tank than canal irrigated areas. The programme was not able to create as much interest among the tank-irrigated area farmers as it did with canal farmers. The main reason seems to be the inadequate fund allocations for taking up minimum required maintenance and rehabilitation activities in tank irrigated areas.

There are clear indications that in many areas the local politics and issues are playing an important role and coming in the way of functioning of WUAs. The positions in WUAs are captured by local influential people. In the study area, a majority of the office-bearers of WUAs belong to rich, upper caste backgrounds and affiliated to ruling party. It is noticed that in some places there is a lack of coordination and cooperation among the members belonging to different backgrounds. But this is not a general phenomenon. In few places it is also observed that members belonging to different backgrounds are closely working together. An important aspect is that though it is the TDP which is responsible for bringing the reforms and creating WUAs, the programme as such is not seen as the programme of the TDP only. The local Congress leadership also seems to extend its support and participate actively in the programme. The three WUA presidents, who belong to Congress, in the study area are as much active as the TDP presidents in implementing the programme.

Thus, in many areas, the WUAs have done commendable work. They have taken active interest in maintenance and rehabilitation works. Works have been completed within specified time and the quality of the works seems to be impressive. As a result of this, in many places there is a considerable improvement of physical structures but they are insufficient for water supply and distribution. The water supply position seems to have improved to some extent, better in canal compared to tank irrigated areas. Hence, the area under irrigation has slightly increased.

But substantial increase in the quantum of area under irrigation which is one of the primary objectives of the reform programme has not taken place due to lack of interest in water regulation functions by WUAs. Despite consistent attempts in generating awareness and subsequent improvement of infrastructural facilities in irrigation systems, concrete persuasion was not taken to regulate equitable distribution of water among the members. It is pertinent to mention that a substantial increase in the quantum of area under irrigation is possible only when the water regulatory functions are promoted effectively. In this regard, majority of members in the study area did not show considerable awareness and motivation towards water regulation. Thus, a systematic and scientific approach to deal with the situation needs to be paid immediate attention.

NOTES

1. In India, introduction of transfer of irrigation management systems at that magnitude has taken place only in the state of Andhra Pradesh. Earlier reforms at such a scale had been carried out in countries like Mexico and Turkey (for details, refer Keith Obillas and J. Raymond Peter et al., 1999).
2. In AP, on an average irrigation sector has received about 42 percent of the total public capital expenditure. This amounts to about 20 percent of the total investment in the state.
3. The data on potential area created and actual area irrigated under surface irrigation vary from one source to another. The data of Revenue Department shows that the

gap is 35 percent in the last decade whereas Irrigation and Command Area Development Department (I&CADD) data show that the gap is only 5 percent. Taking the data of both the departments, which indicated under-estimation and over-estimation, into account, the World Bank report has estimated the gap as 20-25 percent. (for more details see, the World Bank report 'Andhra Pradesh: Agenda for Economic Reforms', p. 38).

4. The budgetary allocation for O&M works in public surface irrigation systems was far below than its actual requirements. The Ninth Finance Commission recommended an O&M allocation of Rs. 324 per ha for 1990-95. In AP the actual allocation in 1990-92 was only Rs. 75 per ha.
5. IRDAS (Institute of Resource Development and Social Management) and SONAR (Society for Natural Resource and Rural Development) are Hyderabad-based NGOs who have been actively working for promoting the farmer's participation in irrigation systems. IRDAS has also been associated with the core group, preparing the state irrigation strategy and implementation programme.
6. The government had initiated a two-year pilot project in 1995 in Sriram Sagar Project introducing Participatory Irrigation Management with the support of two NGOs, i.e., IRDAS and SONAR. Each of these two NGOs worked on 20,000 acres of the command area.
7. The main issue raised at the time of split within TDP in 1995 was the extra constitutional authority exercised by Mrs. Lakshmiparvathi, the second wife of Mr. N.T. Ramarao, which was disliked by majority of the party Legislative members.
8. Though somewhat similar budgetary crisis might have been present in other states when compared to AP, the peculiarity of AP government to opt for economic reforms has a strong base in the political situation and political leadership present in the state.
9. For details see 'Andhra Pradesh Economic Rehabilitation Project, Irrigation Component Preappraisal/Appraisal Mission (November 5-15, 1997)', Aide Memoire, Draft, November 17, 1997 cited in 'World Bank Loan: Boon or Bane', Documents of World Bank and Government of AP, pp. 80-86.
10. Irrigation systems in AP are classified according to size of the command area. Minor irrigation scheme covers less than 2,000 ha. Medium irrigation scheme covers 2,000 to 10,000 ha. Major irrigation scheme covers 10,000 ha and above. There are 15 major, 74 medium and 12,294 minor (tank) irrigation schemes in the state under the aegis of I&CADD.
11. More details on how water charges were negotiated with pressure groups were presented in section I
12. The primary objective of Janmabhoomi being inculcating awareness and motivation among people, which automatically equip people with knowledge to participate in community development programmes which in turn improve their economic and social well-being. Under this concept all important aspects of community development are included covering all important facets of human life. The other programmes in this concept include village education committees for the development and management of primary schools, Vanasamrakshana Samitis (Joint Forest Management groups), women self-help groups (DWCRA) and Watershed

Committees. All these programmes aimed to achieve the targeted results in improving the living standards of people and infrastructural facilities by creating awareness and involving people to shape the programmes in a manner more appropriate to their development. In this regard the role of government has been of a facilitator than provider.

13. Of the five WUAs under major irrigation projects, two are in Kurnool and Cuddapah Canal, two in Nagarjuna Sagar left bank canal and one is in Nagarjuna Sagar right bank canal. Two under medium irrigation project are in Bhahuda Project in Chittoor district. Of the five WUAs under tank irrigation, two are in Kurnool and two are in Nalgonda districts.
14. State-wide data shows 53 percent of the elections for WUAs were unanimous (cited in Keith Oblitas and J. Raymond Peter et al., 1999, p. 26).
15. Several other government programmes refers to watershed, agricultural related programmes in IRDP (Integrated Rural Development Programme). Comparative awareness, interest and participation of people in PIM and above mentioned programmes is stated on the subjective information provided by the respondents. It is also intelligible at the participation of the farmers in the maintenance and rehabilitation works that are taken up by WUAs.
16. During our fieldwork in one village we met the president of a WUA against whom there were several allegations of misappropriation of funds. Thinking that we were government people and had come to inspect the works and accounts he tried to bribe us by offering (through his agent) a bunch of Rs. 50 notes.

REFERENCES

Communist Party of India (Marxist) (1998), *World Bank Loan: Boon or Bane*, World Bank and AP government documents, Hyderabad.

Dr. MCR Human Resource Development Institute (1999), Bhagaswamya Neeti Yaajamanyam: Saaguneeti Samgalu (Reading material for the training of WUAs).

Government of AP, Department Irrigation and Command Area Development (1997) Andhra Pradesh Farmers Management of Irrigation Systems Act (1997), Act and Rules, (Act 11 of 1997).

Government of AP (1998), Reforming the Irrigation Sector for Sustainable Management and Development (policy statement).

Government of India, Planning Commission (1992), Report of the Committee on *Pricing Irrigation Water* (1992-94), New Delhi.

Jairath, Jasveen (1999) 'Participatory Irrigation Management: Experiments in Andhra Pradesh', *EPW,* Vol XXXIV, No. 40, pp. 2823-37.

Keith Oblitas and J. Raymond Peter et al. (1999), *Commencing Irrigation Sector Reforms Through Management Transfer to Farmers: The Case of Andhra Pradesh, India*, draft report.

Mollinga, Peter P. (1998), *On the Water Front: Water Distribution, Technology and Agrarian Change in South Indian Canal Irrigation Systems,* Wageningen, The Netherlands.

Parthasarathy, R. (1998), Reforms in Irrigation Management: Bottoms-up Versus Top-Down Models', working paper No. 104, Gujarat Institute of Development Studies, Ahmedabad.

Sen, Gupta, N. (1991), *Managing Common Property: Irrigation in India and Philippines,* Sage Publications, New Delhi.

Singh, K.K., Syed Turabul Hassan; Sithapathy Rao and Kemuel (1994), 'Promoting Farmers Participation: Action Research in Sriram Sagar Project', in M.V.K. Sivamohand and C. Scott (eds.), *India: Irrigation Management Partnerships,* Booklinks Corporation, Hyderabad.

Venkateswarlu, Davuluri (1988), 'Telugu Desam: A New Regional Party in Power', M.Phil dissertation, University of Hyderabad, Hyderabad.

Wade, R.H. (1987), *Village Republics: Economic Conditions for Collective Action in South India,* Cambridge University Press, Cambridge.

World Bank (1991), *India Irrigation Sector Review, Volume I: Main Report, Vol II: Supplementary Analysis,* Report No. 9518-IN, Agricultural Operations Division, India Country Report, Asian Region.

World Bank (January 1997), *Andhra Pradesh: Agenda for Economic Reforms,* Country Operations, Industry and Finance Division, Country Department II, South Asian Region.

World Bank (May 1998), *Project Appraisal Document on a Proposed Credit for Andhra Pradesh Economic Restructuring Project,* Poverty Reduction and Economic Management Unit, South Asian Region.

8

Designer Participation: Politics of Irrigation Management in Andhra Pradesh

D. NARASIMHA REDDY

The thrust of this paper is that the pre-existing political consensus in favour of farmers' participation in irrigation management is exploited by two dominant forces -- the World Bank and the Chief Minister -- by designing and implementing institutional structures that subserve their interests. The real stakeholders -- the farmers -- are increasingly obsessed with the immediate gains rather than long-run consequences. The irrigation department, especially at the operational level, seem to be gaining lessons enabling a quick readjustment towards inevitable spoils-sharing system. The contractors at the minor WUA level have become invisible like the hidden tenancy, but very much alive and active in carrying on their business. The process of reactionary developments in the form of manipulation of estimates and accounts is already evident in abundance. What is at stake is the transformative role of participatory management and the hope of evolving an irrigation management system that would gear itself towards sustainable development and use of water resources, through the present reforms, is doubtful.

This paper is divided into five parts. The first part deals with the emerging global and national consensus on the concept of PIM, as it is discussed here. The second part refers to the political process of the irrigation reform and the dominant forces behind it in Andhra Pradesh. The third part deals with the agendas of these forces and their implications. The fourth part refers to the mutations at the bottom level. The last is a sampler of questions that haunt.

I

Design from Above

To say that today participatory irrigation management (PIM) is an in-thing in irrigation reforms literature, is an understatement. PIM has come to occupy central place in the irrigation policy of a growing number of countries, in at least two of the continents of the Third World.[1]

However, given the temporal and spatial variations in the nature of initiatives and involvement of farmers in the management of irrigation systems, it is necessary to differentiate the present context of PIM from that of others. There is a binary bottom-up and top-down that is talked about (Parthasarathy, 1998). There could be institutions at bottom only. In any case, for the sake of convenience, we can distinguish broadly three[2] types of PIM: (i) historically-evolved; (ii) grassroots-initiated, and (iii) economically-conditioned.

The *historically evolved* irrigation management systems refer to the experience of local initiatives and practices in the construction and maintenance of irrigation works and distribution of irrigation water as in Dasabandam and Kudimaramat in South India, Banawe terraces and Zenjera in Philippines and the diversion systems of Java and Bali. These were, by and large, smaller systems managed by the farmers or the local community or community-based institutions. In many regions, these systems have decayed or disintegrated. Remnants of them survive in various forms depending upon the local specificities.

Though the historically evolved arrangements referred to as the first category were also based substantially on grassroots efforts, the use of the name *grassroots-initiated* for the second category is more to contrast it from the other, 'top-down' PIM. The grassroots-initiated PIM refers to the relatively recent initiatives that have arisen as a response to certain specific problems of water needs and use. These are arrangements for development and use of small-scale water resources involving either groundwater or surface water or both. These could be autonomous initiatives even within an existing large-scale irrigation system. The *Pani Panchayat* (Water Council) for groundwater use in Maharashtra (Deshpande and Reddy, 1994), the tank irrigation committees in Tamil Nadu (Venkateswamy, 1994) and the 'village republics' (Wade, 1998) under the K.C. Canal could be seen as examples of the grassroots-initiated PIM.

In contrast to the above two types, the *economically-conditioned* PIM has been designed and prescribed as a policy alternative that would improve the performance of the large public irrigation projects. The explanation offered is by now much too familiar. The poor performance and growing gap between potential created and realised is diagnosed as the result of, *inter alia*, neglect of the operation and maintenance (O&M). Low allocation of resources for O&M is because of poor cost recovery. This, in turn, would also result in fiscal burden on the already debt-ridden government of the developed countries. To overcome these twin problems of fiscal burden and poor performance of the irrigation systems, the participation of farmers in O&M, especially from the distributory level down and in the collection of enhanced dues is considered necessary. In a phased manner, it is conceived that the ownership and management could be turned over to farmers and other private investors. This indeed is a conception from above and often linked to political expediency. The discussion of reforms in irrigation management in this paper is concerned with the economically-conditioned and politically-expedient PIM as prescribed from above. Before we deal with actual machinations behind the policy switch in Andhra Pradesh, it may be helpful to have a brief overview of evolution of the policy at the international and national levels.

Early Experiences

This brief review is to suggest as to how PIM presently has come to be equated with economically-conditioned notion. The well-known early experiments towards involvement of farmers in irrigation management under large public works began in South-East Asia. In 1967 and 1968, for example, Water Users' Associations were formed in Thailand in Chaunasut Land Consolidation Project as a part of the North Central Plain Irrigation System (Duncan, 1976). The main objective was to raise adequate O&M funds and to preclude additional demands for O&M on the Royal Irrigation Department (Duncan, 1976: 187). The experiment did develop problems, which offers lessons, to which we shall return later. In 1970s, Philippines initiated PIM by way of two pilot projects in 1976 by externally selected initiators (Bagadion, 1990). By 1979, it was extended to other regions. One of the important considerations favouring the reform was the economic constraint of declining budgetary allocations and growing public debt (Korten and Siy Jr., 1990). In late 1970s, under pressure from the union government, 17 Cooperative Irrigation (Chak) Societies were established under Sone Command area in Bihar (Pant and Verma, 1985). Again, the objective was to reduce the gap between irrigation potential created and the realisation (Pant and Verma, 1985: 39). In the early 1980s, water users' groups on 'chak' basis with about 120 hectares involving 50 to 60 farmers were established under Nang Wai Irrigation Project in Thailand for reasons of augmenting irrigation and improved cost recovery (Kathapalia, 1985). The formation of Pilot Water Distribution Cooperative Society in two villages under Mula Irrigation System in Ahmadnagar district of Maharashtra during 1987-89 with the assistance of an NGO (CASAD) is a well- documented instance (Lele and Patil, 1996). This is a pioneering effort to receive water from the irrigation department on volumetric basis and to distribute the same among irrigators.

The initiative here too is to close the irrigation gap and collect water fees directly. This cursory survey of the 'new' concept of PIM is to emphasise as to how gradually economic efficiency gets equated with less of state involvement and more of farmer or private involvement.

World Bank Policy Shift

At the multilateral institutional level, to be specific, at the World Bank policy formulation level too, there has been a shift in irrigation lending strategy, which incidentally is in tandem with its Structural Adjustment Programme. In the early 1980s, irrigation and drainage projects accounted for 35 percent of lending for agriculture and 10 percent of total Bank lending. And, the success rate is considered to be 80 percent. By the end of 1980s, lending for irrigation and drainage came down to 20 percent of lending for agriculture and 5 percent of total Bank lending. The success rate is considered to be 50 percent in the latter period (Rajagopalan, 1991). The decline in the success rate is attributed to "inefficiency of water distribution and use" and the solution is seen as capacity building of users and of their organisations. Since in many countries, existing irrigation systems have been neglected, rehabilitation projects have become a higher priority on the Bank portfolio than expansion into new projects. "To improve irrigation and drainage projects' O&M, Bank lending has evolved toward low-cost rehabilitation and institutional strengthening, with interventions increasingly national or regional in scope. This new approach focuses on institutional development and the rehabilitation and improvement of existing irrigation system...." (Barghouti, 1991). What goes as implicit in an otherwise explicit statement, is that states which depended on Bank funding for investment in irrigation, are expected to invite private investment into new irrigation projects. And capacity building of farmers' organisations is expected to facilitate such an adjustment in policy. By early 1990s, Bank supported irrigation projects increasingly include provisions for devolving responsibility for O&M to the beneficiaries through Water Users' Associations (World Bank, 1992).

Policy Consensus in India

In the light of the above discussion, it is not difficult to understand that there has been growing consensus, at the policy making level, towards participatory irrigation management (PIM) in India, particularly since late 1980s. This does not include some independent or NGO initiatives, which date back to 1970s. The overwhelming considerations influencing the policy consensus have been the twin-concerns of better cost recovery and closing of the irrigation gap between potential created and utilised through better O&M. It is pointed out that "until the early 1950s, the revenues from water charges exceeded government expenditures for operations and management (O&M) plus investment imputed on investment. By 1989, current expenditures on O&M for public canal schemes exceeded revenues from water charges by an amount equivalent to 2.1 percent of the current agricultural contribution to GDP" (World Bank, 1997b). Another source estimates that major and medium irrigation covers only less than 10 percent of costs through sale of water (Maloney and Raju, 1994: 294).

Under the federal set-up in India, though irrigation has been a state subject, not many states had any legislative provision facilitating the involvement of farmers in the management of irrigation projects. At the behest of incentives provided by the Union Ministry of Water Resources, several state governments, including Andhra Pradesh, enacted Irrigation Utilisation and Command Area Development Act in 1984, which provides for voluntary beneficiary participation in the management of on-farm development. But it is considered to be much too inadequate. In 1985, the Union Ministry of Water Resources introduced an incentive scheme to state government to take up farmers' participation on a pilot-basis at least in one minor project comprising about 1,000-2,000 hectares in each command area and later issued guidelines for the same. The Seventh Plan introduced a centrally sponsored scheme under Command Area Development (CAD), which provided a management subsidy of Rs. 100 per hectare for the first two years, and Rs. 75 per hectare for the third year, for farmers' associations, for taking over management of water distributed below a minor level. The National Water Policy (1987) suggested: "Efforts should be made to involve farmers progressively in various aspects of management of irrigation systems, particularly in water distribution and collection of water rates." The Eighth Plan veered to the view that "eventually, the farmers can take over the maintenance and operation of the system themselves". In 1992, the Committee on Pricing of Irrigation Water, recommended a three-phased movement from wholly government-managed irrigation systems to one of farmers' management (GOI, 1992).

Besides official consensus, there have also been a number of seminars and workshops recommending PIM in India. Even five years back, Maloney and Raju counted 18 such workshops recommending Water Management Associations (Maloney and Raju, 1994: 296-297). In spite of such a wide orchestration in favour of PIM, there has been very limited implementation of the policy. The Draft Ninth Five Year Plan (GOI, 1997) identifies several major reasons for the tardy progress of PIM, which include:

(1) prolonged prevalence of government-managed systems sapped the initiatives of the farmers and made them dependent on government;
(2) non-availability of funds for PIM;
(3) apprehensions in the minds of farmers that under the new system, they might have to incur expenditure on operation and maintenance in addition to increased water rates;
(4) often, lack of homogeneity in the composition of farmers' population and their reluctance to come together, because of differences of castes and classes, to form an association; and
(5) lack of an enabling law for the establishment of WUAs.

The Ninth Plan also refers to several Union Government initiatives in favour of PIM. It refers to the National Conference on PIM in June 1995, which was followed by a state and regional level conferences (GOI, 1997: 515). The Union government suggested setting up of high level groups for the promotion of PIM at the state level and several states set up such groups. What is interesting for the present paper is that Andhra Pradesh did not set up one such high level group! And obviously, PIM was not on the top of the agenda of Andhra Pradesh, not even in 1996.

To return to the causes for lack of progress under PIM, interestingly but understandably, the Draft Ninth Plan does not refer to one oft-repeated cause, viz., lack of political will. Small and Carruthers seem to feel that a major stumbling block for irrigation reforms is the "skeptical and sometimes cynical politicians". They suggest: "Politicians will need to be convinced... that direct cost recovery from enhanced user feels that are retained on the scheme to finance operation and maintenance costs is economically efficient, equitable, fiscally efficient... and, potentially at least, organisationally feasible" (Small and Carruthers, 1990: 213). If one pursues the question as to what crystallises 'political will', one of the possible answers is that it is often a crisis, a crisis in irrigation or even more likely a political crisis. A crisis situation is seen as creating flux and space for change (Mollinga, 1998: 241).

In the light of the above discussion, the emerging policy consensus on PIM in India may be summed up as follows:

(1) In the large and medium public irrigation systems, it is desirable to have PIM, particularly for O&M at the level below the distributory.
(2) It is desirable to recognise the diversity of conditions, specificity of needs, and to begin by pilot projects and to build expansion based on experience.
(3) The basic components for the PIM to take-off are: proper motivation of the farmers, resources for rehabilitation of the irrigation system, appropriate legal framework and political commitment.

However, even at the risk of repetition, the emerging consensus referred to above *does not pay attention to the alternative forms of PIM for smaller or minor works* but is exclusively addressed to larger works.

II
The Designers and the Design: Andhra Pradesh PIM

The PIM experience of Andhra Pradesh is of very recent origin. Only a few studies are available (Parthasarathy, 1998; Jairath, 1999; Venkateswarlu, 1999; Oblitas and Peter, 1999). Since the concern of this paper is more on the nature of the PIM in AP, the forces behind it, their interests, for one, looking for the details of the scheme, the above-cited sources are helpful. There are state sponsored advertisements in leading national and regional newspapers, which refer to the Water Users' Associations in the state and proclaim: "Andhra Pradesh has become the role model today, not just for the country, but for the whole world."[3] Until 1997, there was hardly anything in Andhra Pradesh that would draw much attention in terms of PIM. It did get attention when Robert Wade proclaimed certain command area villages as 'Village Republics' when farmers in these villages evolved voluntarily water management systems under and in the vicinity of Kurnool-Cuddapah Canal (Wade, 1987). But in any case, such grassroots initiatives have no place in the PIM advertisements of the present. In Andhra Pradesh, the occasionally referred to Pochampad experiment with 'Pipe Committees' lasted only as *warabandi* committees for a few years during 1979-84 (Singh, 1985). According to the Draft Ninth Five Year Plan (GOI, 1997: 514), Andhra Pradesh hardly had 32 Water Users' Association (WUAs) covering hardly 17,500 ha[4], while Kerala while had 3,432 WUAs covering more than 137 thousand ha and West Bengal had 10,000 WUAs extending over 37,000 ha. But, there is not much information or news, nor advertisements are seen about the latter two states. In the neighbouring Karnataka state, the government appointed an Expert Committee in October 1995 to study, *inter alia,* Farmers' Association and a high level Working Group was constituted in June 1996 to initiate action on the recommendations of the Expert Committee (Mollinga, 1998: 262). None of such initiatives were in evidence in Andhra Pradesh, even till the later half of 1996. Then how is it that Andhra Pradesh PIM is attracting so much of attention and that too so suddenly? What happened to irrigation management in Andhra Pradesh since 1997, so as to make such sensational claims? How and why did it happen? There is a claim that Andhra Pradesh has followed a 'big bang' approach in PIM instead of 'incrementalist' or 'gradualist' approach (Oblitas and Peter, 1999). What are the consequences of the 'big bang', is a question to which we shall return later. Presently, how do we explain the so-called 'big bang' is the policy?

The Chief Minister and the World Bank

The answers to these questions take us to the politics of the present regime and the forces behind and at the behest of it. The present Chief Minister, N. Chandrababu Naidu, who delights in being described as the CEO of Andhra Pradesh, assumed office by usurping power from N.T. Rama Rao (NTR), his father-in-law and the founder-President of Telugu Desam Party (TDP) in September 1995.[5] Of interest to us here is the change in the leadership resulting in a paradigm shift from 'welfarist' NTR regime to 'developmentalist' Naidu regime. In itself, TDP when it came to power for the first time in 1983, breaking the monopoly of the rule of the Indian National Congress (INC), is credited to have brought about a 'regime change' (Srinivasulu, 1998). The 'regime change', *inter alia*, meant more political space for the backward communities and the 'new rich', both of which did not get adequate share in power in the dominant-caste based politics of patronage practiced under the INC. The TDP ruled from 1983 (with a short interregnum) till 1989 when it lost power to the INC, but NTR came back to power in 1994 with an overwhelming majority, on the promise of restoring subsidised rice scheme without any dilution, introduction of total prohibition and supply of subsidised power to the farmers. But within the TDP, a growing section which perceived the expenditure on these programmes and subsidies as unproductive and unsustainable, were resentful of the 'welfarist' approach and were more inclined towards reformist and market-oriented 'developmentalist' approach. But none, including their unchristened leader, Naidu, could dare challenge NTR. They bided time and occasionally made their position discretely clear. For instance, when Chandrababu Naidu, the then Finance Minister, presented a deficit budget in March 1995, his reply to the opposition's criticism of the budget was that it should be seen as a surplus budget by excluding the 'loss' of revenue because of prohibition and additional burden of expenditure due to subsidies on rice and power to farmers. He did not defend them as investment in people and succor to farming community, as NTR would.

When Naidu finally could dare and displace NTR in September 1995, his top priority was to establish himself as a 'development-oriented' leader. Though politically he had substantial majority, what he needed was to bolster his own image as 'development-oriented' and the financial support to demonstrate his commitment to development, which in his perception, is building roads, bridges, flyovers and the like. In November 1995, the state government announced a New Industrial Policy with ten thrust areas with an objective of achieving 20 percent annual industrial growth in the state and attracting Rs. 1,00,000 crore of investment in the next five years. The Chief Minister offered tax and power tariff reliefs to new industries. During the six months that went past the announcement, the results were not very encouraging and therefore, he had to look elsewhere for succor. And his eyes were set on the lender of the last resort or the only resort -- World Bank. But to get the favours of the Bank, he had to pass through the test of commitment to the reforms. By June 1996, there were White Papers. The one on the 'State Finances' and another on the irrigation sector are of relevance here. Apparently, the objective was to put to debate the World Bank-inspired reforms of reducing the public expenditure at the state level too and by raising user charges to recover the costs of public utilities and services.

In the last week of June 1996, the Chief Minister returned from New Delhi, obviously with a briefing from the World Bank, and announced that he would take concrete steps to put the state finances on even keel as soon as possible. And he did take steps in July. Since the exigencies of elections were a distant four years away, he saw no imminent risk in raising the price of subsidised rice and reducing its quantity, imposing fresh taxes on a number of items, massive hikes in power tariffs and reintroduction of the controversial turnover tax. These together were expected to net an additional revenue of about Rs. 1,200 crore every year to the exchequer. And this was done in spite of the repeated promises during the national parliamentary elections that there would be no compromise on subsidised rice being supplied at Rs. 2 per kg, subsidised power to farmers and total prohibition -- the three programmes associated with the former chief minister. These promises did help the TDP to put up a respectable performance in parliamentary elections, even in the absence of charismatic NTR. In this context, the decisions to raise more revenue by raising taxes and power tariff, though it was still a modest rise in the case of farm sector, and to reduce expenditure by increasing the price of subsidised rice, were bold and politically risky steps. Still, similar courage was not mustered in the case of lifting of prohibition, which had to wait for some more time. The present TDP regime had to take such risks just to demonstrate to the World Bank that it no longer wears the welfarist mantle of NTR and, on the contrary, was too eager to toe the reformist 'developmental' path, if only the Bank comes to its rescue.

How desperately the present regime was in the need of World Bank's assistance could be seen from the fact that in October 1996, when the visiting World Bank president, James D. Wolfenson, declined the Chief Minister's invitation to visit the state capital, the latter went all the way to New Delhi where he managed to get some time of the Bank president to make a presentation on the 'developmental' agenda of the state. The Chief Minister presented his agenda of building up a package of infrastructure facilities, including the reforming of the power sector, improving irrigation projects and information technology, together involving Rs. 7,000 crore and pleaded for immediate Bank assistance of Rs. 1,635 crore. He tried to convince the Bank President of his commitment to reforms by drawing attention to his difficult and unpopular but determined decisions in reducing subsidies and raising power charges. The release of a series of White Papers on state finance and other factors in June 1996 and the announcement of a 'three-fold increase in water charges' for irrigation, in August 1996, were also part of the preparation for the presentation. The Chief Minister pleaded that though he could raise additional resources to the tune of Rs. 2,100 crore in his short stint, he was constrained to meet the demands of infrastructure, because only 10 to 12 percent of the revenues could be devoted to capital expenditure.

The October 1996 meeting of the Chief Minister with the Bank president was a critical turning point in aligning of interests[6]. It was an alliance of interests of the Chief Minister, who was desperately looking for the Bank's resources, even if it were to be loans, in stabilising himself as a political leader capable of delivering 'developmental' needs of the state, and the interests of the World Bank in finding an ally in the Chief Minister for the pursuit of its reforms agenda at the state level[7]. One is likely to expect that the state government, which was harping on 'developmental' programmes, would have a long-term agenda and the World Bank would be interested in enforcing the short-term conditionalities. Strange as it might seem, the Bank's agenda has been aimed at redesigning the institutional structures of the state in the long-run in favour of privatisation, while the state government's so-called long-term agenda does not seem to go much beyond perceiving future in terms of, often unachievable, high rates of growth. To this, we shall return at the end. Presently, let us see how the opportunistic alliance between the state government and the Bank shaped the PIM in Andhra Pradesh.

The Consensus on PIM in AP

The consensus on irrigation reforms at the state level was built up by drawing attention to certain incontrovertible facts. First is the huge investment and poor performance of the irrigation system. In spite of a cumulative investment of Rs. 5,250 crore made on major and medium irrigation projects during the plan period till 1995-96, the performance in terms of expansion of the area irrigated and the returns to meet the maintenance are far too inadequate. Still, there was continued large flow of investments even in the Eighth Plan (1993-97), which amounted to Rs. 2,500 crore, or 24 percent of plan expenditure. There has been a growing gap between the irrigation potential created and the utilisation, and in the 1990s, it further deteriorated. Worse still, the actual area irrigated under public sources has been declining during 1990s. The net area irrigated from public sources has declined from 3 mha in 1988-89 to 2.3 mha in 1993-94 and the gross area declined from 3.9 mha to 3.1 mha during the same period.

Second, there has been extensive under-funding of O&M works in public surface irrigation. According to one estimate, the actual O&M allocation is about one-quarter of the funds needed for adequate upkeep of the network, which in turn, adversely affected the performance of the system. "The Ninth Finance Commission recommended an O&M allocation of Rs. 324 per ha. for 1990-95. Actual allocation in AP in 1990-92 was only Rs. 75 per ha" (World Bank, 1997a: 39).

Third, the poor allocation to O&M is the consequence of inadequate cost recovery. In A.P., water rates, which are the only source of revenue from irrigation, hardly account for 20 percent of the required O&M provisions.

These factors explaining the under-performance of the public irrigation systems are stylised as a 'vicious circle' of deteriorating irrigation performance due to poor O&M provision due to inadequate revenue yield due to low water rates (Oblitas and Peter, 1999). Institutional reforms in favour of PIM involving water users' associations in the tasks of operation and maintenance and cost recovery, given adequate support structure, would result in better system performance and adequate cost recovery. The institutional intervention thus described is stylised as 'virtuous circle' (Oblitas and Peter, 1999).

The need for irrigation institutional reforms of legally creating and involving Water Users' Associations (WUAs) in the management of public irrigation systems is hardly contested. There has been a near unanimity among all the political parties on the need for PIM. But the question still remains as to why no earlier government, including the one in which Naidu was the finance minister, took any steps to realise this reform? What are the interests of the main forces behind the present reforms in irrigation management?

III
The Hidden Agendas

The Bank Agenda

The Bank's engagement with the Andhra Pradesh government, in irrigation inclusive of conditions on participatory management, begins with the Third Andhra Pradesh Irrigation Project with an estimated investment of Rs. 1,890 crore. The share of the Bank and IDA loan is estimated to be about Rs. 1,300 crore or 70 percent of the cost of the project. The third AP Irrigation Project (APIP) reflects two of the important elements of the policy shift of the World Bank. One is not to lend to new irrigation projects but only to strengthen or rehabilitate existing projects. The other is the involvement of farmers' organisation in the management. In fact, the third APIP is nothing but a rehashed version of the second APIP, which was funded by the bank. That the second APIP, which after almost eight years (1985-93) of bungling, had ended up as one of the notorious cases of failure, need not detain us here.[8] The major components of the third APIP include: (i) completion of Srisailam Right Bank Canal (SRBC) network to facilitate irrigation of 65,000 ha, and (ii) rehabilitation of 2,53,000 ha of ayacut under Sriramsagar Project (SRSP). Besides, it also involves providing of agricultural support services, rehabilitation of project affected persons, environment engagement, dam safety and project management and rehabilitation. The Bank proposed as a part of the project, Water Users' Association Promotion Programme, which was expected to develop with the involvement of Irrigation and Command Area Development Department (I&CADD), 472 autonomous WUAs in SRSP and 150 WUAs in SRBC. Since it was a beginning, it was conceived as a phased process-oriented strategy, which could serve as a precursor to similar development WUAs in other irrigation projects. The Bank's appraisal explicitly recognised the social and institutional diversity of different regions and made the following suggestions:

(1) The programmes for WUAs establishment in SRSP and SRBC… would vary in detail because of basic differences in scheme type… local establishment practices and perceptions.
(2) Recognition of the need for social preparation, socio-anthropological studies, motivation camps and meetings in each village over a period of six to nine months to promote WUAs and to integrate local indigenous farmer organisations like tank irrigators within the WUA framework.

(3) ...WUA representatives would be selected through the existing village structures and other functional organisations.

The Third APIP also launched two pilot projects on PIM covering an area of 16,000 ha. For the Bank, which has been notorious for applying a package more or less uniformly and with little research into environments into which they are to be introduced (Wade, 1982a: 172), this was indeed an elaborate and down to earth perception of the need for gradual and well-prepared strategy to launch WUAs. But when the AP government came out with a bang, a statewide uniform system, the Bank had no objection in releasing the loan for third APIP for 1997-98. For, the Bank transferring management to WUAs in only a first step in reducing the involvement of the state and a right step towards providing an environment to attract private initiatives. We shall return to this soon. Presently, let us understand the Bank's interest in the third APIP, which is succinctly summed up in the appraisal report: "Involvement in the project at this juncture would strongly support GOAP's irrigation sector reform programme and *enable the Bank to maintain an irrigation policy reform dialogue with India's second largest irrigation state*" (emphasis added).

The other much more comprehensive and decisive engagement of the Bank with the institutionalisation of WUAs is through the irrigation component of the Andhra Pradesh Economic Restructuring Programme (APERP). APERP is an ambitious Bank-sponsored programme involving an investment of Rs. 3,280 crores ($830m) and covering six sectors, besides fiscal reform programme. APERP aims apparently at providing resources to meet priority needs in human development and in the maintenance of economic assets affecting the rural poor. The irrigation component, which accounts for almost one-third of the investment, is a part of the economic asset maintenance programme. Beginning with 1998-99, it will cover an area of 3.7mha by way of minimum rehabilitation and recurrent maintenance. The basic resource support to establish WUAs on reliable irrigation assets comes from the irrigation component.

It is through the irrigation component of the APERP that the Bank pursued its own agenda by linking the formation of WUAs to certain long-term objectives of the Bank. Halfway through, the negotiations on the irrigation component of APERP, the Bank made issuance of irrigation sector policy by the AP government conditional to the completion of the appraisal of the irrigation component of the APERP. Considering the importance of the issue, it may be necessary to explain the sequence of events and the brazen methods adopted in pushing through the decision on issues of critical importance to the long-term interests of the people of the state.

A draft of Andhra Pradesh Irrigation Sector Policy was prepared by the Bank in September 1997 and passed on for review by the AP government. It was discussed between the I&CADD and the visiting mission of the World Bank during November 1997 and a draft was finalised with "changes largely in details and some presentational matters". The visiting Bank mission suggested that the government should complete the process of cabinet review and issue the irrigation policy statement within a month. Besides the repugnant nature of the methodology of almost forcing the state government to issue a policy statement on irrigation, let us see what exactly is the nature of the policy statement in which the Bank was interested. The AP government finally issued in May 1998 a sector policy statement called "Reforming the Irrigation Sector for Sustainable Management and Development".

The following are the short-run and long-run objectives set for WUAs in this Bank-inspired policy statement. "The short-run objective is that WUAs and the Irrigation and Command Area Development Department (I&CADD) should become financially autonomous for O&M revenues and expenditures. They would generate their own revenue from water charges and entirely finance O&M from their revenues, thus giving them the financial and cost-quality consciousness critically lacking in the past. WUA members will also contribute to interest earning reserve fund accounts, to progressively build up financial capacity to invest or undertake improvements. New investment will continue to be financed by government, *but with users or prospective users contributing through appropriate cost-sharing arrangement*" (emphasis added). So much for the short-term goals. Now, let us see what the WUAs, according to the Bank, would be doing in the long-run.

"In the longer-term, further change may be appropriate as experience is gained and opportunities develop. In other countries such as USA, Australia, Chile and Mexico, where WUAs have taken significant responsibilities and here the private sector has developed as a provider and investor in irrigation and water provision (direct private investment, growth thorough capital market), the *government agencies have adjusted their roles...*" (emphasis added). The Bank is obviously pleased because it could get its own agenda of expanding the space for private sector incorporated as a part of the Irrigation Policy Statement of AP government.

Babu's Agenda

Considering the question as to how Andhra Pradesh, which was hardly known for any initiatives in participatory irrigation management, suddenly emerged as an example since 1997, we have tried to seed the links between the leadership's desperate need for resources and programmes that would help build public activity to project the developmental face that would attract the attention and support of the people. In this search for sources of resources, the World Bank looms large. In the search for ways of blowing programmes of even very low populist potential into larger proportions has been there ever since Babu assumed office. For, the World Bank, the Chief Minister familiarising the concept of

WUAs at every public meeting, launching of massive training programmes and workshops often addressed by the Chief Minister represented a major commitment by Andhra Pradesh. The unprecedented step to form WUAs across the state is an action only undertaken on this scale in recent years by countries such as Mexico and Turkey, international success stories in irrigation reform. For the Chief Minister, it was yet another opportunity to build his popularity as the politician who minds 'development' as his top priority. Babu could see statewide PIM as a powerful political opportunity in building up cadres through favours and to an extent performance in improving irrigation systems. If it succeeded, his vote banks would swell. Further, here is yet another feather, the WUAs, which he could claim in his 'Janmabhoomi' cap. There is little wonder that WUAs prominently figure in the manifesto of the TDP.

The interests much larger than involving the participation of the farmers is evident from the design and the breathless pace with which the programme of WUAs was ushered in the state. Sometime in early February 1997, it seems to have been decided that the farmer participation should be designed as a statewide programme. This was contrary to the earlier decision in 1996, in the context of the third APIP, which stated that "...the GOAP would like to develop a new approach to water management. The new approach would be based on participatory irrigation management through the creation of Farmers' Organisations. It is proposed to implement this new approach first under the proposed AP III Project in the SRBC and SRSP commands...". What brought the switch from such as idea of phased, or what is pejoratively referred to as 'gradualist', approach to one of 'big-bang' in the programme is obviously a desire to turn it as an aid to high political visibility. The explanation that the 'big-bang' decision was taken in the interests of regional equity[9] is an earnest administrative explanation but doesn't prevent us from seeing the more important political interest involved. In early February, a circular was issued by the I&CADD to all the Chief Engineers of all irrigation works informing them of the impending legislation, requesting them to provide information delineating zones of 2,000 acres (minor irrigation) and 800 acres (major and medium irrigation) on hydraulic basis and to provide the information by 10 March, 1997. And in March 1997, the A.P. Farmers' Management of Irrigation Act got passed and 10,292 WUAs were constituted by June 1997. This, by no means, is a minor political constituency for any ambitious politician.

Regarding the pace of implementation, while the World Bank has been all praise for the unprecedented decision of the AP government in raising water rates, in getting the Act passed, in announcing a policy commitment, in getting the WUAs constituted, all within the space of one year -- June 1996 to June 1997, the Chief Minister has been repeatedly referring to one or the other visiting missions or review groups of the Bank, in his weekly high-tech teleconferences with the district authorities in reviewing progress. In this process of mutual interest and mutual appreciation between the Bank and the Babu, what actually is happening to participatory management and the participating farmers is an important question. Even the assessment of the participation seems to be more as a part of the design that adds more to political mileage – it apparently may appear programme-support propaganda. But the huge advertisements that have been accompanying the programme are vulgar, false and exaggerated.[10]

IV
PIM/Machinations

The Farmers

In the legislation and constitution of the WUAs in the state, the basis is hydraulic, not social. The delineation of WUAs or the Territorial Committees is by hydraulic measure, not by socio-economic structure. This is not by accident but by design – a design mutually acceptable to the Bank and AP government. In the entire Bank-driven conception of water management, the objective function is efficiency, not equity. The questions are, given water resources developed in a territory, how to bridge the gap between the potential and the realised. How to extract the cost, not who is paying the cost. There is no room for consideration of farming classes or castes. All farmers are homogeneous buying units. This should ensure through their participation, the quality of the product, viz., water supply and should pay for its cost, if not the economic price. In this ideological framework within which the PIM legislation is made and WUAs are constituted, there is no room for questions of representation by small-marginal farmers, weaker sections or women. Given the socio-economic framework within which the programme is embedded, it is politically highly risky. And therefore, though the constitution of the programme has no inbuilt space for it, by design, the class-caste-gender questions are raised and answered.

Since even the World Bank cannot assume a social vacuum, it was constrained to find a social justification, which was asserted in the APERP Project Appraisal Document (pp. 24-25): "The main beneficiaries of the project would be farmers with holdings averaging less than 2 hectares (about 1.1. ha in the Coastal Region, 1.7 ha in the Telangana Region, and 2.1 ha in the Rayalaseema Region). The project would lead to a substantially more equitable distribution of water, benefiting particularly tail-end farmers who are generally in the poorest group. Based on Participatory Rapid Appraisals (PRAs) done during appraisal, the first impact of the WUA formation process would also have a positive impact on equity. WUA presidents tend to be from the same social and economic background as the majority of farmers (predominantly small to medium farmers and without bias according to caste). The WUA committee elections, based on territorial representatives for different areas in the WUA command and use of secret ballots, has resulted in balanced area representation and election of leaders. This is an important change from previous community development efforts, where representatives tended to be large landowners. Another positive feature has been the election of a few women as WUA leaders, but this is still not common. To focus on the greater integration of women in WUA management, a sociological study on Opportunities for Women's Participation would be mounted in early PY2."

The Bank's elaborate explanation that "WUA presidents tend to be from the same social and economic background as the majority of farmers and will be without caste-bias" is to give an impression that there is nothing wrong in the present constitution of the WUAs, which does not take into consideration the socio-economic differentiation that prevails at the ground level. This is sought to be perpetuated by the numerous advertisements and pamphlets, which carry information collected from a survey of the presidents of WUAs at state-level workshop (Oblitas and Peter, 1999). The results of the survey show the following class-caste classification of the presidents:

Farm Size	*Percept of WUA Presidents* (%)*
Marginal Farmers (upto 2.5 ha)	42
Small Farmers (2.5-5 ha)	31
Large Farmers (5 ha)	23
Scheduled Tribes	2
Scheduled Castes	3
Backward Castes	31
Other Castes	60

* 4% non-response

Source: Oblitas and Peter, 1999: 26-32.

First, the above statement blows off the Bank's assertion that there will be no caste bias; while the survey cannot hide caste in its tabulation, it tries to perpetuate the myth that most of the presidents will be from the small-marginal farmers by tampering with the widely accepted and used classification that marginal farmers are those with less than one hectare, small farmers with 1 to 2 ha, semi-medium 2 to 4 ha, medium 4 to 10 ha and large 10 ha. A proper retabulation would show that small-marginal farmers are only a minority of WUA presidents. The field information reveals that the speed with which the WUAs were constituted and elections held, in most of the villages, the impression was that it was one more committee in the series of committees coming to villages of late, that there would not be much resources and therefore, the really powerful have put-up their candidates either from small-marginal sections or from the weaker sections. Only much later, when the resources started flowing, they realised the importance of the office and of course have been making quick compensating arrangements.

Diverse Conditions of WUAs

Further, trying to assess the nature of the leadership or performance of WUAs without considering their location -- whether these are under major and medium works or under minor works, whether in drought-prone region or in the better rainfall region -- would lead to more of obfuscation rather than better understanding. While the condition under each project may vary, the systems, mechanisms of operation, regulation, and arrangements for distribution are likely to be, by and large, similar under all major and medium irrigation projects. But minor work like tanks are spread all over the state and are likely to have extensive differences based on their historical and spatial specificities. A uniform command approach may not be appropriate. This needs significant attention given the fact that almost 80 percent of WUAs are under minor irrigation works like tanks. The politics of 'big-bang' may cause lot of dislocation to the existing institutional arrangements and to avoid any adverse effects, urgent attention is needed to look into their specificities and provide for appropriate changes.

Further, the 70,000 minor works under Panchayat Raj may not be exclusive hydraulic systems without any link to the other storage systems. These links should be studied and appropriate integrated participatory system should be developed.

Participation

It is needless to say that farmers' participation in the maintenance of all types of irrigation systems existed in the past and continued in several informal ways. The participation was extensively in terms of voluntary labour and under conditions of water-stress, it took even the form of monetary contribution. Water Councils of *Panchayats* and Watermen (Nirgantis) with diverse payment arrangements continued in various forms. Besides maintaining works and water distribution, these arrangements also helped to preserve village as a community. The field experience on the working of the new WUAs suggest that the flow of funds and along with that the rules and commands run the risk of seriously undermining the community orientation through depersonalised bureaucratic procedures. There is need for deeper grassroots level research as to how the new WUAs could be integrated with existing positive community spirit, particularly in the countryside under the minor irrigation works. This is more so in the context of the looming ground water crisis, which is likely to soon emerge as the major social problem.

There is a dangerous tendency to make the WUAs function the commands from above. Participation has to be nurtured through grassroots initiatives.

Turning the WUAs into 'Janmabhoomi' like image-building political campaigns would be counterproductive and turn participation into a farce. By now, the notoriety of 'Janmabhoomi spirit'[11] is quite well-known in the countryside. Under the Janmabhoomi, 20 to 30 percent of the expenditure on local works is to be met by *shramdan* (voluntary free labour) or by local contribution in cash. Very rarely does it happen. Most often, the estimates of the works are jacked up by 20 to 30 percent, the work is completed with 70 percent state funds and 20-30 percent *shramadan* or local contribution is shown in accounts. This exercise in manipulation has reached a level of perfection in most of the villages in Andhra Pradesh. The Chief Minister, at one of his weekly Thursday tele-conference with the Collectors in May 1999, desired that the WUAs should conduct general body meetings once in 15 days, but the Collectors pleaded their inability to organise such meetings which depended on the cooperation of the WUAs members. In the next teleconference, the Chief Minister directed the collectors to organise general body meetings of all the WUAs throughout the state on banks of canals on 30 and 31 May, 1999. And it was done with a big bang, mobilising the entire official machinery to urge people to attend the meetings. Perhaps, that was the first general body meeting for many of the WUAs after the June 1997 elections. And one hopes that it would not be the last.

The Irrigation Department

The pioneering zeal of a band of officers and engineers in the Irrigation and Command Area Development Department (see Oblitas and Peter, 1999), which seems to have been providing logistic support to the 'big bang' of Babu, down the line in the irrigation department, there has been resigned compliance. Though many of them have been actively participating in the training and execution of the WUAs, there are many more, who have been reticent. If the irrigation department functionaries at the WUAs level, are reduced to be 'accountable' to the WUAs, what will happen to all those proverbial 'savings on estimates' and 'saving on ground' gratifications (see Wade, 1982) for which the department is well known? There has been suppressed displeasure among several functionaries but it could not be expressed as a protest. But, this did express itself in the form of a pen-down strike by the Irrigation Department (ID) official in July 1999 as protest against the death of the Chief Engineer, allegedly because of the uncouth behaviour of the Chief Minister towards him. But interaction with the presidents and the ID officials show that the 'savings on estimates' and 'savings on ground' on their return. Still, there is friction in some areas on the delaying of estimates by the ID officials. There are also disputes on the demand by the WUA presidents that the *lascars* in their area should be under their control and not under the ID engineering staff as it has been the practice.

Contract System

All minor works under WUAs are supposed to be carried out by the managing committee itself and contracts should be avoided. This is a very naive proposition in the absence of any other local institution like a labour cooperative or farmers' cooperative. Even small works involve some equipment, which cannot be hired easily at the village level. Sometimes, prior investment may have to be made. Some works need lot of supervision. In many villages visited, apparently, WUA works are carried on by the managing committee itself, but in reality, contracted out. Contracts and minor contractors have become hidden under WUAs. It may be necessary to investigate the implications of these developments.

V
Questions?

The case for PIM management cannot be reduced into narrow economic considerations only. The diversity of the socio-economic conditions of the irrigators, as much as the irrigation contexts, should guide the designing of PIM. Water resource development should be seen in totality of the surface and groundwater resources and their rapidly changing balance. It is obvious that the PIM introduced in Andhra Pradesh has been designed in a manner that would subserve two main factors, the World Bank and the Chief Minister, but leaves the sustainability of the programme subserving the interests of the needs of people in doubt. Therefore, there is need for a detailed research and analysis into the following aspects:

First, what are the implications of the Bank's short-run interest in trying to force the state to recover entire O&M cost within three years? Second, is the Bank's long-term agenda of designing the state's irrigation policy on the lines of USA, Australia or Chile, where private sector has developed as a provider and investor of irrigation, necessary or desirable in a state like Andhra Pradesh? What are the consequences of the Bank's imposed irrigation policy of the state, which declares that government will invest in new projects provided the users contribute to cost sharing, at a time when the state has used only one-fourth of the state's total surface water potential of 8.5 million ha? Third, what are the consequences of the straight-jacket uniform management practices imposed on all irrigation systems? What are the dangers of mixing-up of political publicity with the meetings of WUAs and official mobilisation of members even for general body meetings at the village level. Fourth, can the WUAs evolve as transformation (Michener, 1998) institutions where people identify with resource conservation and use? Would it end up as a grandiose showpiece behind the veneer of which manipulative accounting, duplicity in functioning and hidden contracts and rent-seeking would flourish?

These and many more questions that would surface need to be answered once the noise of the 'big-bang' subsides.

NOTES

1. There is extensive and growing literature on PIM in Asia and Latin America. But, there is hardly anything about the developments in Africa, except some reference to capacity building in Ghana or poor realisation (only 15%) of potential created (World Bank, 1991).
2. There could be alternative ways of classifying PIM. But the three-way classification is more to minimise the overlapping notions of PIM and to emphasise the specificity of the third type, viz., economically-conditioned.
3. See for instance, the advertisement issued by the Engineer-in-Chief (Irrigation and Command Area Development Department) in *The Hindu*, 22 April, 1999.
4. These 32 WUAs turn out to be the two projects under Sriram Sagar Project as a part of the condition for the World Bank loan for 'Third Andhra Pradesh Irrigation Project', carried out by the two NGOs, IRDAS and SONAR, each carrying out a project.
5. We wouldn't have touched upon his personal relationship, had it not been mixed up with the politics and public stakes of the state. Apparently, the differences between the present Chief Minister and his predecessor were attributed to the former's resentment at the political power wielded as kind of extra-constitutional authority by the wife of the former Chief Minister. Besides personal ambition, the reasons for unseating the former CM are more to do with the simmering governance and policy differences that became sharper in the months before the act, which was variously described as treachery, ingratitude or deceit.
6. It was a turning point in terms of Bank lending too. By October 1997, the Bank announced a package of loans to the tune of $2 billion, a feat unknown to any state in India.
7. Though economic reforms were introduced in India in 1991, the efforts of the World Bank to get the reform agenda, like reducing fiscal deficit, eliminating subsidies, raising user charges, etc., implemented at the level of the states through the Union Government reached *almost a dead-end*. The Union Government, because of its dependence on fluid coalition politics, was much too weak to push the state governments to fall in line with the Bank-induced reforms.
8. The Second APIP was launched in 1985 with a World Bank loan of $177million. The main components of the project were: (i) rehabilitation of the existing 2,34,000 hectares of Srisalam Project (SRSP); (ii) extension of SRSP ayacut by another 4,000 ha on the Kakatiya Canal; (iii) development of 65,000 ha of new command area under Srisailam Right Branch Canal (SRBC); and (iv) retrospective actions to rehabilitate project-affected persons. At the end of eight years, and a loan disbursement of $118 million, there was not a single additional hectare of irrigation added, whatever might be the funding with the figures of rehabilitation of the ayacut

under SRSP (see for details, World Bank, *India's Third Andhra Pradesh Irrigation Project: Staff Appraisal Report*, 1997).

9. The administrative explanation is given in the draft implementors' report (Oblitas and Peter, 1999).
10. See the following sample:
    Vulgar: "Just like the thunder thighs gave way to the sveltes at the box-office, the large-dam multipurpose projects have given way to small tanks and ponds.... Overcoming political rhetoric (sic), the present government is committed to reversing the decision-making process, of superimposing yojanas from above (sic)... The Water Users' Association (sic) is a political realisation coupled with a statesman-like decision in the process of decision-making. The political philosophy embedded in 73rd and 74th constitutional amendments of taking power back to the grass-root level and decentralising its executions is based on strong commonsense". GOAP full-page advertisement in *The Hindu*, May 23, 1999.
    False: "....Investment in irrigation infrastructure has been a priority of the state. An amount of Rs. 64,000 crores (!) has been spent so far during plan period...." Advertisement in *The Hindu*, April 9, 1999.
    Exaggerated: "....Sriramsagar Project had been recording an ayacut of 95,000 acres... With the coming of the farmers' organisations, the ayacut in the first year rose to 2,37,000 acres and this year, has risen to 3,45,000 acres". Advertisement in *The Hindu*, April 19, 1999.
11. The GOAP, under the present leadership, has launched a populist programme called 'Janmabhoomi', which literally means 'motherland'. This is supposed to be an umbrella programme without any direct investment, development or welfare components in it, but most of the programmes including centrally-sponsored programmes are projected as part of this.
    In brief, the 'Janmabhoomi' programme is supposed to be aimed at what is called 'regeneration of values along with fundamental transformation of the work styles of people'. Such a transformation is seen as a basic requirement if the benefits of development are to reach all equally. Towards this, the programme enunciates that all institutions should model on the basis of the following fundamental principles: people's partnership, unbiasedness, equality, transparency, accountability, search for new solutions and relentless effort. And then follows the elaboration of the sermons in the tonnes of booklets on 'Janmabhoomi'. It is more like a religious propaganda but the difference is that it is state-funded. And the Bank too does not seem to have any objection to this type of public investment.
    The point of relevance here is that the PIM in state is also put under this umbrella. Apparently, there is nothing wrong in such ideals. But, what is observed is that the ideals enunciated are so lofty and completely devoid of the behaviour of those preaching them that people come to regard it as yet another political gimmick with which they have nothing to do. And continue to grease the hand of the powers that matter either to get their ration card or release of water or whatever. There is a danger that 'Janmabhoomi' duplicity would befall on the WUAs too.

REFERENCES

Bagadion, B.U. (1989), "The Evolution of Policy Context: A Historical Overview" in Korten, Francas F. and Robert Y. Siy Jr. (eds.) op. cit.

Barghouti, Showki (1991), "Agricultural Diversification and the Role of Irrigation" in World Bank (1991) op. cit.

CPI(M) (1998), *World Bank Loan: Boon or Bane?: Documents of World Bank and Government of A.P.*, Hyderabad.

Deshpande, R. S. and V. Ratna Reddy (1994), "Social Dynamics and Farmers' Society: A Case Study of Panipanchayat" in R.K. Gurjar (ed.) *New Dimensions in Water, Environment and Ecology*, Vol. IV, Printwell, Jaipur.

Government of India, Planning Commission (1992), Report of the Committee on Pricing of Irrigation Water, New Delhi.

Government of India, Planning Commission (1997), *Draft Ninth Five Year Plan 1997-2000*, Vol. II, New Delhi.

Jairath, Jasveen (1999), "Participatory Irrigation Management: Experiments in Andhra Pradesh", *EPW,* Vol. 34, No. 40.

Kathpalia, G. N. (1985), "Farmers' Organization in the Irrigation Systems of Asia and the Pacific Region" in Kanwar, J.S. (ed.), *Water Management: The Key to Developing Agriculture,* New Delhi, Agricole Publishing Academy.

Korten, Frances F. and Robert Y. Siy Jr. (eds.) (1989), *Transforming A Bureaucracy: The Experience of Philippine National Irrigation Administration*, Connecticut, Kumarian Press.

Lele, S. N. and R.K. Patil (1996), *Farmers' Participation in Irrigation Management: A Case Study of Maharashtra.*

Maloney, Clarence and K.V. Raju (1994), *Managing Irrigation Together: Practice and Policy in India*, New Delhi, Sage.

Michener, Victoria, J. (1998), "The Participatory Approach: Contradiction and Co-option in Burkina Faso", *World Development,* Vol. 26, No. 12.

Mollinga, Peter, P. (1998), *On the Waterfront: Water Distribution Technology and Agrarian Change in a South Indian Canal Irrigation System,* Wageningen, The Netherlands.

Muthyam Reddy. K. (1999), "Water Users' Associations in A.P.: Some Issues", Department of Economics, Osmania University, Hyderabad (Mimeo).

Oblitas, K. and J. Raymod Peter, "Commencing Irrigation Sector Reforms Through Management Transfer to Farmers: The Case of Andhra Pradesh, India", Draft Report, I&CADD, GOAP/World Bank (Mimeo).

Pant, Niranjan and R.K. Verma (1985), *Farmers' Organization and Irrigation Management,* New Delhi, Ashish Publishing House.

Parthasarathy, R. (1998), *Reforms in Irrigation Management: Bottoms-Up Versus Top-Down Models,* GIDR, Ahmedabad.

Singh, K.K. (1985), "People's Participation in Command Area Development: Experience and Research Issues" in J.S. Kanwar (ed.), *Water Management: The Key to Developing Agriculture,* New Delhi, Agricole Publishing Academy.

Small, Leslie, E. and Ian Carruthers (1990), *Farmer Financed Irrigation: The Economics of Reform*, Cambridge, Cambridge University Press.

Srinivasulu. K. (1998), "World Bank, Economic Reforms and Welfarism: The Case of Andhra Pradesh", Department of Political Science, Osmania University, Hyderabad (Mimeo).

Venkataswamy, R. (1994), "Success Story of a Farmers' Organization in a Tank Irrigation System in Tamil Nadu" in R.K. Gurjar (ed.) op.cit.

Venkateswarlu, D. (1999), "Politics of Irrigation Management Reforms in Andhra Pradesh". Paper submitted to International Conference 'Long Road to Commitment' held at Hyderabad, India, 11-14 December, 1999.

Viswanathan, Rajagopalan (1991), "The Global Problem of Land and Water Constraints" in World Bank (1991) op.cit.

Wade, R. (1982a), "The World Bank and India's Irrigation Reforms", *The Journal of Development Studies*, Vol. 18, No. 2, January.

Wade, R. (1982b), "The System of Administrative and Political Corruption: Canal Irrigation in South India", *The Journal of Development Studies*, Vol. 18, No. 3, April.

Wade, Robert (1987), *Village Republics*, Cambridge, Cambridge University Press.

World Bank (1991), *Developing and Improving Irrigation and Drainage Systems*, Technical Paper 178, World Bank, Washington, D.C.

World Bank (1992), *Water Resources Management in Asia*, Vol. 1, Technical Paper 212, Washington.

World Bank (1997a), *India: Andhra Pradesh: Agenda for Economic Reforms*, Country Operations, Industry and Finance Division, Country Department II, South Asia Region.

World Bank (1997b), *India: Third Andhra Pradesh Irrigation Project: Staff Appraisal Report*, Washington, D.C.

9

Contradictions of a Supply Side Approach: Participatory Irrigation Management in Andhra Pradesh

JASVEEN JAIRATH

This article attempts a brief overview of the developments in the irrigation sector in Andhra Pradesh (AP) during the last two years, with particular reference to the AP Farmers' Management of Irrigation Systems (APFMIS) Act of 1997 (ACT). While noting the positive impacts and constructive activities stimulated as a result of the reforms to institute participatory irrigation management (PIM) on an extensive scale in the state, the article raises first, the issue of sustainability of these initiatives. Second, it questions some of the claims of enhanced economic benefits following from the reforms, e.g., increased irrigated acreage and yields, etc. Third, it points to the equity implications of the reforms and questions the probability of equal water distribution under the existing pattern of land and water use. Fourth, it focuses on the limitations of the data categories along which most official data is recorded. Finally, the financial viability of Water User Associations (WUA) is examined, in light of the proposed objectives of the reforms.

Following from the above, the underlying thrust of the argument is as follows: Given the political constraint of neutralising the head-tail inequity in access to irrigation, discrimination against tail-end water users will continue. They will therefore desist from paying their water fees. This will seriously threaten the financial self-sufficiency and, therefore, sustainability of WUAs. The single major motivation of the heightened activity in the irrigation commands currently being witnessed, is based on the heavy inflow of funds through the network of WUAs. Discontinuation of external funding coupled with the likely failure to generate resources from within the WUA network will threaten the sustainability of these organisations.

The argument is not for giving up the reforms but carrying them forward to include the realm of water use within its fold. Thus, it is necessary to wean the head farmers away from water guzzling crops such as rice. This will enable increased release of water for downstream users. The productive use of supplies thus received creates the possibility for users to pay their financial dues. Conscious interventions have to be made to actualise this possibility, given the poor record of payment of water cess even in the head reach areas -- which have typically received more than their due share of water. Confining the reforms to supply side initiatives, on the other hand, reinforces the extant contradictions which inhibit sustainable reproduction of participatory institutions.

It is recommended that market and non-market mechanisms (of agricultural extension) should be used to influence choice of low water consumption crops, water-efficient technologies and adoption of practices which enhance water productivity. Technology transfer from lab to the field has to be facilitated to demonstrate the practical viability and economic attraction of low water consumption cropping practices. Technology transfer and constraints to technology diffusion need to be coopted as prominent items on the research agenda.

The article is partly based on fieldwork conducted in areas of Nagarjuna Sagar Project (NSP) and Sri Ram Sagar Project (SRSP) commands as well as secondary data collected from irrigation and revenue departments and other sources. Distributaries approximately at the middle of the main canals (NSP-Left Canal under NSP and Kakatiya Canal under SRSP), i.e., Wazeerabagh Distributary (WD) and D-89 respectively, were selected for more intensive surveys. A sample of 27 farmers from each distributary was selected from head (H), middle (M) and tail (T) of the distributary located at head, middle and tail of the minors. At each nodal point the attempt was to interview a large, medium and small farmer (i.e., three farmers). Therefore, a total of nine farmers each at H, M and T were selected, i.e., a total of 27 per distributary and 54 for both the distributaries. Random surveys to elicit farmer responses to PIM were also conducted.

Overview

It had been noted that "existing usage of water is inefficient, wasteful, inequitable with tail end deprivation being almost universal. Further, widespread deterioration of infrastructure with little or no participation of farmers has aggravated the low utilization of water of irrigation commands" (note on the APFMIS Act, 1997). The area irrigated under some of the major and medium projects was shrinking, inspite of secular increase in capital expenditure (corrected for inflation) and increasing Maintenance & Operation (M&O) expenses (Jairath, 1999) (see Table 1 for evidence for NSP and SRSP projects).

Table 1

Year	Area irrigated (lakh acres)	
	NSP	*SRSP*
1990-91	22.33	3.23
1991-92	19.08*	3.33
1992-93	22.50	3.53
1993-94	21.50	2.23*
1994-95	22.00	3.35
1995-96	17.56*	3.23

* Water not released for second crop to take up canal repairs.

Source: Irrigation Sector: A Factual Note, GOAP, 1996.

This was well acknowledged within the government circles, in particular for its implications for realisation of irrigation potential created.

> "The state of Andhra Pradesh has invested large amounts of its scarce financial resources to create irrigation schemes. The net irrigational potential created thus far 6.4 M ha. In spite of this large effort, statistics show that only about 4.3 mha. or 67% of the potential created is utilised thus leaving a gap of 2.1 mha. or 33% unutilised. Some of the main reasons for this gap are the *non-compliance of farmers to the designed cropping pattern,* (emphasis added) the poor conditions of the irrigation systems, and the lack of operational plans. In addition to this gap, water distribution within the command areas is often neither reliable nor equitable with large differences in water availability between the head and tail end of the irrigation canals.
>
> Irrigation & CAD Department, has attempted to remedy this situation through various measures among which are the rehabilitation of the distribution networks, the introduction of rotational water supply based on structured systems and the promulgation of the AP Irrigation Utilisation and Command Area Development Act, 1984 authorising the creation of Command Area Development Authorities (CAD) and the formation of Pipe Committees. Though some improvements were noted, much of the disorders noted previously continue and none of these initiatives have really proved to be fully *sustainable* (emphasis added). The failure is mostly due to a top-down approach to irrigation planning and management without adequate consultation with farmers" (Action Plan, GOAP, 1995).

Within this backdrop, the AP Farmers' Management of Irrigation Systems Act was passed in the Legislative Assembly on 27 March, 1997. The main aim of the legislation was to provide a legal backing for implementing measures to institutionalise the participation of irrigation users in the operation and maintenance of irrigation systems. Major stated objectives were as follows: to ensure equitable and reliable supplies, bridge the gap between irrigation potential created and utilised through preparation of suitable operation plans, bring about social/water audit in irrigation and enable the department to withdraw from

M&O in order to concentrate on the reservoir management and development of new systems.

The thrust was on decentralising irrigation management, thereby according a greater responsibility as well as greater powers to bodies comprising elected representatives of water users. Involvement and control by the users at the local level, it was believed, would constitute an incentive to improve operational performance. The entire exercise would lead to improvement in quality and cost efficiency of irrigation management leading to increased profitability of irrigated agriculture, often in spite of increased cost of irrigation to the user. The process was perceived to be a cost saving venture for the government, wherein the released resources could be used elsewhere.

The proposals for operationalising the above ideas envisaged the creation of an institutional structure of irrigation water users with Water User Associations as its basic elements (see Jairath, 1999, for details). Following these proposals and passing of the Act in 1997, elections for nearly 10,000 WUAs were held on 17 June 1997 (one president and technical constituency members as per the WUA command). Subsequent to the elections, extended training was provided to government officers and presidents of WUAs to sensitise them about PIM, its features and necessity. Training programme for all the WUA presidents and members was conducted in June 1999 at a cost of Rs. 2.5 crores (DIPR, 1999). As a result of the reforms 22,171 works at a cost of Rs. 107 crores have been taken up and completed (DIPR, 1999). During the first year activities taken up in canal irrigated areas took the form of silt clearance of channels, followed by strengthening of embankments, small repairs of pipe outlets and drop structures, plugging of breaches, etc. The emphasis during the second year had shifted to the fixing of shutters at the various offtake points. This is significant, as existence of shutters is a precondition for any attempts to institute water regulation.

Creation of these institutions has succeeded in providing a forum for expression of the collective demands, problems and concerns and channeling the communication to the government in a systematic manner. In place of anarchy, confusion, and randomness there is a relative order in getting the message across. Most farmers, as revealed by the field survey, preferred to deal with the WUA office-bearers rather than the irrigation department officials, as was the case formerly. Whatever reservations the individual farmers may have had regarding the degree of effectiveness of the WUAs, they were unanimous in welcoming the establishment of these organisations. While earlier personal patronage seemed to be the only avenue for getting a hearing at the higher echelons, now there did exist institutions manned by office-bearers, where a relatively anonymous farmer could go for redressing his water related difficulties. How effectively they functioned is a separate issue. Organisation of elections for the formation of 10,000 WUAs on a single day for the entire state was a stupendous task. Smooth and fair conduct of elections by Irrigation Department (ID) and the local administration was commendable for its organisation, planning and execution, specially considering the sheer scale of elections and the fact that they were conducted for the first time. Nearly all the farmers surveyed in the field were satisfied with the mode of elections. It was indicative of the political will to ground the programme well.

As a result of instituting WUAs, irrigated acreage has shown a distinct and sharp increase in certain commands of major and medium projects. An official release by the irrigation minister noted that 5.12 lakh acres of ayacut has been *additionally* brought under cultivation as a result of better maintenance work due to the reforms. The state had a total ayacut gap of 15 lakh acres. In all 23,011 works had been completed by March 1999 at a cost of 135 crore. Paddy production was reported to have gone up by 10 percent and yield had increased from 30-35 bags per acre to 40-45 bags per acre (DC, 1999).

According to the recent statement by the irrigation minister, the maximum gap bridged in irrigation potential was in respect of SRSP project where irrigation potential created was 6,59,000 acres, average area irrigated during past three years was 1,15,00 acres and ayacut transplanted in the current year was 3,66,000 acres. Next came the Nizamsagar project with gap bridged being 74,000 acres, irrigation potential created being 23,000 acres, average ayacut irrigated during the past three years being 1,14,000 acres and ayacut transplanted during this season was 1,85,000 acres. A total of 17,000 acres of ayacut was bridged under Vamsadhara project, 50,000 acres under NSP-LC, 53,000 acres under NSP-RC, 10,000 acres under Tunghabhadra High Level Canal, 11,000 arces under Low Level Canal, 2,000 acres under Rajolibandh Diversion Scheme, and 23,000 acres under Kurnool Cuddapah Canal (Minister's Statement, 1998). As argued below, while there has been a significant increase in area irrigated, the official figures are clearly exaggerated.

From the fieldwork done it was seen that the head of the distributary continues to have comfortable water, the middle experienced some improvement, the tail continues to suffer erratic and low supplies, in spite of water 'touching' the tail end on this distributary for the first time in the last 15 years. For D-89 maximum increase in area irrigated is at the tail end WUA but only for irrigated dry crops. The intervening WUAs have lower proportion of increase in ayacut irrigated. If wet crops are also taken into account the increase in area is highest at the head WUA, i.e., 68 percent. In this case the use of increased water availability has been used to shift from irrigated dry to wet crops, i.e., to rice cultivation. This is corroborated, for example, where for kharif 1997 and 1998 show an increase of nearly 600 percent of total area irrigated. A ratio of 40 percent wet and 60 percent irrigated dry was reversed to 55 percent wet and 45 percent irrigated dry, indicating a shift to rice cultivation. For rabi also there is a mild shift to rice from 27 percent to 29 percent. Same trend is evident in NSP while kharif wet area shifts from 73 percent to 83 percent, for rabi there is a dramatic shift from 2 to 82 percent.

Given the unreliability of the irrigation statistics, much weight should not be added to absolute numbers. But that there is a shift to wet crops with increased water availability during the post-reform period was also confirmed during field visits as discussed below. The higher proportion of irrigated dry crops in case of D-89 is explained by the existing pattern of cultivation prior to the reforms where more than 70 percent of rabi cultivation was of dry crops. The situation has not changed much for rabi but for kharif there is a distinct bias towards wet crops. To some extent this can be explained by higher rainfall during these months also. But whether the entire shift can be thus explained is an open question. This shift to wet crops underestimates the area increases that is *possible* with additional water released since water required per acre is higher. Overall, therefore, we notice the *fact* of increase in area irrigated, which is however overestimated, and a proportional shift to wet crops which is mainly rice in this region.

Critical Assessment

Increased Irrigated Acreage

To begin with, the claim of increase in irrigated acreage requires critical examination. That the area under irrigation has shown a marked improvement under some of the projects where PIM has been implemented effectively has been widely claimed by the irrigation department as per evidence discussed above.

Field visits to the sample areas also verified that water had in fact reached further downstream in some of the channels where silting, clearing, etc. had been undertaken. While this is understandable and plausible, the extent to which this increased flow could contribute to increased irrigation comprising requisite number of wetting, i.e., adequate irrigation is the moot question.[1] The concept of area irrigated has to be clearly defined and there has to be a commonly accepted understanding. Is the area receiving single multiple or requisite wettings. There is thus no indication of the *quality* of irrigation in terms of quantum of water available and the degree of certainty of supply. The quality of irrigation accessible to the user is, however, crucial for production decisions of the farmers and the eventual profitability levels realised. The ability and inclination to pay the water cess depends on this to a large extent.

This is a critical point to which the government agencies are yet not sensitive. Therefore, the reported increase in area irrigated has to be interpreted with caution. It is possible that there has been a qualitative and distinct improvement in the irrigation water availability after the organisation of WUAs and the initiatives undertaken to clean and clear the channels, repair the drops, etc. This is reflected in water travelling further downstream as noted earlier. Data from the sample farmers does not reflect increased area irrigated after PIM reforms to any significant extent even for the upper and middle reaches. However, this can be explained by the small size of the sample. The more random response survey conducted among farmers in the two regions distinctly reflected their perceptions of increased irrigation as a result of repair works done in the area, but only upto the middle reach villages/farmers. The tail end farmers continued to be dissatisfied with the water accessible to them. Feedback survey conducted by the ID during the various meetings of WUA and DC presidents and office-bearers also confirmed increase in irrigated area (Feedback, 1998, 1999).

However, the *extent* of resulting additional area irrigated is overstated based on the following reasons: area where water is barely touching gets counted as area irrigated and areas in canal commands which are rain-fed, specially at the tail end, where water does not reach at all, also get entered as area irrigated by canals. A second source of data on irrigated acreage is the revenue office. In this case mis-reporting of area irrigated exists due to a strong bias on part of the revenue personnel for under-reporting the area irrigated. This leads to lower recorded acreage and thus the need to deposit smaller amounts of revenue (which might have in fact been collected) in the treasury. It is possible that this recording of area has increased of late. The increase in area irrigated is thus partly a reflection of increased reporting and not actual increase.

The second point to note is that whatever the increase, it is a one-time rise in area irrigated. There is little likelihood of sustainability of increase at this rate. The percentage increase in area irrigated will be highest during the initial stages, since the original state of maintenance was very poor. The result will therefore be very marked and distinct. Subsequent improvements are likely to induce incremental increase in irrigated acreage.

While the irrigated acreage cannot be expected to increase indefinitely, the point of initial quantum increase not being sustainable is that most of the increase possible by reducing wastage had taken place and further avenues of increasing area irrigated through channel improvement at the terminal level are limited.[2] Subsequent to waste reduction -- and this by itself may not be able to irrigate the designed command -- further ability to meet the water demands of the area left over will depend on water releases due to better regulation and redistribution of water from head to downstream areas within the designed command.

This raises the issue of adequacy of the designed command. Are the assumptions of cropping pattern that go into delineating a given command legitimate and reasonable or should they be reviewed to redefine the command to fit the area actually being cultivated even if only for wet crops.

The latter approach would rationalise the status quo and exclude the tail-enders from their right to water. This would amount to legitimising the discrimination faced by them. In addition, by reducing the area within a command, the calculation of cost of irrigation development per acre would rise. As of now this cost is highest for canal irrigation as noted above. Such a situation would call for a review of the investment decisions in favor of irrigation by canals compared to other surface and ground water sources of irrigation, which are technically feasible in the area.

A more viable approach would be to explore technologies which entail alternative patterns of water use over not only existing command but possibly beyond that as well. In contrast to the approach above, the command could be redefined to *enlarge* the area sought to be irrigated. The advantages in terms of spreading irrigation benefit over larger area and number of cultivators, along with resulting higher total output, would serve the objectives of growth and equity simultaneously.

Even if we take the existing command as a starting point and accept the increase in irrigated acreage as a result of recent reforms, following questions arise: Where did this extra quantum of water come from? Was it from a reduction of wastage during transit or from a redistribution of water, which was formerly being appropriated by the upstream farmer or both? There is as yet no evidence from the field visits or from the official records that there has been any redistribution. This is a politically sensitive issue and is being avoided for understandable reasons. Clearly it is the surplus or the wastage water which has been partially mobilised for productive use. And this is no doubt a desirable development. But whether this is sufficient by itself to enable further increase in irrigated acreage is by no means certain. The hunch is that once the scope of the above noted source of area increase is exhausted, further increase will depend on the ability of the government to impose redistribution of water strictly according to the principle of water distribution in proportion to the land cultivated. This is likely to be an unpopular move and has built-in confrontation elements. Conflicting interests of tail-head farmers may not be that easy to resolve. But that is also where the real challenge lies. Particularly in view of the fact that PIM in AP is a top-down program, with hardly any grass root efforts to mobilise the community. Neither has there been any sensitisation of the users to water as a limited natural resource with multiple uses and entitlements of the entire community to use. There is need for a social discipline in the exploitation of a common property resource such as water since the externalities of its use bear on the entire community. Monopoly by a section of the community, with adverse implications for the others requires to be checked by the government as the largest regulating authority.

On the whole, we can infer that while irrigation department figures tend to overestimate the area irrigated, the revenue figures tend to underestimate the same. The real value lies somewhere between these limits. There is overwhelming evidence of increase in area irrigated but its estimation as of now is way off the mark. While the revenue department collects data village-wise, the ID collects according to minors, distributaries, etc. The two do not overlap. The data referred was taken for the WD as a whole from the local division office of irrigation department, while year-wise data from revenue department was collected for the specific villages with the command of WD and then consolidated for the major as a whole. The WD passes through three mandals of Miryalguda, Dharmacherla and Tripuravaram and irrigates a total of 25 villages. Some of these villages are irrigated by more than one major. The corresponding area has thus to be deleted by approximation. Due to the complexity and scattered availability of the data, corresponding calculation for D-89 could not be carried out. Accessing data from the local revenue offices is always a daunting task. With increased computerisation of the records, hopefully data retrieval will become less cumbersome with time.

It may be mentioned that evidence of irrigation at the tail areas might be true of certain head to middle reach distributaries of SRSP. But it is certainly not as widespread as claimed in the media, which is projecting an unrealistic picture. Although WD is a large major drawing a discharge of 600 cusecs compared to the designed discharge of 500 cusecs, and is closer to the head reaches of the main canal than the tail end, two of its tail-end villages of Irkigudem and Wadapalli show nil irrigation from WD. Since there is no supplemental irrigation through wells, borewells, tanks, etc. the entire area cultivates only one crop of rainfed dry crops. It is however possible that the single irrigated dry crop gets entered in the irrigation department figures as 'irrigated dry', thus inflating the cultivation in tail areas due to canal water as noted above. The foregoing illustrates the need to subject the available data to detailed scrutiny before arriving at hasty generalisations. It also underscores the need to give serious attention to monitor the developments according to modified categories, which capture the trends with greater accuracy and adequacy.

Yield and Cropping Pattern

Second issue is the impact of PIM on the yields of the major crops and fluctuations over time. A higher proportion of water-insensitive crops in the annual production package will reduce the fluctuations in the total output due to variations in available irrigation water, which may be inevitable due to climatic factors. PIM is expected to influence total productivity and per acre yields by enhancing controlled water application as well as inducing changes in crop combination due to water rationing.

Increase in per acre yields, especially of rice, is an issue that is occasionally touched upon while highlighting the achievements of PIM, although not with equal force. It is very difficult to comment on the validity of the hypothesis that crop yields have increased as result of PIM. In fact this claim is not forcefully made even in the case of IMT in Mexico, which is recognised as a success story of

PIM. Impact assessment study for the Alto Rio Lerma region states that there has been no impact of the irrigation transfer program in the area on economic and agricultural productivity, including physical yields and even area irrigated (Johnson et al., 1997). Yield changes are a function of a complex set of variables including natural-climatic, technological and timelines of input availability, and cannot be obviously related to reforms in the irrigation sector per se. Secondly, it is probably too early to comment either way on the direction of impact on the physical yields. Field data from both canal irrigated areas do not indicate any trend of increase in the yield of rice over the last two years inspite of the potential to increase, given the average yield of about 2,500 kgs per acre, which is below that of Punjab and Haryana.

An interesting observation during field visits from canal commands points to lower rice yields in head-reach areas which are right next to the canals, compared to other areas which have surplus water. Possibly this is caused by seepage of canal water in addition to regular irrigation of the rice crop. This points to diminishing returns to water application and ironically low water productivity *inspite of and because of surplus water*, in the absence of adequate control and drainage structures. However, data collected does not bring out this point. Since this was quoted frequently and independently during discussions with farmers and division engineers, there is need to explain the anomaly. Especially since average yield of the head-reach farmers in the selected sample is marginally higher compared to the downstream yields for SRSP and NSP areas. It is possible that seepage in fields close to canals (from half to 10 kilometres) did depress the yields by about 5 bags per acre (1 bag = 70 kgs), other conditions being similar. Beyond this stretch, however, the impact of surplus water, which has drain outlet, shows up in higher yields in lands beyond the seepage belt. It is the locking in of the water along canal banks that is responsible for depressing the yields in that region.

Significant correlation of yield variation of rice with size of land holding was observed, indicating the precarious position of the small landowner. These variations would be expected to get worse as one moves away from the irrigation source. However, field observations do not bear this out. This could be explained by access to alternative irrigation sources. It may be noted that rice can only be grown in the downstream areas on the condition of access to wells/borewells/ tanks or possibly in the low-lying areas close to the natural drainage such as a river bed, as was the case of some areas of the tail-end villages under D-89. The expected trend therefore partially gets neutralised. Use of fertilizer and pesticides are other critical determinants of yield. However, no correlation of per acre use of these inputs was observed with land size and location relative to the irrigation source. The data on input use was the average for the farmers interviewed for the recent past. There was no mention of any increase in the use of these inputs after the reforms. Farmers mentioned that use of pesticide often increases the biomass of the output without increasing the seed (i.e., the yield of the grain). Another indicator of production decision determined by irrigation was the area devoted to rice during rabi season as it is entirely irrigation dependent cultivation, unlike the rice cultivation during kharif, in which case expectation of rainfall plays an important role. Although no overall pattern correlating proportion of rabi rice cultivation with land size and location was observed from the field data, it was noticed that tail-end farmers without wells were unable to cultivate rice during rabi, which was the sought after crop. This could mean considerable hardship for the small farmers, as they were dependent on rice for their food security. The main source of irrigation, i.e., canals in the area support rice cultivation during rabi only in the upper reaches, confirming the discriminatory water availability according to location. This underscores the need for equitable distribution of water. How far PIM has succeeded in redressing this imbalance is the pertinent question.

The more important issue regarding irrigation impact on yield was its stabilising role. Better management of the system would contribute to greater stability by ensuring timely and predictable supplies. Although it is not possible to overcome the fluctuations in natural climatic factors, controlled irrigation can often mitigate the adversity. However, no such impact was in evidence. It is also a little too early for such impact to become visible.

The only shift in the cropping pattern induced by the reforms is to reinforce the cultivation of wet crops, like rice. The farmers are inclined to shift to rice wherever possible. The interesting question is that this is justified by the lack of any feasible alternative which has comparable ease of cultivation and economic returns. Rice is popularly known as a lazy man's crop in the area. When questioned about the readiness to shift to less water consumptive crops if given a feasible alternative, the response was positive. Another argument put forward by the farmers in favor of persisting with rice cultivation is that the soil gets addicted to rice and it is not possible to switch over to dry crops usually during a period of water scarcity. This is however only a conditional constraint applicable to low lying areas and wet soil areas. Another reason for preference for rice was that it is a food crop. However, that does not rationalise its production as a cash crop as well. While these arguments may be reasonable under some specific conditions, they are not infallible; what is more definitely constraining is the externality of rice cultivation in one field for the neighboring fields, especially if they are downstream of the rice field. The excessive watering required for rice percolates underground and the seepage may extend laterally. This prevents the viable cultivation of irrigated dry crops in the neighboring area. Independent decisions taken by cultivators can limit the choices open to the others in the vicinity in an immediate sense. Feasibility of dry crops is thereby constrained due to this factor. Some regulation of the cropping pattern and/or interventions to promote the cultivation of less water intensive crops are required to be made by the overall governing authority.

What the above does point to is the need to explore technical alternatives to rice cultivation that can be transferred to the field under actual conditions prevalent in the area. Transfer of technology from the lab to the field through practical demonstration along with an active awareness campaign is required. Conscious effort is required to wean farmers away from rice cultivation, which can facilitate equitable water distribution by demonstrating the viability of low water consumption cropping pattern. The latter has to be made attractive to the head-reach farmers, at least beyond the seepage zone, in order for the government to tackle the issue of unpopularity of any measure which aims at reduced water allocation to those who are accustomed to unrestricted supplies (and who could constitute a strong lobby). Mere appeals to their sense of justice are unlikely to motivate them to relinquish their favorable hold. The anticipated resistance by a section of the farmers will also be a disincentive for the political party in power to press hard for the redistributive reforms.

The cropping intensity changes are likely to be witnessed with increase in irrigated area, and to that extent only. As of now, most of the irrigated areas under canal head and middle reaches are double cropped, except the ones under annual crops. Tail-end areas without wells have lower cropping intensity. Evidence of two tail-end villages on WD show the cultivation of only single rainfed crop, as there are no wells, tanks, or any other supplemental irrigation scheme in the area. However, if the cropping pattern changes, the cropping intensity will change in proportion as the crop are less water consumptive and therefor cover a larger area. No definite clues or trends were noticed from the field data or the secondary data with regard to cropping intensity changes across projects. Low cropping intensity in the downstream areas under canal irrigation can partly be explained by annual crops such as turmeric, cotton, etc. In tail areas under WD, land was often left fallow during rabi, which did not have the benefit of low-lying areas. Limited water under canals and lower cropping intensity further underscores the need to use water for extensive cultivation to benefit the entire area. Water shortage is thus *artificially* created when limited water resources are monopolised by water guzzling crops such as rice, restricting the area benefited by irrigation.

To sum up, therefore, we can say that there is no basis for attributing an increase in per acre yields of rice (as a representative crop in the area) to the institution of irrigation reforms. Yield variations continue to be relatively higher for smaller farmers across the canal commands. PIM has induced a shift towards rice cultivation. Lower cropping intensity in the tail areas is related to water scarcity and this situation has not changed. Cropping intensity is expected to change in proportion, as the new cropping pattern is more or less water intensive.

Double crop of rice at the head continues to co-exist with water scarcity at the tail reaches.

Water Efficiency and Regulation

Third critical issue with respect to PIM is the control and regulation of water supply to individual users. This constitutes the main challenge for most public irrigation projects. One of the primary motivations for introducing PIM in the state of AP was to ensure equitable water distribution by instituting systems and rules of water distribution under major, medium and minor irrigation. Given the total quantum of water available, as determined by natural climatic factors, the time spread and volume of water available to users depends on the technological system and rules of water distribution. Typically water control is affected at four major points on the irrigation system.

First, it is at the stage of release of water from the reservoir. This is consequent on the decisions of the irrigation agency (the department) which manages the entire system, in accordance with criteria agreed upon depending on the priorities of the governing institutions. Technologically and managerially this remains more or less beyond the domain of PIM activity. The next point of control is at the offtake points from the main canal. Installation of control structures to regulate flow into different channels (distributaries) is unproblematic to the extent that it remains under the control of the irrigation department. Tampering of structures by water users at this point has been the typical experience in the past, specially during low rainfall years. Once damaged, their repair is taken up, if at all, after considerable time lag, during which the system continues to decay. It is not surprising therefore that repair of the offtake points on the main system has been a major thrust of activity. Further downstream, there is another set of channels (minor) which take off from these distributaries. Some control mechanisms at the head (take-off points) of these minors are also provided. Finally pipe outlets are provided on these minors, which let out water to the field channels. These outlets are usually free flow outlets without any control mechanism. The individual water users access water by cutting openings on the field channels to allow water flow to their fields. Water use beyond this point is entirely under the control of the users. Control of water at the first and the last points does not lie within the domain of the reforms. Releases from the reservoir are no doubt subject to the expected demand based on the cropping pattern of the eventual water users. However, technological and managerial control remains under one agency, without any coordination problems experienced further downstream.[3] Neither are there any such impediments at the lowest level of the system as it is entirely under the control of the individual user. It is mainly at the level of the offtake from the distributaries, minors and to some extent from the main canal, that control of water distribution constitutes the problem area.

Technical structure like shutters had been installed at these points during the initial stages of the projects for controlling the flow into these inlets. The flow could be reduced or totally cut off depending on the water available and the requirements of equal distribution within the command areas. This was done through instituting a system of rotation of water supply, which means a sequential delivery of water to different channels in a predetermined order as opposed to simultaneous supply to all the channels taking off from the parent canal. While the latter could be feasible during surplus water availability, the former was a means of coping up with water scarcity, whereby allocations during lean period would be reduced in equal proportion between the entire command. Institutional functioning of the rotation system, however, requires complementary technical function of the control structures. Tampering with the shutters, offtake structures, etc. has been the observed pattern of user behaviour in the past, which has seriously jeopardised the efficacy of attempts to regulate water distribution. Unauthorised withdrawals of water by upstream farmers using engines for pumping water, further accentuate the unequal distribution, which is to some extent in-built in the delivery system due to increased losses en route. This had resulted in continuation of the head-tail discrimination in access to irrigation water from public sources, due to systemic and technical features as well as the violations of rules of water distribution.

Control Structures

The existing method of water regulation and distribution, as originally planned, entailed an assessment of the water availability in the reservoir at the beginning of each cropping season. This determined the water releases during the season in the main canal. Flow in the main canal continued throughout the season (usually 3 to 4 months). Flow for kharif starts in July and ends in December. For rabi season it starts from January to April-May of the following year. April to July is called the closure period during which most of the major repair and maintenance works are carried out. Water is distributed to the command areas by means of the distributaries, and through them onto the minors. Shutters had been installed at these offtake points from the main canal as well as from the distributaries, as per the original project design for both NSP and SRSP. The purpose of the shutters was to control the flow of water to different areas by deliberate design as prioritised at the beginning of the season. However, most of these regulating structures had been tampered with or damaged. There is thus, no effective control mechanism for directing the flow of water on a need basis. The head reach farmers of both the main canal and distributaries first appropriate as much water as they require and only the leftover flows downstream. This leads to the typical problem of water scarcity at the tail-end areas even when there is surplus water available to the head-reach farmers. During the year 1998-99, this led to an ironic situation of the farmers taking two crops of rice at head reaches of WD while the tail-end farmers on the same distributaries had severe problems in raising low water consumption crops, i.e., cotton, chillies, etc. The line diagrams of the two distributaries in the sample areas show the dysfunctional state of the control structures just before PIM. Out of 60 shutters installed at the beginning, none were functioning, and 36 were reconstructed as offtake sluices with shutters. The same is contemplated for SRSP area in the following season.

Rotation

Rotation of supplies among the distributaries in the main canal of SRSP stage 1 has manifested itself in a distinct improvement of supplies to the downstream areas. While the distributaries till about the middle could be supplied water earlier, water was reaching the tail-end distributaries for the last two years. However, the distribution below the distributaries has not recorded any corresponding improvement.

The tail-end areas of the sample distributaries continue to be deprived of their share of water. There was no effective way of prevailing over the head villages even through the mediation of distributary committee. Supplemental irrigation by wells, which were recharged by the surrounding low-lying area, was the mainstay of the tail areas of D-89 distributary. The farmers mentioned having made a few attempts to monitor the water flow from head areas to their village, but were unsuccessful, as the village just upstream of their village intercepted the supplies. Problem such as these dissuaded any payment of water dues for the last more than 10 years. The minor reaching their field, which had been cleared of weeds, bushes, etc. at government expense, was full of weeds again. There was no incentive to maintain the channel in the absence of expectation of receiving any water. One time investment in such a situation also goes waste.

Secondly, even if the system of rotation works ideally it has an in-built discrimination against the tail-enders due to time taken for water to reach the tail after resuming supplies. The scale bias in transit losses for tail-enders is widely observed in case of most surface irrigation projects.[4] Rotation, therefore, accentuates the uncertainty and irregularity for downstream areas.

Quite apart from the locational disadvantage in receiving water, there is the disadvantage based on the size of land holding. Getting water in a lumpsum wets a larger area compared to the same water divided into smaller units of land and entering each piece of land separately. This problem was not found to be that acute among the small landholders of the distributaries located at the head reaches of the canal. However, as we go downstream along the main canal as well along the distributary, the locational and land size disability vis-à-vis access to water reinforce each other, with its adverse consequences for agricultural productivity and financial returns.

Field Responses

As of now there does not appear to be any change in the ability to *control and regulate* the supply of water. There are technical constraints, which are planned to be overcome by installing shutters during the summer closure period of canals. However, there is yet no assurance that these technical artifacts will not meet the fate of their predecessors. Particularly in light of poor awareness and motivation levels noticed among the head reaches farmers, as discussed below. The issue requires not merely technical solutions but, more importantly, a political will and social mobilisation effort to effectively ground rules of equal water allocation. It will be possible to handle the real challenge of equal water distribution between areas and within areas, only when it is possible to restrict the free and unbridled access to water by upper reach farmers. The question of witnessing any change in the direction of *redistribution of water* was posed to groups of farmers and the departmental engineers at the field division offices. The near unanimous response from them was in the negative. Field officers were very categorical about their inability to control or restrict water consumption at the head reaches. The problem continued to persist after the reforms and they did not foresee any likelihood of its mitigation as a result of any provisions implied in the reforms. Water-intensive cultivation of paddy crop, they felt, had to be discouraged wherever possible, to reduce water demand at the head reaches and enable extensive cultivation with same quantum of water. In case where the option of switching away from rice is technically feasible, it is possible to cultivate a higher acreage with less water consumptive crops with the same quantum of water.

Increased Water Availability

No doubt significant quantum of water has been released for additional irrigation as a result of physical works being undertaken to repair the channels. This is manifested as flow of water reaching further downstream, relative to earlier years. Many farmers in the NSP-LC mentioned that water reached the channel at the tail end after 15 years as a direct result of silt clearance. Instances such as these were cited quite frequently during the course of fieldwork. This has also been reported in official statistics as increase in area irrigated as discussed in the preceding sections. However, as noted above, this *increase in water availability is due to water saving measures and not due to redistribution of water from head to tail.* There is a quantum jump in the additional water availability as a result of technical works, which facilitate smooth water flow and reduce transmission losses. But, this is a one-time jump and not an ongoing one. Further works are going to be manifested only as incremental water increases. Therefore, further drastic increases in area irrigated are not expected to occur with the existing practice of water distribution and its use. Since this has clearly not provided satisfactory irrigation to the entire command, there is need to either redefine the command or institute control mechanisms effectively. The latter will necessarily entail a change in the cropping pattern across the length of the distributaries.

Water Wastage

Water wastage in the head-reach areas of the canal-irrigated areas was very prominent. Excessive watering of the rice crop led to relatively lower yields in areas close to canal banks and discharge of excess underground water as a flowback to the river or drains. No direct estimation of this wastage was available from the department. But absence of control structures and continuous flow of water preempted a close monitoring of water let into the field. Rice is known as a lazy man's crop in so far as the cultivator can take it easy once the transplanting is over. All that was required was the submergence of the field. The tendency is to always over-water as a security, rather than take the risk of under-watering. Specially as water was easily available at the head and in plenty. Ironically, it was present in areas where the tail reaches were being starved of their legitimate due. The flow was often of the order of 20 to 30 cusecs that compare with that of a minor.

This indicates that there is potential for additional mobilisation of water resources by controlling wastage and increasing area irrigated without affecting the differential and discriminatory use of water by head-reach farmers. Schemes to utilise about 20 percent of the flow were in operation in the NSP-LC area. These are small lift schemes where water is diverted from drainage streams to irrigate areas further away.

An indirect method of estimating the wastage is to look at the variation of irrigated area per unit of water or what is known as the 'duty' in irrigation parlance. Of course this index has limitations in so far as it does not capture the changes in cropping pattern. If we look at the variation in total area divided by the total water supplied at the head of the distributary over different seasons/years, then a sharp rise should be evident during 1998 and 1999, over the previous years. While duty varied from 3 to 5 prior to 1997, it showed a sudden jump from 7 to 10 from kharif season onwards. While the irrigation data does show a marked increase in the duty calculated for equivalent wet irrigation, from close to 40 and 50 in the earlier years to nearly 70 during 1998-99. This can also be explained by the good rainfall during that year and the fact that area irrigated in the canal commands by rainfall also counts as canal irrigated area. The duty based on revenue data also points to marginal increase during 1998-99 over the earlier years. In fact, the values are closer to those of 1991-92, not only for the duty, which is a ratio, but also for the areas irrigated and water supplied. Possibly the situation reflects a turnaround in the technical efficiency of the system due to the physical improvements carried out in the main system as well as distributaries.

Again, it is not the absolute numbers that are of significance. Rather it is the fact and direction of change that is to be noted. This increase in duty reflects increased efficiency of water use due to reduced wastage as a result of works like silt clearance, construction of drops, etc., all of which facilitate smooth flow of water.

Conclusion

The hypothesis regarding PIM resulting in a greater control over water use and increasing the equity of distribution is not validated from the field investigations. Whatever increase in irrigation has come about, is a result of reduced wastage due to revamping of the supply network of irrigation channels. This has its limits and may soon be saturated as a source of additional irrigation. In spite of the possibility of installing technical structures for controlling and regulating water supply in the near future, it is highly unlikely that a genuine *redistribution* of water will be brought about as it is bound to be politically unpopular. This fear is reflected in the farmers insisting on punitive action for illegal extraction of water. But redistribution is necessary for ensuring equity in distribution of water in proportion to land cultivated as per the professed criteria of the government. The present system of water distribution exists by default in the absence of any effective regulatory mechanism, resulting in disproportionate withdrawals of water by head-reach farmers amounting to a discrimination against the downstream farmers. In addition, there is the likelihood of even the new control structures being tampered with. It is not clear how the reform provisions will prevent this. Further, it is extremely unpractical to install control structures at all the minor heads and pipe outlets. What then might be the social or community measures to share water equitably according to well-defined and accepted rules is the question, which is *not* being posed.

As noted above, redistribution of water implies equal distribution on per acre basis, as stipulated in the government policy. This is necessary for ensuring equal opportunity for all the landholders to enhance productivity on their lands with consequent income flows. This is imperative from equity considerations as it influences the income and employment opportunities that are possible. It may, in addition, lead to higher productivity levels at the macro level. For instance, if farmers are given a certain limited ration of water, they are forced to spread it across their cultivable area by growing a combination of wet and irrigated dry crops. Less water will irrigate same area, which may currently be devoted to rice due to uncontrolled access to water. The water thus released will be used to cater to the lands of other farmers on a pro rata basis with similar cropping pattern, that is the farmers who were hitherto denied irrigation for their entire area. This will lead to additional productivity on gross area brought under cultivation. The point is that as of now the head-reach farmers make their production decisions with only land as the scarce resource. The proposition being hypothesised is that water too should enter their calculation as a scarce resource for efficient utilisation. Secondly, water constraint at the level of individual producer will result in extensive use of water, greater land coverage, and higher quantum of agricultural produce in addition to generating more employment. This would be achieved without compromising the economic returns of even the head-reach farmer. The macro and micro objectives of growth and equity would complement each other. The above is possible only if water control and regulation rules are instituted effectively. Warabandi and water pricing are some of the methods of water regulation tried elsewhere. However, their indiscriminate applicability in a different environment cannot be taken for granted. We have drawn attention to the limitations of water pricing as a means of inducing efficient water use elsewhere (Jairath, 1999).

Institutional Changes

The major institutional issue, vis-à-vis the irrigation reform that excites curiosity and evokes interest, is the nature and extent of participation that the reforms have succeeded in realising, with particular reference to the initial objectives. The fundamental understanding of the malaise afflicting the performance of the public irrigation projects was the managerial inefficiency. The technical, financial, water-use efficiency and productivity problems associated with irrigation projects were rooted not so much in lack of resources or know-how. Rather, it was the poor overall performance in spite of significant resource allocation. While shortage of finances is often cited as a limiting factor, the argument does not carry weight given the high degree of investments done in irrigation sector during the successive plan periods, representing 23 percent to 36 percent of total plan outlay amounting to a total of Rs. 5,250 crore (*Factual Note: Irrigation Sector*, 1997, GOAP).

In order to overcome the above hurdles, various attempts had been made in the preceding decades to provide know-how through extension coupled with credit support at the grassroots level by programmes such as CADA, without however much success on a sustainable basis. The Irrigation Department (ID) had been unable to maintain either the main system or the terminal network of supply channels. Water distribution rules and regulation were flouted more as a routine rather than exception and had lost all meaning. Within this backdrop, and motivated by external circumstances and pressures (abstracting from them for the present), the irrigation reforms -- with a focus on users participation in the M&O of the irrigation system -- were instituted. The underlying expectation was first, to systematise water distribution and second to improve the terminal level of supply channels by decentralising their upkeep and maintenance through shifting its responsibility to organised groups of users. This, as argued earlier, was motivated not out of altruistic concerns but as being perceived to be a more efficient management strategy. Since the water users are spread out in space it is practically convenient to allow them to handle micro level distribution of water among them, while the ID is entrusted with the responsibility of providing water in a lumpsum at head of the channels. Devolution of responsibility had to go hand-in-hand with the devolution of authority to take decisions. This is the crux of participation that is envisaged, i.e., the facility to take decisions regarding the *distribution of water* and the *maintenance* of distribution network of water channels. With this in mind, organisations of water users were formed and the rest is history.

Two prerequisites for such a participation to be viable are one, the availability of financial resources with the WUAs on a sustainable basis for M&O and two, the ability of WUAs to *implement* the rules of water distribution which are democratically agreed upon. To begin with, financial grants have been allocated to WUAs and DCs for initiating the long overdue and urgently required repairs to the irrigation structures. As noted earlier, significant and commendable work has been completed and is continuing as a result of this intervention. Although it is easy to find faults with progress of activities being undertaken, the fact that a wide range of constructive activity which is *need* based is being carried out simultaneously over a vast area on a scale never witnessed before, should not be lost sight of. What is even more striking is the fact that this has resulted from a government-initiated program, which has its usual limitations of bureaucratic functioning.

Participation of the WUAs (that is the president and/or managing committee *only*) in the execution of the works is evidently manifest. We discuss below the question of participation by the *general body* separately. From the field observations it was found that all the works undertaken were identified by the elected bodies and also executed by them without the mediation of contractors or any other intermediary. The interaction was directly between the WUA and competent authority, who is from the ID. There were well-defined channels of communication between user representatives and the irrigation agency, i.e., ID. That this communication network functioned effectively is also evident from the feedback obtained by the ID, where a majority of the presidents responded positively to their interaction with the department personnel. Compared to the prior anarchy and confusion in approaching the department for their problems, there now existed 'a single window' – the competent authority. He too had been given clear instructions and training about his role vis-à-vis the WUAs. The institutional structure of PIM has provided a forum for the users and the agency to come together and discuss irrigation issues, which was absent earlier. DCs have brought together farmers' representatives from head and tail regions on a common platform. This is facilitating at least the *positing* of issues of discriminatory water availability across regions and its implications, especially during years of water scarcity. Whatever difficulties some WUAs experienced in accessing help from the department were more in the nature of teething troubles and could be expected to sort out in due course of time. Similarly the flow of funds to the WUAs was by and large satisfactory, if not perfect. In fact, it is this financial flow that singularly underlies the momentum that has currently been generated.

The second pre-condition, i.e., the ability of WUAs to institute water distribution according to well-defined rules, has been relatively difficult to achieve. This can be explained in the context of excessive preoccupation with paddy cultivation in the region, which engenders correspondingly heavy demand of water by upstream farmers leaving little surplus for the downstream farmers. The latter in turn default on their payments of water dues. Most often they are not assessed for payment of water tax. This response threatens sustainability of the reform initiatives. Government records also corroborate this. "Farmer in the head reaches of major and medium irrigation schemes are drawing water far in excess of their allocation and as a consequence, water is not flowing to tail areas. The problem is particularly acute in new commands which are based on irrigated dry cropping pattern and where farmers are resorting to paddy cultivation" (Factual Note, 1997: 8). The availability of financial resources as of now was not problematic. Clearly, external flow of funds cannot continue indefinitely. Ultimately, WUAs are conceptualised as financially self-sufficient entities for ensuring sustainability of the PIM framework. In order to achieve this, the water cess raised from within the commands of individual WUAs is proposed to be handed over to them eventually for managing their total administrative and work related expenses. As short-term measure to facilitate the transition, the government proposes to continue external funding, but only in proportion to the water cess collected in the command of respective WUAs. This is visualised as an incentive for WUAs to ensure better compliance with payment of water dues.

We have noted earlier the widely relevant tradition of defaulting on payment for water charges. How effectively, if at all, the WUAs counter this practice constitutes the critical challenge for institutionalised 'participation'. How and to what extent are the WUAs going to discharge their responsibility of ensuring the payment of water cess is yet to be seen. What aspect of reforms has contributed to the acceptance of WUAs by the general body as authoritative institutions empowered to insist on payment of water dues – something that is a departure from previous practice? During fieldwork we have noted the reluctance of some WUAs to undertake this task as they run the risk of facing social ire. The extant system of collections through Village Administrative Officers (VAOs) has been wrought with problems. There was very little in the field to suggest the emergence of a trend to mitigate the situation. Evidence illustrating continuing default on payments even in head-reach areas is discussed below.

This raises the issue of relationship between WUA managing committee and its general body. How representative are these organisations? What is their social and political base? What is the extent of control/respect that they command within their jurisdiction? On this depends their ability to execute their role as fee collectors. It may be noted that so far none of the local elected bodies are charged with function of collecting taxes, and if they are, it is seldom successful (see UNICEF (1997) for problems associated with tax collection by Gram Panchayats in Ranga Reddy district). To a certain extent it can be argued that the ability of WUAs to provide and manage effective irrigation might enable them to command social respect and legitimacy, which in turn would earn them the social recognition to be able to ground fee collection institutions. This in turn depends on how effective are the WUAs in *distributing* water.

We have discussed at length the problem of controlling and regulating water flow among individual cultivators. Although it is agreed that water is to be distributed in proportion to the land cultivated, extensive violation of the norm was evident. This practice had entrenched itself to the extent that extra withdrawals of water by head-reach users for paddy cultivation had become the norm and was no longer even considered as 'unauthorised'. Over time this practice had acquired a sort of social acceptance. While differential water access was prominent *between* jurisdictions of different WUAs, it was noticeable even *within* jurisdiction of a single WUA. Illegal water drawls continued to persist after the reforms. WUAs were unable to control such malpractice to any significant extent. Therefore, while the channels for carrying the water were technically shaped up, the water they carried continued to be erratic.

The above situation reflected weak control of the WUAs over their command area. This is not surprising, given the top-down nature of the reforms and very little awareness at the grass root level. In fact, most of the WUAs were single person dominated organisations, owing allegiance to one or the other political parties. From a brief survey done on the educational, political and economic background of the presidents of WUAs, it was found that each had distinct political affiliations, had some elementary education and were reasonably well-off. There was a marked difference between the president and the other members of the managing committee in terms of their social and economic status. They were by and large subservient to the president. Most of them had very little information about the activities of the WUA or interaction with neighboring WUAs or about the meetings of the DC. The awareness of the general body members was even less. Some farmers at the head-reach areas had no idea about the program. Although they did recall the event of elections to WUAs, it was a distant memory. Only 30 percent of the farmers surveyed knew the name of the president, and hardly anyone knew the names of managing committee members. Very few general body meetings had been held. And even the ones that were held were poorly attended, with the quorum being 'managed' by circulating the register for signature the following day. Exceptions were certainly noticed, of extremely enthusiastic and well-known presidents of some WUAs and DC presidents, e.g., the woman WUA president from distributary D-89 on SRSP project. The character of WUAs is similar to the Panchayati Raj institutions. Their effectiveness will depend on the extent to which a grass root level awareness is generated regarding the entitlements of general body members, and to the extent that general body members are organised to exert pressure for better services from these organisations. In other words, it is only grass root mobilisation for the activation of basic democratic processes that can lead to a greater responsiveness of WUAs to members of its general body. Wider dissemination about the PIM program and its potential for enhancing efficiency, stability, incomes and productivity through innovative communication strategies is a necessary though not sufficient condition for grounding these reforms.

However, mere publicity is not enough to realise greater effectiveness in institutionalising these reforms. Their practical viability depends on the ability to remove the root cause of out-of-turn water drawls, so as to bring about better adherence to rules of water distribution on per acre basis. As discussed earlier, the need of excessive drawls of water are based in the widespread paddy cultivation prevalent in the area. Weaning farmers away from this cropping pattern, wherever possible, by providing technological alternatives which are practically feasible and economically lucrative, holds the key for arriving at a better control of water use by the general body.

Effective control of water distribution is likely to facilitate higher revenue collections. As discussed above and illustrated by the field data, payment of water dues has typically been problematic. Reasons are two-fold – one, default due to lack of pressure to pay in spite of getting water and two, refusal to pay due to poor irrigation service provided or accessible. Added to this are the lower remittances by the local revenue department personnel. Efficient functioning of the system on a sustainable basis will have positive impact on the financial management by providing an incentive to pay. Financial inflows in turn will make possible the continuous maintenance of the system requiring fewer quanta of funds because of regular upkeep, as argued below. This is a critical requirement since the external funding support available at present cannot continue indefinitely.

From the foregoing discussion we can infer that the WUAs have performed very successfully as institutions for executing activities for improved maintenance and repair of the channels by accessing funds from the government. However, as regulatory institutions in charge of water distribution on equitable basis their performance has been quite weak and ineffective. This bears adversely on their ability to generate resources through collection of water cess. In the coming years this would seriously constrain the reproduction of WUAs as viable organisations. Technological interventions to provide realistic alternatives to high water consumption crops need to be undertaken to enable extensive spread of irrigation benefit rather than concentrate its benefits to a few. The latter would negate the probability of sustainable reproduction of WUAs, thereby threatening their institutional viability.

Financial Sustainability

Financial inefficiency experienced typically with most irrigation projects, was another major motivation for the rethinking which led to the organisational reform in the irrigation sector. It is a well-acknowledged fact that low financial efficiency and poor financial returns at various levels characterised most irrigation systems projects, especially the large-scale ones. First, inefficiencies are noted at the level of creation of irrigation potential. Second, at the level of M&O there is the problem of insufficiency of funds as well as inefficiency of the utilisation of whatever financial resources that are available. While more than 70 percent of these resources are spent on work charged establishment comprising salaries, infrastructure, etc., there is little left for actual execution of the works. Added to that is the poor *quality* of works done with financial implications for repeated maintenance and greater financial requirement. Finally, there is the whole issue of *poor collection* of irrigation revenues as water cess from the users. It is well-known that the recovery rates are very poor. The latest figures for Miryalguda mandal, whose villages are at the head of WD, show a maximum of 64 percent (of the total demand) collections during rabi season of 1998-99 only. The collections during kharif were nil. The total demand collected for the year works out to barely 10 percent. And this is the case in the heart of 'rice bowl' of the country!

Looking at the accumulated picture of collections and balances due, we find that the arrears amounting to more than Rs. 2 crore against most recent collections of less than Rs. 38 lakhs, i.e., more than 84 percent of water tax is due as of date from this mandal. The farmers in this region certainly cannot claim to suffer from water scarcity. Department sources have noted that until the 1950s irrigation development used to be financially profitable. Revenues from water charges levied on beneficiary farmers were larger than government expenditures on operation and maintenance and imputed interest on initial investment. Over time, however, this situation has eroded. The difference from current revenues from major and medium irrigation projects and non-pan recurring expenditure on these projects has exceeded Rs. 115 crore in 1995-96 (see Table 2 below):

Table 2
Revenue From and Expenditure on Major and Medium Projects

(Rs. crores)

Description	*1992-93*	*1993-94*	*1994-95*	*1995-96*
Revenue	20.44	21.84	25.11	19.06
Non-plan expenditure (excluding interest borrowing)	139.30	150.70	138.28	133.20
Plan expenditure	579.14	826.06	778.31	648.93
Revenue as % of non-plan expenditure	14.67	14.49	18.15	14.31
Expenditure on staff (plan + non-plan)	152.16	153.29	181.48	193.99
Revenue as % of staff expenditure	13.43	14.24	13.83	9.82

Source: Irrigation Sector: A Factual Note, 1996, GOAP.

The resource crunch was accentuated by the rise in per hectare cost of new investments. The cost of providing canal irrigation at 1996 prices is estimated at Rs. 50,000 per acre. Accounting for interest on this investment, the annual per acre cost works out to be Rs. 8150. The capital investment cost per acre is as follows (Irrigation Sector: A Factual Note, 1996, GOAP).

Major Projects Rs. 50,000
Medium Projects Rs. 40,000
Minor Projects Rs. 35,000

Pre-reform estimates of fund requirements for maintenance of canals, dams and other structures of major and medium projects is worked out at Rs. 120 per acre. As against this, the water charges were fixed at Rs. 40 to Rs. 60 per acre for different crops (see Table 3 below). These costs did not include establishment and capital cost of the projects. Receipts of revenue from water rates were far below the recurrent expenditure incurred on maintenance and operation of irrigation schemes. As noted by the department, the above criteria of charging fees were barely sufficient to pay for the salaries of the work charged establishment, let alone the capital costs (Factual Note, 1996).

Table 3
Per Acre Water Rates Fixed

Nature of the crop	*Rates of water tax per acre in respect of water sources under (Rs.)*	
	Category I	*Category II*
First or single wet crop	60	40
Second and third wet crops	60	40
First crop irrigated dry	40	20
Second and third crop irrigated dry	40	20
Dufasal crop in fasli year	120	80

Source: Irrigation Sector: A Factual Note, 1996, GOAP.

The present water rates do not even cover the M&O expenses, let alone yield a return on huge investments made. On the basis of cumulative investments of

Rs. 5,250 crore made in irrigation projects till the end of 1995-96 and assuming an interest rate of 8.7 percent on government borrowing and 1 percent depreciation on these investments, plus actual O&M expenses, the cost per acre of canal irrigation to government works out to be Rs. 1094. This is in sharp contrast to the water rate of Rs. 40 to Rs. 60 per acre paid by the farmers (Factual Note, 1996:6).

Water Rates: Fixation and Collection

Raising the fees by itself is not sufficient to increase irrigation revenues. We have discussed elsewhere the limitation of water pricing as a lever of efficient water use (Jairath, 1999). In fact, the steady increase in the official water rates has not resulted in more prudent use of water. The critical question is the *extent of compliance* with these rates. Evidence from the field points to large variation in the payment pattern. Defaulting on payments of irrigation fees consistently by tail-enders from the field sample, was rooted in their poor access to irrigation from the main source. As noted above, there was a categorical and explicit refusal on their part to pay the water tax till the irrigation service was improved. Fee collection at the tail end continued to be marginal even after the reforms. Thus claims of providing assured irrigation to the tail areas are not tenable. Even farmers in the head-reach areas, which have been getting good water consistently, appear to have defaulted on their payments. We have already discussed the evidence from Miryalguda mandal in this context. The water rates have been revised upwards by more than 100 percent after PIM in AP. Data from MRO's office in NSP area, on the water charges collected, do indicate marginal increase in the absolute quantum of revenue collections. But when this is compared with the revenue expected as per the demands, the amount collected at higher rate per acre may correspond to a smaller proportion of area cultivated or fewer persons paying their dues. Added to this are the collections made but not deposited. Disaggregated data is required to come to a firm conclusion. It is not clear, therefore, whether the higher fee collected reflects the fee for larger acreage or higher rates charged for a smaller area. This does point to the caution required to interpret the statistics. And simple *claims of higher revenues* are not necessarily reflective of higher rates of fee collection. It can in fact *conceal* lower rate of fee collection. The point is that these issues cannot be sorted apriori but have to be analysed in-depth for which disaggregated data is required.

The second cause of low revenue collections is the mis-reporting of revenue which has actually been collected. It works as follows: if a farmer owns 'x' acres of land, he pays revenue for 'x--a' acres only on the understanding that it is not entered in the official records. Such an arrangement is mutually beneficial to the VAO (local collecting authority as representative of the revenue department) and the farmer. While the former pockets the collection, the latter has to part with fewer funds. This would also depress the financial returns for the WUA, as these are proportional to irrigated area. Therefore, the reason for low returns is three-fold -- one, due to under-reporting by revenue department data collectors, i.e., VAOs, two, due to noncompliance by water users as result of poor water service delivery and consequent lack of motivation to pay their dues and three, due to default on payments in spite of getting water. Recently the department has initiated a policy of conducting a 'joint azmaish', i.e., a combined assessment by four persons, one each from the department of agriculture, irrigation and revenue and the local WUA office-bearer. This is a welcome development and is gradually being implemented in more and more areas under canal commands. However, there is a resistance to the grounding of this method of assessment of area irrigated by the local revenue person as it threatens their ability to manipulate statistics. The field engineers cited many instances of a stalemate between the two departments. In such situations the irrigation and revenue departments maintain separate records and the ID does not sign the joint azmaish. It is however the revenue figures which are entered in the official records by the Bureau of Economics and Statistics and is considered the authoritative figure by the Chief Planning Officer for the purpose of development planning. There is thus a significant discrepancy between the data on area irrigated, from the two sources.

From the perceptions of the farmers surveyed, it was found that there was not much resistance to paying higher charges for water. This was particularly true of farmers from among the medium to small landowners who were at the tail end. The important condition of payment was the dependability of the irrigation service. As of now, although expectations are raised, there is little likelihood of the tail areas getting the water due to them. Payment of water cess from them will correspondingly be negligible. Increasing water rates will by itself provide little relief to the public exchequer. This is probably a matter of time, as water rates are a very small percentage of total cost of production (less than 5%).

High water tax is not a serious issue with the farmers at present, at least in the relative sense. The weak pressure to pay can partly explain this. From the cost of production data collected for a small sample of farmers, it was found that water cost, as per the new rates, amounted to maximum of 5 percent of the total annual paid out costs of production. This is of the same order as for the Mexican program of PIM, where the corresponding figure varied between 3 percent to 5 percent (Johnson, 1997). The monthly cost of water per acre would work out to be not more than Rs. 20. Thus it is not the 'high' rates per se which might depress the revenue collections and eventual sustainability. Rather it is the effective delivery of water to the tail areas -- a categorically stated objective of the PIM reforms. This in turn provides the basis of insisting on better compliance to terms of payment. Farmers were very categorical about payment of water fees only if they got good service. Few were also critical of the new enhanced rates as not being justified by any increase in productivity. For the majority, however, it was not the high rate but poor irrigation that was a matter of concern.

Maintenance and Operation (M&O)

Discussion on poor M&O in the literature are purely in financial term -- with the physicality and timeliness of maintenance not accorded its due recognition for its impact on the requirement of financial resources as well as utilisation of the existing/available resources. A welcome departure from this was the collaborative study conducted by IIMI (Colombo) and IIM (Ahmedabad) on the impact of irrigation management transfer to the WUAs. With reference to maintenance, they find that inspite of similar average annual expenditures on maintenance of the transferred and non-transferred minors in their sample, the performance of the transferred minors was distinctly better. The explanation for the difference lies "in the way maintenance was managed.... the irrigation department prefer to desilt and clean whole minor at one time every few years rather than clean and desilt small portions each year.... Also ID employees tend to spend little time on any particular canal or canal section, resulting in relatively low knowledge of detailed maintenance needs" (Maharashtra IMT Impact Assessment, 1999). This reinforces the point made below about better maintenance possible with same quantum of funds or similar output with smaller funds. That is, it is the inefficiency of use of given funds that determine the quality of maintenance. This is not to discount the need for minimum financial resources. The root cause of poor levels of maintenance lies more often in the inefficient utilisation of this resource. The impact assessment study of the Mexican program of PIM also underscores the noticeable improvement in maintenance of the system consequent upon the transfer of management to the users. Not only has the quality of maintenance improved, it has been achieved with fewer staff and at a lower cost. The focus of maintenance has shifted from main canal to the terminal end of the system. These effects were corroborated by the perceptions of the farmers (Johnson, 1997). In AP, the recent attempts are focussing on all round maintenance since condition of the entire system was in a bad shape. While the ID is managing the main system rehabilitation, the institutional network of farmers' representatives is reconditioning the terminal network beyond distributaries.

Maintenance can broadly be characterised under three heads: breakdown maintenance, preventive maintenance and upkeep maintenance. If routine maintenance is done regularly to keep the system in good shape and continuous monitoring identifies any weaknesses that are developing, it can be checked in the initial stages and major breakdowns can be prevented. This has significant cost saving consequences for the maintenance budget. If however, the system is allowed to deteriorate to the extent of absolute breakdown, then there is no choice but to go in for high cost repair and maintenance options. The technological options/choices for maintenance are a function of ongoing maintenance. Insufficient allocation of funds for such routine maintenance therefore turns out to be very expensive in the long run. The second issue is of utilisation of available funds for ongoing maintenance by the irrigation agencies with the usual bureaucratic constraints and inefficiencies. In this context, handing over the management to locally elected bodies of the users is a very powerful mode of enhancing financial efficiency and ensuring tremendous improvement in quality of works done. It is not surprising therefore that the most impressive and immediate outcome of the reforms is in the field of works undertaken for repair and rehabilitation, which is physically apparent even to a casual visitor.

Ample evidence is available for this from field visits during the first phase of repairs being carried out on the minors and distributaries. The main works are the following: removal of silt, clearing of weeds and vegetation, repair of structures and sluices, closing of breaches, repair of canal banks, repair of drops, etc. The entire exercise is managed by the interaction with the competent authority (CA), who is a department engineer, and the WUA office-bearers. Right from identification of works to be undertaken, preparation of estimates by the CA and release of funds in a phased manner by the P&A department after sanction by the Superintending Engineer. The department gives monitoring and technical support and conducts periodic inspections. The government has allocated Rs. 100 per acre as maintenance grant to WUA, and an additional amount to DCs. Therefore this year a total of Rs. 200 per acre has been allocated for maintenance to the PIM bodies. Money is reaching and works are going on in full swing and concurrently. This is not only more effective in terms of total amount of work output but also has lower financial leakage and more efficient use of the funds. The users have a direct stake in better quality work. Some farmers did make cynical comments about the contractors becoming WUA presidents or these presidents operating like contractors. While this may be true, it is still better to have a contractor from the same or nearby village as he is likely to be more accountable to the farmers and has a vested interest in doing a better job compared to a total outsider. Secondly, since the positions are elected, the members of WUA always have the option of not re-electing the errant candidates. The situation may not be ideal in these respects but does constitute a distinct improvement over the earlier condition.

The greatest obstacle to the smooth functioning comes from the Pay & Accounts Section of the department. There is a widespread feeling among the farmers that they expect bribes for speedy processing of the papers. This is also confirmed by the regular feedback conducted by the ID (Saddassu Feedback, 1998-99). In fact, there was considerable evidence of cordial relations with the CA – the nodal officer dealing with the WUA officers. Occasional instances of conflict came to the fore (TH, 1998). By and large, however, interaction with the CAs did not emerge as a serious problem, even when the farmers were deliberately questioned about this.

While recognising the positive impact of PIM on maintenance of the irrigation channels which were in a bad shape, there arises one basic question, i.e., can the system survive if the continuous flow of funds comes to halt? It is proposed in the ACT and subsequent government pronouncements that the entire water fee collected would be diverted to the WUAs as their maintenance grant. Even the fee collection responsibility would be handed over to them. As of now the fee is collected by the VAO along with the land revenue. Field survey points to a tremendous *resistance* to fee collection responsibility by the WUA presidents and TC members. Their argument is as follows: being elected representatives they have no authority to impose the payment of water charges due to fear of incurring their wrath in case farmers refuse to pay. They have no way of ensuring compliance. If they use any pressure they run the risk of earning unpopularity among the users, which will be reflected in their being thrown out as office-bearers. The farmers were therefore very reluctant to accept this responsibility. Whether their apprehensions are valid or not – can only be tested after actual implementation. Secondly, if it is proposed to finance all future maintenance through the water fee collected, what happens when the rates of collection (nearly Rs. 200 per acre, which are equal to what is being spent now) are not adhered to. Already there is lot of resistance to the proposal of farmer contribution for WUA funds. How is the financial self-sufficiency proposed to be ensured such that the level of maintenance can be sustained? The question of *financial sustainability of WUAs* in the absence of external inflow of resources is very critical and needs to be faced squarely.

Although continuation of external funding is contemplated for both the new and ongoing schemes, it still leaves open the question of financial viability of the entire irrigation organisation. While there is a justification for some form of subsidy for the capital expenditure for irrigation investment, as is the case all over the world including developed countries, atleast the recurring costs of service are required to be met through revenue generation. It may be pointed that success of the Mexican program is critically dependent on its financial self-sufficiency. Johnson's study of the irrigation management transfer in Mexico points out that "....water user associations have proven capable of operating and maintaining the modules, even upto sizes in excess of 50,000 hectares, and water fees collected by the users have not only supported the module O&M activities but have also funded most of O&M activities by CAN staff at the main canal and water source levels. This is in sharp contrast to the situation that existed when the systems were heavily dependent upon government subsidies and consequently were deteriorating rapidly due to lack of stable funding" (Johnson, 1997). The fee recovery from the modules was sometimes more than 100 percent, i.e., they often paid the fee in advance of the required time, which was also before delivery of the irrigation water.

Table 4

Sources and Financial Outlays for Two-year Action Plan (1998-2000) for Irrigation

World Bank (AP III APERP, Hazard Mitigation)	Rs. 1990 crores
OECF Japan (KC canal modernisation)	Rs. 550 crores
Rural Infrastructure Development Fund (NABARD)	Rs. 500 crores
Accelerated Irrigation Benefit Program (AIBP)	Rs. 542 crores
Bond Money	Rs. 370 crores

Source: White Paper on Irrigation, 1997.

That the same maybe possible in case of Andhra Pradesh, by charging water rates which do not exceed 5 percent of the cost of production of the farmers, point to the realistic feasibility of self-sufficiency in terms of M&O expenses. Absence of this will continue to result in a drain on the public exchequer. This would put pressure for cuts elsewhere or possibly a weakening of M&O leading to deterioration of the system, which means going back to square one.

Concluding Discussion

To sum up the foregoing analysis, we have noted above the upsurge of activity in the irrigated commands initiated as a result of implementing reforms to institute participation of water users in the management of irrigation systems. More than 10,000 organisations have been created with clearly specified mandate, responsibilities and powers in the legal enactment. There is overwhelming evidence for the repair and rehabilitation work done throughout the state and spread over different projects. Heavy inflow of funds has been injected through these organisations, which have infact been utilised for what they were earmarked. Dispersed works of smaller scale and concurrent activity in this direction has served to discourage leakage that is typically associated with the conduct of physical works in the irrigation sector. The establishment of the entire structure of participatory institutions with enabling capability to initiate action is indeed commendable for a state-initiated programme and speaks volumes about the political will and determination.

It would indeed be a tragedy if this dynamic initiative for revolutionising the management of water use for agriculture is shortlived. The experience is being watched the world over and the lurking apprehension at the back of everyone's mind is: How sustainable will the reforms be? Will they meet the fate of many a development program with ambitious design and start, which have eventually entered the annals of history? Or those that have withered down to the mundane after the heat and dust have settled down. The expectations and stakes are high.

Disappointment generates cynicism that takes a heavy toll of faith in participatory approaches. It is a risk of paradigm failure – one that the entire community of stakeholders cannot afford to take. For the process of participatory approaches to acquire the recognition, legitimacy and patronage, has been a long and uphill drive; there is a heavy stake in making it work. This working has to be self-reliant. Artificial sustenance based on life-saving measures in terms of external loans violates the tenets of participation. The key question therefore is -- how do we ensure the self-reliance of these atomic WUAs, which are the building blocks of PIM institutions in Andhra Pradesh?

One of the necessary conditions -- though not sufficient by itself – for sustainable reproduction of WUAs as viable entities is their financial self-sufficiency. We have pointed above how critical this was for the successful entrenchment of PIM institutions in Mexico. That this is problematic under the present system of an overwhelming priority of using water for paddy by the cultivators in almost all the commands has also been argued above. *Monopolisation of water* (by head-reach farmers) *inspite of increased availability* as a result of improvement in physical structure, confines the benefits to relatively smaller number of farmers. This discourages the payment of water cess by the downstream as a matter of right, while the head-end farmers are also defaulting on their payments inspite of getting more than their share of water, even during water scarcity period. This reduces financial inflows required for recurring expenses to maintain the system in a functional state over the years. This will set into motion the cycle of poor upkeep leading to loss of functionality and consequent dampening of expectations of a section of water users from the system. Such a situation weakens the hold of participatory institutions.

Rice is a highly water-intensive crop. It is the only attractive option in soils with high moisture content in many areas. It also constitutes the chief food crop in the region. While growing rice may be necessary on a portion of the command, its extensive cultivation as currently practiced cannot be rationalised. In fact, this has been made possible only due to the total breakdown in rules of water distribution and regulation that has ensured unbridled access to water by water users at the head of main canal, distributaries and minors. Unless the PIM reforms can bring about significant control over water distribution, not only is the situation likely to continue, paradoxically enough it may worsen. That PIM induced activity can be positively detrimental for the tail-end distributaries, was a fear expressed by one of the field engineers in charge of tail area distributaries. The repair and streamlining of head-reach distributaries had enabled the withdrawal of larger discharge by them. This cut into water supply scheduled for downstream channels. He had noticed a positive *decrease* in water supplies reaching the tail as a result of improvements effected in the upstream distributaries. While this may not be a serious concern during high rainfall years, the consequences during the drought year (1999) were serious.[5] Enforcing compliance to rules of water distribution by the establishment will be unpopular and cannot be expected to happen on account of political imperatives. Social pressure of WUAs on each other does not have the 'teeth' required for insisting on adherence to the code of water sharing agreed upon. Examples of cooperative action by rural communities have been noticed only at the village level, i.e., *within* villages and only under special circumstances of perceived costs and benefits (see Wade (1984) for a discussion on social cooperation for management of irrigation water in rural areas of Andhra Pradesh). There is hardly any evidence for inter-village cooperation based on community initiatives. It is therefore unrealistic to expect a sudden spurt of social initiative to force the head-reach WUAs to bend to the common will.

Financial incentive is sought to be offered to the WUAs for enhancing water fee collection in their commands. This will be futile in the absence of effective water regulation. Secondly, even where the water is available, there are limitations of the control of WUAs over its general body members. We have earlier pointed out the weak links of the user representatives with the electorate. This restricts their ability to effect higher collections as well as monitor unauthorised withdrawals of water from the channels that must pass through their lands for carrying water to downstream cultivators. Moreover, higher collections are an incentive for the members of WUA, as they control the funds, not so much for the members of general body. No doubt, the latter too have a vested interest in the ongoing maintenance of the system, but in the absence of a strong grass root level awareness and mobilisation drive, the individual cultivator might like to buckle on making his payments if he can get away with it. Lack of financial transparency at the WUAs level will further weaken the ability of office-bearers to exact payment of full dues.[6]

Given the above scenario, it is imperative to include the realm of water use within the fold of PIM reforms or other policy intervention. Effective strategies are required to be worked out for weaning the farmers away from cultivation of water-intensive crops such as rice. AP has been called the 'rice bowl' of the country. Living up to the image can be a costly blunder. There are already reports of over production of rice in the country, which are rotting in the FCI godowns (Omvedt, 1999).[7] Clearly, the macro considerations dictate a shift to less water consumption cropping package.

The point is to influence micro level decisions of an individual producer to favour less water demanding crops and practices by interventions geared to induce changes in his environment. The above calls for market and non-market interventions. We have noted the limitations of purely price-oriented policies. The successful impact of price interventions depends on the non-market environmental factors that allow a producer to avail of price concessions. For example, price support for inputs and outputs for stimulating the cultivation of low water-intensive crop should be accompanied by corresponding provision of technological extension, credit, input availability, access to output markets, etc. The important point is that these inputs should converge on him *concurrently* for a focused support. Support that lacks coordination in time and space entails highly inefficient use of resources.

The final argument is for carrying the process of irrigation reforms – which are necessary and inevitable – beyond the narrow confines of a supply side approach to include the critical realm of *end use regulation of water*. It is thus to take forward the process, which has been initiated, to its logical conclusion by bringing within its purview the modes of on-farm utilisation of water, the cropping pattern and agricultural practices. Using market and non-market interventions, attempt should be made to influence choice of low water consumption crops, water-efficient technologies and adoption of agricultural practices which enhance water productivity.

This has corresponding implications for research agenda for the PIM fraternity. In the context of AP, one of the key questions to be investigated is the reason underlying the strong preference for water-intensive cultivation and resistance to less water demanding and/or high economic return cropping pattern. Issues of technological change in agriculture and impediments to technological diffusion and absorption need to be explored.

NOTES

1. For example, on Wazeerabagh distributary while the DC president reported 100 percent irrigation in the post-PIM period, it was found that areas where the water was barely touching were also counted as areas recently irrigated.
2. As discussed later, there is some scope for increasing irrigated area by reducing water wastage through outflow of drainage water from paddy fields in upper reach areas
3. During discussions with field officers, instances of political interference to release water on a priority basis to specific distributaries were often cited as significant pressures which constrain the ability of division office
4. See Jairath (1984) for a discussion with reference to canal irrigation in Punjab (India). Case of Wazeerabagh distributary on NSP-LC also illustrates the point. This channel is closed for two days in every fifteen days in order to allow water to feed the downstream distributaries. It has a balancing reservoir towards the tail end of the distributary. When the water stops at the head it continues to flow for two to three days towards the end. But since the flow is progressively reduced this is used to fill up the reservoir and water does not reach further. When the distributary again receives the water after two days, it takes three days to fill up and start irrigating the furthest point possible. The gap of irrigation for the tail-end areas therefore becomes nearly five days.

5. This was corroborated during the field visit to Wazeerabagh distributary, closer to the upper reach, which was drawing a discharge of 600 cusecs as against its designed share of 500 cusecs, to cope up with the water shortage due to low rainfall.
6. This is exactly the case with gram panchayats. Although it is the lowest elected body and is entrusted with authority to implement local taxes, e.g., housing tax as a part of resource generation, the collections are negligible.
7. Estimated losses of food grains according to ministry of food and supplies are about 10% of the total production or 20 million tons per year... most of these losses take place in storage in the godown of FCI. According to Kirit Parikh, an expert on food security and PDS, the government stocks 11.7 million tons of rice against a requirement of only 4 million tons per year (Omvedt, 1999).

REFERENCES

Action Plan, 1995, *Action Plan for Completion of Irrigation Projects*, Irrigation and CAD Department.

APFMIS, 1997, *Andhra Pradesh Farmers' Management of Irrigation Systems Act*, GOAP, Hyderabad.

DIPR, 1999, From a Drop to an Ocean, *The Hindu*, 24 May.

Factual Note, 1996, *Irrigation Sector: A Factual Note*, GOAP, Irrigation and CAD Department, Hyderabad.

Feedback, 1998 and 1999, Analysis of WUA President's Feedback obtained in the Sadassus in Hyderabad and Vijaywada, April, CAD, GOAP.

IIM and IIMI study (Brewer, J. et al.), 1998, *Irrigation Management Transfer in India: Policies, Processes and Performance*, A Collaborative Study in Indian Institute of Management, Ahmedabad, India and IIMI, Colombo, Sri Lanka.

Jairath, J., 1999, *Status Report of Irrigation in Andhra Pradesh*, Centre for Economic and Social Studies (CESS), Hyderabad.

---, 1999, Participatory Irrigation Management in Andhra Pradesh: Experiments in Andhra Pradesh, *Economic and Political Weekly*, Vol. XXXIV, No. 40, October 2.

---, 1999, *Participatory Irrigation Management in Andhra Pradesh: An Assessment*, ongoing research project at CESS, Hyderabad.

---, 1984, Technical and Institutional Factors in Utilisation of Irrigation: A Case of Public Canals in Punjab, 1985, *Economic and Political Weekly (EPW)*, Vol. XX, No. 13, March 30, pp. A-2 to A-10.

Johnson, S.H. and others, 1997, Impact Assessment of Irrigation Management Transfer in the Alto Rio Lerma Irrigation District, Mexico, *Research Report No. 15*, International Irrigation Management Institute (IMMI), Sri Lanka.

---, 1997, Irrigation Management Transfer in Mexico: A Strategy to Achieve Irrigation District Sustainability, *Research Report No. 16*, International Irrigation Management Institute (IMMI), Sri Lanka.

Minister's Statement, 1998, Gap in Irrigation Potential Bridged, *The Hindu*, 16 December.

Note on the APFMIS Act, 1997, *Note on Andhra Pradesh Farmers' Management of Irrigation Systems Act*, for District Collectors' Conference, Irrigation and CAD Department, Hyderabad.

Omvedt, G., 1999, Rotting Food, *The Hindu*, 23 October.

----, 1999, 5.12 Lakh Acre of Land Benefited from Irrigation Works: Tummala, *The Hindu*, 15 March.

---, 1998, Irrigation Official Suspended, *The Hindu*, 25 May.

UNICEF, 1997, *Income and Expenditure Patterns of Gram Panchayats*, UNICEF and PRED, Hyderabad.

Wade, R., 1987, *Village Republics*, CUP and Orient Longman.

IV

Training and Gender

10

Capacity Building for Participatory Irrigation Management in Andhra Pradesh

B. GURU PREM

In most developing countries, the irrigation sector has always been a top-down government-administered sector, with very little or no participation of the farmers and users. It is assumed that only the state was capable of handling large irrigation projects. Recent experience has proved that government-operated systems are poorly maintained and inefficiently operated. Yet, some of these systems showed significant improvement when their management was transferred to the farmers in some of the countries like Mexico, Turkey, Philippines, etc.

Prior to British rule, irrigation in India was normally created through diversions from rivers and other natural water sources, which were completely managed by farmers. Some of these systems are operational even today which include the PHAD system in parts of western India, the canals built by Vijayanagar Empire (present Karnataka) and several thousands in southern states of India (Karnataka, Andhra Pradesh and Tamilnadu).

After India attained independence in August 1947, paramount importance was given for attaining self-sufficiency in food and the state governments were encouraged to expeditiously formulate and develop water resource projects for specific purposes like irrigation, flood control, hydropower generation, drinking water supply, industrial and various miscellaneous uses. As a result, a large number of projects comprising dams, barrages, hydropower structures, canal networks, etc., have come up all over the country in successive five-year plans.

India is a country largely made up of small and marginal farmers, thus the strategy in planning and managing water resources assumed greater importance.

Andhra Pradesh Scenario

Andhra Pradesh (AP) is the fifth largest state in India, with geographical area of 27.68 mha and a population of about 80 millions. The economy of the state is predominantly agrarian. The state is endowed with three major rivers, namely, Godavari, Krishna and Pennar, which have a total drainage area of 70 percent of the total land of the state. In addition to these, there are nine inter-state rivers and 28 minor rivers. The water potential of the state is about 7.78 million hectare metres (2746 TMC).

There are 15 major projects, 74 medium projects and 12,294 minor irrigation tanks under the control of irrigation department, and 70,000 minor tanks under Panchayat Raj Department. The total irrigation potential created is 4.80 mha against which only 2.84 mha is being irrigated, leaving a gap of 1.96 mha.

Genesis of CADA

It was a matter of great concern to the government to bridge the gap ayacut (1.96 mha), and improve the conditions of the irrigation systems. During 1984, the government had promulgated the AP Irrigation Utilisation and Command Area Development Act, with a view to remedy the above situation by introduction of Rotational Water Supply (RWS). However, the programme could not achieve the target set, as the farmers neither enjoyed the freedom nor could make decisions due to lack of strong legal framework, lack of financial autonomy or any specific responsibilities.

Subsequently, several seminars, workshops, awareness programmes were conducted in the state to pave way for farmers' participation in the irrigation water management. The first seminar was inaugurated by Chief Minister N.T. Rama Rao on 4 April, 1988 at Water and Land Management Training and Research Institute (WALAMTARI), Hyderabad. Another seminar was conducted at Sriramasagar Project (SRSP) command on 11 March, 1989 on Farmers Role in Irrigation Management for Major Projects in Andhra Pradesh.

Pilot Projects

In the early nineties, to vest the management of irrigation system in farmers, a pilot project was initiated to demonstrate the utility and practicability of Water Users' Association (WUAs). A minor was selected in Sriramasagar project (5L of D-64) on a small scale, covering 500 ha. This was followed up with a large-scale pilot programmes covering a command area of 20,000 ha under six different distributories under SRSP command during 1995-97. This was taken up by two local non-governmental organisations (NGOs), viz., Institute of Resources Development and Social Management (IRDAS) and Society for Natural Resources and Rural Development (SONAR).

The farmers in the command were helped by these NGOs to form into WUAs, for taking up minimum maintenance works and for ensuring free flow of water up to the tail reaches in the system. They were also helped to organise water distribution in their jurisdiction, resolve disputes, if any, and adopt their own crop-planning, etc. The WUAs were registered as legal bodies to provide the required identity. To supplement the efforts of WUAs, a Distributory Committee (DC) was formed federating all WUAs under this distributory.

As a result, an increase in ayacut of about 50 to 80 percent was achieved with the same quantity of water supplied as earlier and without any substantial investment on rehabilitation or repairs of the system.

These results were widely discussed and considerable confidence generated among the farming community, creating a conducive environment for the promotion of PIM and formation of WUAs in the state.

Follow-up Action

The first proactive step to address the problem was taken up by the government with the preparation of a white paper in June 1996 containing a diagnostic of the state's irrigation sector. Extensive discussions were held with farmers, District Collectors and officials of irrigation department through seminars and workshops spread over a period of six months, from July to December 1996.

To empower the farmers in the management of irrigation systems, the Chief Minister has initiated an enabling environment among the people's representatives and farmers. Conferences were organised in all 22 districts of the state followed by several rounds of dialogues with farmers and political parties by the Chief Minister Chandrababu Naidu, on the benefits of PIM, during July to December 1996.

A conference was held with the leaders of all political parties of the state legislative assembly to discuss the provision of draft act in January 1997. This has favoured and created a political will.

Delineation of Hydraulic Areas

The government had appointed seven consultants for discussions with farmers and local irrigation officials to delineate the hydraulic areas for formation of WUAs.

Legal Framework

The above enabling environment coupled with political will had culminated in giving a legal framework to the introduction of Participatory Irrigation Management (PIM) on a big scale. In May 1996, a factual note on the irrigation sector was laid on the table of the AP Legislative Assembly, highlighting the performance of the irrigation sector and investments made.

The AP Farmers Management of Irrigation Systems Act, 1997 came into being with effect from March 1997. It is a revolutionary legislation and first of its kind, which exclusively empowers farmers as decision takers and implementors in the management of irrigation systems. Consequently, 10,292 WUAs and 174 Distributory Committees were formed on democratic basis during June 1997 and November 1997 respectively.

WUA: A Body Corporate

The act recognises every Farmers' Organisation (FO) as a body corporate with a distinct name, having perpetual succession and a common seal. FO has the capacity to enter into contracts and it shall sue or be sued in its corporate name, represented by its president/chairman.

Elections of WUAs

Elections were conducted for all the 10,292 WUAs on democratic basis on 17 June, 1997. In November 1997, elections to 174 DCs were conducted. Action was taken to create project level committees by federating all the distributory committees under the projects.

An incentive amount of Rs. 50,000 was given to 5,481 WUAs for having elected their presidents and members of the committee unanimously. An advance of Rs. 30,000 from water charges to be collected in future was given to 4,224 WUAs, which contested in elections.

Capacity Building

Capacity building is a continuous process of developing the awareness, knowledge, skills and attitudes that have been defined as necessary to achieve the objectives by the most cost-effective means available.

The basic elements of capacity building can be listed as following:

(a) increasing awareness of the farmers about PIM,
(b) creation of legal awareness about PIM, and
(c) institutional reforms and strengthening of PIM.

Farmers Awareness

The walkthrough surveys, general body meetings, executive committee meetings, social audits, which are being conducted from time to time are instrumental in increasing the awareness about PIM. It helps the stakeholders in identifications of works, prioritisation, execution and better monitoring of works. Thus, strengthening the bottom-up linkage in further consolidation of the WUAs.

Legal Framework

The laws and regulations of PIM are fundamental and hence a compendium comprising government orders related to PIM has been published and made available to officials and farmers, in order to achieve transparency of the system. It is also imperative to learn from international experiences, viz., Mexico, Turkey, Philippines, etc.

Institutional Reforms

The institutional reforms can be broadly categorised as technical and social.

(i) *Technical skills*: For capacity building, the institutional reforms on technical front are being organised by conducting massive training programmes, providing funds for works, and strengthening the infrastructure using the latest technology.

(ii) *Social skills*: The mindset is being changed by conducting walkthrough surveys, general body meetings, social audits, etc.

Launching of Massive Training Programmes

A series of training courses and orientation programmes for farmers' organisations and government staff was undertaken (Insert-A).

During July 1997, for the first time, officials of I&CAD, revenue and agriculture departments participated in a workshop to determine the thrust of the training programmes. The detailed programmes for training of trainers were prepared based on the recommendations of the workshop. The first round of training was given to all elected WUA presidents in two-day district-level workshops during October 1997. Training modules have been progressively developed in 18 subject areas (Insert-B) for the capacity building of WUAs and irrigation officials.

A number of other training programmes involving all WUAs and including an Indian Network on PIM/World Bank Institute/Ford Foundation sponsored workshop on training of trainers during April 1998, have also been conducted subsequently. During September 1998, a one-day orientation workshop for I&CAD department accounts officers and engineers was held to discuss new procedures of accounting in accordance with the implementation of APFMIS Act, 1997.

Training of Trainers

During March-April 1999, eleven intensive four-day training programmes were conducted for the training of trainers to become the district-level trainers. Five officers each from I&CAD, revenue, and agriculture departments were trained as trainers for the districts. A "Trainers Handbook" and a "Manual on Orientation in Training" were brought out for the guidance of the trainers.

Training Kit

A special "training kit" has been designed for training farmers with flip charts and a manual prepared in local language.

During July-August 1999, the training programmes were conducted in all the districts at selected places for training of 65,000 presidents and TC members of WUAs. About 55,000 presidents and TC members were trained during this period. These trainees were given a manual on WUA activities and details of the provisions of the act with the rights and responsibilities of the WUAs.

Cooperation from Line Departments

(1) *Irrigation Department*: A competent authority of the cadre of Assistant Engineer (AE) was notified for three or more WUAs in the case of major irrigation system, one AE for every four WUAs in medium irrigation system and in the case of minor irrigation system one AE for 15 to 20 tanks. These irrigation engineers provide WUAs with all technical support in the preparation of estimates, execution of works etc.

(2) *Agriculture Department*: The officials of the agriculture department provide the WUAs with the advanced technologies in maximising the agriculture production, providing proper marketing facilities, and also enhancing value addition of products. The department has taken up Agriculture Intensification Programme with a layout of Rs.113 million to enhance the capacities of departmental personnel and farmers at large.

(3) *Revenue Department*: The co-operation of revenue department is also mandatory in providing command area maps, collection of water cess, identifying increased ayacut area and other related information necessary for effective functioning of the WUAs.

Revision of Water Charges

For the sustenance of WUAs in the long run, irrigation water charges were raised three-fold on April 1, 1997, effective from 1996-97 rabi season. Ten percent of the amount of water charges collected were allocated to gram panchayats and the balance 90 percent were to be shared by farmers organisations (WUAs, DCs and PCs) as under:

WUAs	50 percent
DCs	30 percent
PCs	20 percent

In addition, the act empowers WUAs to levy special fees for achieving objectives of the act, collect contribution, raise income from the properties within the system jurisdiction (FO).

Andhra Pradesh Economic Restructuring Project

The AP government has taken full cognisance of prevailing deteriorating situation in irrigation sector and the imperative need for reorientation. Several steps have been taken by the government to help further implementation of reform programmes and to create the capacity building for sustainability of WUAs. The government has taken the assistance of World Bank (Rs. 12,986 million) for this project.

The major objectives of the project are:

1. To improve the performance of all major, medium and minor irrigation schemes through the financing of a minimum rehabilitation programme and operation maintenance costs on a declining basis.
2. To operationalise the WUAs and their committees at distributory and project levels, through awareness camps and training programmes for irrigation officials and WUAs.
3. To support institutional reforms within irrigation department, by introducing state-of-art communication and computerised Management Information System (MIS).

The AP Economic Restructuring Project supports the following components, which are operationalised:

Minimum Rehabilitation Works (Rs. 4,549 million)

For deferred maintenance and special repairs to head works, canals, drains and hydraulic structures, de-silting of canals and drains, etc.

Recurrent Maintenance Works (Rs. 6,518 million)

For works such as annual desilting of canals, drains, annual maintenance of canal lining, cross regulators and their gates, drop structures, cross drainage works, etc.

Scheme Improvement and Farmers Turnover (SIFT) (Rs. 1,202 million)

For modernising a limited number of medium irrigation schemes and a part of major irrigation schemes on a pilot basis, to serve as a demonstration on how overall scheme performance could be substantially improved to facilitate management and turnover of O&M responsibilities to PCs, DCs, and WUAs.

Agriculture Intensification Programme (Rs.113 million)

For increasing the irrigated agriculture productivity and water use efficiency in all major, medium, and minor schemes. This includes dissemination of agriculture technology through training the farmers, and on-farm demonstration through pilot projects, etc.

Institutional Support (Rs. 603 million)

For planning and monitoring of irrigation activities by a computerised MIS and extensive communication system to facilitate flow of information. Also for human resources development of WUAs through farmers' training units, study tours, etc.

The expected outcome of these measures would be: (i) full recovery of water charges, (ii) full financing of irrigation O&M, (iii) improved performance of irrigation schemes, and (iv) increased irrigated crop production.

Government Funding

During 1998, the government had released the following grants to WUAs for taking up maintenance and rehabilitation works:

	(Rs. in crores)
Repairs and maintenance, Rs.100 per acre	107.23
Minimum rehabilitation	58.98
Additional grant for minor irrigation schemes	48.00
Functional grant to SRSP and NSP	64.00
Field channels in SRSP	3.89
Minor drains under hazard mitigation project	22.00

This has generated tremendous enthusiasm among WUAs and maintenance works were taken up on a large scale at the rates which are 10 to 20 percent less than what has been done by irrigation department through contractors. The WUAs have taken up 22,171 works at a cost of 141.53 crores.

During 1999, the government had released following amounts to all WUAs for taking up the maintenance and rehabilitation works successively for the second year:

	(Rs. in crores)
Minimum rehabilitation (major and medium)	165.84
Operation and maintenance (major and medium)	32.45
Operation and maintenance (minor)	30.95
Irrigation works under AP hazard mitigation	67.75

Additional Measures taken for Capacity Building

The AP government has taken up several other measures for capacity building for WUAs, like these discussed below:

Wireless Communications

For providing a speedy communication between controlling officers and field level officers and WUAs, the government has sanctioned VHF wireless sets under the jurisdiction of Collector of each district, to the Superintending Engineers, Executive Engineers and Deputy Executive Engineers incharge of O&M of irrigation systems. This wireless comprises of mobile and fixed sets.

Publication of Magazines

For dissemination of information and to create awareness among farmers' organisations, the government has accorded an amount of Rs.78 lakhs towards publication of monthly state-level magazine and district newsletters from all district headquarters.

WALAMTARI is publishing the monthly magazine under the title "SAGUNEETI SANGHAM". About 55,000 copies are circulated among all the presidents and territorial constituency members of WUAs, field engineers and others in the entire state. Similarly, all the District Collectors are publishing district newsletters under different captions, for circulation among WUAs in each district.

This is enabling the WUAs to be familiar with latest developments in the irrigation sector, besides dissemination and exchange of information.

Administrative Expenses of WUAs

To meet the day-to-day expenditure for maintaining their offices, the government has sanctioned the following grants towards administrative expenses of each FO:

WUA (minor irrigation)	2,500
WUA (medium irrigation)	5,000
WUA (major irrigation)	10,000
DC	25,000

Technical Publications

Several publications (in English and local language) were issued to the WUAs on the provision of APFMIS Act, the rights and responsibility of WUAs, etc., during the state and district level conferences. A compendium of all government orders was also published for use by irrigation officials and WUAs.

Teleconferences

For redressal of grievances and difficulties in execution of works by WUAs, the Chief Minister has conducted weekly teleconferences from state headquarters and district headquarters with the field officers and WUAs and reviewed the progress of works by WUAs and issued instant orders for speedy execution of works. Now, the government have launched video conferencing, as a further step in hi-tech communications.

State-Level Conference

During the first year of constitution of the WUAs (on 27 April, 1998), a state-level conference was conducted at Hyderabad with all 10,292 WUAs, for direct interaction with WUAs on their problems. The Chief Minister had chaired the meeting. There was an overwhelming response from all the presidents of WUAs and the feedback on SWOT analysis was taken during interaction meeting. Consequently, the following orders were issued in pursuance of the policy decisions taken:

- Fishing revenue to the WUAs;
- WUAs term extended from three to five years; and
- Lascar and work inspectors brought under the control of WUAs.

Regional Conferences

Similar conferences at regional level were also conducted at Rajahmundry and Hyderabad. A regional conference was conducted at Rajahmundray, East Godavari district on 27 December, 1998 covering the coastal districts of the state. During the same month, another regional conference (31 December, 1998) was conducted at Hyderabad covering other two zones, namely, Telengana and Rayalaseema. People's representatives, ministers, secretaries, WUA presidents and farmers had participated in direct interaction with WUAs and the Chief Minister had chaired the meetings. Various issued were discussed on the problems of WUAs along with the action plan for 1999-2000.

District-level Conferences

During April 1998, district-level conferences were conducted in all 22 districts, wherein thousands of farmers, WUA presidents and members had participated in the interaction. Handouts were distributed to all farmers explaining the provisions of the APFMIS Act and the rights and responsibilities of WUAs.

National Conference on PIM

During January 1999, Fourth National Conference was held at Hyderabad and the Chief Minister had inaugurated the conference. World Bank officials, delegates from all states in India and WUA presidents, government officials, etc., had participated in the seminar. The delegates were taken for a field trip for interaction with the presidents of the WUAs.

Women's Role

Indian women have traditionally an equal participation in agriculture activities. However, their involvement in decision-making is insignificant. Efforts are being made to increase the capability of women members of WUAs by means of effective training programmes. With greater participation by women, they would have potential contribution to make in terms of:

(a) their greater access to accept less privileged groups (schedule castes, schedules tribes and poorer sections);
(b) their propensity for consensus seeking; and
(c) their greater recognition about necessity of water in domestic sphere, environment and water quality

Though the APFMIS Act, 1997 is gender-neutral, it is encouraging to note that among 10,292 WUA presidents, 98 are women presidents and about 1,097 are TC members. To further strengthen WUAs, facilities may have to be provided to women for their empowerment.

Role of NGOs

The massive process of farmers' empowerment needs considerable skills in social engineering, and irrigation department is the agent of change throughout the process. But, for further consolidation of these transformations, NGOs' participation is very essential and imminent. The NGOs facilitate:

(a) increase in social awareness;
(b) motivation and identification of leaders;
(c) imparting training; and
(d) providing linkage with various WUAs and building strong network.

Conclusion

The AP government has initiated a major shift in its approach to governance – from being a major operator to a prudent facilitator, from a regulator and controller of economic decision-making to an enabler of market-based development, and from a top-down administration to decentralised decision-making through people's participation.

The introduction of parallel committees in rural areas, viz., Watershed Committees, Village Education Committees, Vanasamrakshna Samithis have paved the way for generation of new leadership coupled with strong political will, which in turn established WUAs as most appropriate rural institutions for efficient water management and agriculture development.

The reforms are continuing and the steps taken so far represent only the beginning of a broader reform programme requiring a number of years to further develop and, nevertheless, AP's reform programme is already attracting interest, both within India and internationally.

11

Role of Women in Participatory Irrigation Management: A Study in Andhra Pradesh

C.VIJAYA SHYAMALA and C.SITHAPATHI RAO

To promote Participatory Irrigation Management (PIM), Water Users' Associations (WUAs) have been constituted in Andhra Pradesh through the Andhra Pradesh Farmers' Management of Irrigation Systems (APFMIS) Act, 1997. The primary objective of the PIM programme is empowerment of farmers in managing the irrigation system on their own and bringing about social justice and equity in sharing the resources.

Women constitute around 47.5 percent of the total population and a majority of them, accounting for nearly 75 percent, belong to the families of small and marginal farmers. Women constitute nearly 50 percent of the total agricultural labour force and their contribution to agriculture is estimated to be between 54-66 percent, depending upon the cropping patterns of the areas and the technologies involved. Maria Mies (1987) observes, based on her field studies of three villages in Telangana region, that approximately eighty percent of operations necessary in the agricultural production process feature women labour. When measured by the labour time thus spent, it amounts to 70-80 percent of the annual labour time.

The 1997 Act provides that every WUA shall consist of all the water users who are landowners in a water users' area. It is an established fact that very minute percentage of women are landowners. Nevertheless, it was found that about 98 and 830 women have emerged as presidents and members respectively of the WUAs in the normal electoral process. It becomes pertinent, in the light of this intervention, to understand the role of women in PIM as it is an unexplored area and can help develop future strategies.

This study, thus, is aimed at throwing light on the existing role and status of women in irrigation management in Andhra Pradesh and exploring the opportunities for increasing the women's participation in the WUAs and the PIM programme.

Methodology and Sample

A total of 32 women [14 WUA Presidents and 18 Managing Committee (MC) members] were randomly selected from seven districts, viz., Medak, Warangal, Krishna, West Godavari, Chittoor, Mahabubnagar and Karimnagar representing the three regions of the state. Interview method through open-ended questionnaire was used to gather the background details, their role in the WUAs and their perception on women's involvement. Also, group discussions were held with both men and women in the concerned villages of the WUA to gather their views on the role of women in PIM.

Background Particulars

The background particulars of the women presidents and members was gathered mainly to understand and assess if there is any correlation between these factors, viz., age, caste, education, etc., and their selection to the posts, effective performance, etc.

The following table gives the background details of women presidents and members surveyed:

Table 1(a)

District	*Total Sample*	*Head of the Household*	*Occupation*		
			Housewife	*Agriculture*	*Non-Agriculture*
Medak	6	2	2	3	1
Warangal	8	4	5	3	-
Krishna	3	-	2	-	1
West Godavari	3	-	3	-	-
Chittoor	8	4	3	5	-
Mahabubnagar	2	2	1	1	
Karimnagar	2	2	1	1	
Total	32	14	17	13	2

It can be observed that:

- More than 70 percent of the sample fall in the economically active age group of 25-50 years.
- About 50 percent of the sample belong to the forward castes. The rest being Backward Classes (BCs) and Scheduled Castes (SCs).
- About 50 percent of the sample are illiterate. Among the literates too, it was found that majority have studied upto primary or secondary school (maximum upto seventh class) only.

Table 1(b)

District	*Total Sample*	*Age Groups(years)*			*Caste*				*Education*				
		<25	*25-50*	*50+*	*SC*	*ST*	*BC*	*OC*	*Illiterate*	*Literate*	*Level of Education**		
											1	*2*	*3*
Medak	6		4	2	2		3	1	5	1		1	
Warangal	8		7	1	1		3	4	4	4	1	3	
Krishna	3	1	2					3	1	2			2
West Godavari	3	2	1					3	1	2		1	1
Chittoor	8		6	2	4		2	2	4	4	4		
Mahabubnagar	2		1	1			1	1	2				
Karimnagar	2		2				1	1	1	1		1	
Total	32	3	23	6	7		10	15	18	14	6	5	3

Note: * 1=Primary School, 2=Secondary School, 3=Intermediate and above.

- About 50 percent of the sample are primarily housewives. However, some of them have said that they sometimes need to supervise the agriculture operations. In the balance 50 percent, 41 percent of the women perform agriculture operations and the rest 9 percent have non-agricultural occupations.

Election of the Women Presidents and Managing Committee Members

About 84 percent or 27 women office-bearers were elected unanimously. In majority of the WUAs, the election process was engineered so as to avail the Rs.50,000 incentive announced by the government for those WUAs whose office-bearers were elected unanimously. The social and political dynamics of the respective villages were found to be the main influencing factor in the selection of the candidates for the posts of presidents and members.

It was observed that majority of the women did not voluntarily enter the WUAs but were pushed into it to function as 'token' members to serve the self interests of their male member, who is either a contractor, a political party activist, an influential person in the village, etc., but cannot directly obtain the position as the land is not in his name.

Among the 14 women presidents surveyed, it was found that in 12 cases the husband/ father/son is the interested party but since the land in the WUA area is registered in the particular woman's name she was pushed to occupy the position. The two exceptional cases are, one of Padma in Warangal district and the other of Sammakka in Karimnagar district, who have voluntarily contested and won the election, both of them having earlier experience in the watershed committee and the village panchayat respectively.

In the election of MC members, in 60 percent of the cases it was observed that while giving representation to the different castes, it was also thought fit to have one woman member like in the gram panchayat. Why this particular woman was selected depended again, like in the remaining 40 percent cases, largely on various factors. About 60 percent of the total MC members surveyed came into the position due to the male member's influential position in the village. In the remaining cases it was observed that the main reasons were, no male member coming forward in the Territorial Constituency (TC) area, being the only woman owning land in the command, etc. About 87 percent of the MC members have said that they were informed of their selection at a later date.

Role Performance

The women office-bearers are holding their positions for the past two years. Having gained entry into the WUAs, the next issue is their level of performance, i.e., whether these women are able to carry out their roles and responsibilities independently, and what are the constraints, if any.

Although all the women office-bearers covered are aware of the WUA's main activities as undertaking maintenance works and equitable water distribution, with regard to the actual performance of the tasks, male members invariably are playing the major role. It was found that excepting for two women presidents, none of them have had any earlier experience of public life.

It was gathered that the meetings are entirely organised and conducted by the proxies, in about 57 percent of the cases the women presidents are not even physically present. It was also found that 80 percent of the women MC members are not attending the meetings, in majority cases the decisions are taken informally by the president and other male members, the women are asked to give their signatures at a later date. In about 10 percent cases the women members are openly discouraged by the men in attending the meetings. An illustrative case is of one woman president in Krishna district who was advised by the distributory committee chairman that she need not take the trouble of attending the meetings and her husband was allowed to represent her in the meetings.

Excepting in two cases, none of the women presidents surveyed are personally implementing the maintenance works. Execution of works, financial transactions, etc., are all being performed by the male members; 64 percent women are totally ignorant of the funds received, expenditure incurred, works undertaken, etc. They simply give the answer, "he only knows everything, I do not know". This is because execution of works requires much interaction with irrigation officials, contractors, other menfolk in the village and it also requires extensive moving about on the system, supervising the quality, etc., which are all considered to be tasks that can be performed only by men in the existing social system. An illustrative case is of a woman MC member in a particular WUA being refused the charge of maintenance works of the system in her TC area by the president. She was vehemently asserting her right to prove her capability when the president expressed doubts about her capacity to get the works done.

Among the two women presidents personally participating in the execution of the works, one is C. Padma, president of Oora Cheruvu WUA, Lingal Ghanpur, Warangal. Prior to the WUA, Padma has undertaken construction of checkdams as member of watershed committee. As president WUA, she has personally executed the maintenance works. Her dynamism and sincerity has earned her a good reputation and an organisation called World Vision has approached her for implementing their activities in the village. The women in the village are appreciative of her efforts and the women from DWCRA groups are approaching her for support in their various activities.

The other is Sammakka, president WUA No.3, D-89, Sriramasagar Project, Karimnagar. She personally participated in undertaking the maintenance works and she also personally pursues the release of funds from the government.

Training programmes were extensively conducted for the WUA presidents initially at WALAMTARI, Hyderabad and later at the district level. About 21 percent of the women presidents attended all the training programmes accompanied by the male members in their family, 37 percent attended the training programmes given at the district level and the rest 42 percent have not attended any training programmes at all. The reasons stated for not attending were, husband representing her, long distance and lack of proper transportation, inability to leave the domestic chores, not getting information at the right time, etc. Only one training programme was conducted so far for the MC members at the mandal level and about 90 percent of the sample respondents said they have attended the training.

All were appreciative of the facilities provided and the content of the training programme. However, about 50 percent of them expressed the view that more intensive special training be given to women presidents and members at a nearby convenient location to orient and encourage them towards their responsibilities. One woman president when asked why she was not playing an active role, surprisingly her husband answered by saying that "no one came and talked to her like this", implying that she needs much encouragement through intensive counselling.

About 50 percent of the total sample covered have had some formal education; three of them in fact have gone to college also. However, it cannot be said that these educated women are performing better than the others. Although literacy should enable them to read the material given during the training and effectively perform, majority of the women have not even read the material given to them, as they are not actively involved.

Perceptions of Women Presidents and Members on Role of Women in PIM

That women have no role whatsoever in irrigation management has been quite evident from the present study. Excepting for a negligible section, all the women contacted have responded negatively on the question of their participation in irrigation management. They all agreed however that they perform various other irrigated agricultural operations like transplanting, weeding, threshing, harvesting, etc.

None of the women contacted were clear as to what could be the role of women in Participatory Irrigation Management. Although all of them are aware of the general benefits that accrue to their families through the WUA, none of them could perceive any direct benefits to the women in the command area or how women development could be linked up with WUAs.

Although the basis for selecting women MC members in majority of the cases surveyed was to have a woman member in the committee as in other local institutions like the gram panchayat, none of the women members surveyed could state in what way they could represent women farmers. They also said that so far they have not had any interaction with other women in relation to the working of the WUAs.

Analysis and Conclusions

The women have occupied the present positions in WUAs at the behest of their male members, who obviously were motivated by their self-interest of acquiring the de facto powers. When the women were probed as to why they were not carrying out the activities themselves majority of them said, "we are not used to going out. He only takes care of everything". Some of the men, particularly of some castes, have also said, "our women do not go out". Some women also expressed that they fear being criticised by other women in the village if they play an active role outside the home.

Thus, the traditional patriarchal social system, which has prescribed certain norms of behaviour for women, seems to be the major influencing factor for the inactive role of the women presidents and MC members.

The study also clearly brings out that women are generally not involved in the direct irrigation activities. They also do not perceive any direct and immediate benefits to the women agriculturists, although all of them agree that benefits will accrue to the family in terms of more assured irrigation and increased production. This obviously raises the question as to their stake in the irrigation management and participation in the WUAs.

Performing about 80 percent of the irrigated agriculture operations and deriving benefits from the irrigation system as a part of the farm household, women are obviously stakeholders in the irrigation system and in its effective performance, even though they do not necessarily carry out the irrigation activities.

The ultimate objective of the WUAs is poverty reduction through increasing agricultural productivity by securing social justice and equity. The water users' associations are the only farmers' organisations currently providing a forum for major decision-making in the irrigated agriculture and many linkages are proposed to be developed in terms of support facilities like training and extension, credit, marketing, etc.

Agriculture being a joint farming activity involving the entire household, the women's role in the decision-making at the family level cannot be ignored. Developing their skills and knowledge to participate better in performing the various irrigated agriculture operations and also in effective decision-making obviously enhances the effective functioning of the WUAs and the realisation of its objectives. While it is recognised that majority of the irrigated agriculture operations are performed by women, not involving them in the water users' associations is not only a great injustice to them as it sidelines them from the developmental process, but also detrimental to the progress of the society. The need for gender equity and empowerment thus makes it pertinent that women are involved in the WUAs.

The APFMIS Act, however, has been discriminatory, practically denying majority of the women agriculturists a role in participatory irrigation management, by making landownership the criterion for membership in the WUAs and not making any special provision for women to be involved.

Although few women have entered the organisations in the normal process, they form a very negligible percentage of the women to be represented. The issue then is how can a larger number of women be involved in the WUAs.

Interventions Needed

Agriculture being a joint family affair in India, the roles of women and men in performing the various agricultural operations are complementary to each other. Decisions are mostly taken jointly and interventions to bring about development through participatory decision-making should view the family as a unit and jointly empower both men and women. For this, the following measures should be taken:

1. Women agriculturists in irrigated agriculture need to be empowered along with the men through the WUAs. This is obviously possible if larger number of women are associated with the WUAs. While representation in the decision-making bodies is an important part of empowerment, involving women merely for the sake of involvement through provision of reservations is not going to be helpful. The present study clearly is in the existing social context, a reflection of the danger of 'tokenism' being promoted if 'reservations' is considered to be the strategy for increasing women's participation. It has also been observed during the field survey that the women landowners constitute less than one percent of the total landowners in each WUA, which makes adequate representation a impractical proposition, even if reservation is provided. There is a need to explore the various opportunities for involving women in the water users' associations in the context of the existing legal and social constraints.
2. Empowerment of women agriculturists in other respects like developing their skills and knowledge in improved agricultural practices, promoting appropriate technology, increasing their awareness levels, etc., through training and extension are the most pertinent issues. This would build up their confidence, enhance their participation in decision-making both in the household and family. This intervention will empower and enable them to participate on equal level in the various activities of the WUAs instead of being represented as nominal members.
3. The social norms and values are mainly discouraging women to participate actively. However, about half the sample asking for more intensive training is a clear reflection of their interest to be actively involved. The role model cases also exist to prove that women can personally carry out their various

responsibilities. To ensure that the present women presidents and members evince interest in the WUAs and inspire other women to participate in the irrigation management, there is a need to activate them. Since majority of them belong to economically active age group, intensive training for changing their traditional inhibitions and attitudes and building their confidence to accept the challenge will enable them to function independently.

REFERENCES

Jasveen Jairath, Dharma Sammakka: Pioneer of the Women's Role in *INDIANPIM, Newsletter*, IndiaNPIM, No.5, April-June 1999.

Kitty Bentvelsen, PIM and Gender: Examples from Macedonia, *INPIM Newsletter*, 7 April, 1998.

Margreet Z. Zwarteveen, Gender Aspects of Irrigation Management Transfer: Rethinking Efficiency and Equity, Internal Irrigation Management Institute, Food and Agriculture Organization of the United Nations, Rome, 1995.

Margreet Z. Zwarteveen, Gender Issues, Water Issues, discussion paper, June 1993, International Irrigation Management Institute, Colombo, Srilanka.

Mies Maria, *Indian Women in Subsistence and Agricultural Labour*, New Delhi, Vistaar Publication, 1997.

Shashi Kumar M., Lalita K. and Gopal K.S., Women and Development Profile of Andhra Pradesh, THINKSOFT, Hyderabad, 1995.

V

Beyond Andhra: Replicability and Sustainability in India

12

Power in Motion: A Critical Assessment of Canal Irrigation Reform in India

PETER P. MOLLINGA

Irrigation reform policies have in most cases been conceived by government agencies, often in collaboration with or at the instigation of multi- or bilateral funding agencies. There is now an international and there are many national debates on irrigation reform, finding expression, among other things, in the establishment of organisations like INPIM and IndiaNPIM.[1] Though these are NGOs, their genesis is closely associated with the same funding agencies and governments. INPIM emerged within the World Bank, Washington, D.C., IndiaNPIM from the Ministry of Water Resources, India. Civil society organisations and water users have rarely been involved in irrigation reform policy formulation, though some exceptions to this are discussed below. This may be considered as a contradiction given the label of Participatory Irrigation Management (PIM), that is now given to the reform exercises. Participation of irrigators and other water users seems to refer to the end-state of the reform process, and to some extent to its implementation, but not to its design.

This observation on the non-participatory nature of irrigation reform policy formulation introduces the central theme of this paper: the introduction of PIM as a process in which the relationships between the government and its citizens, water users in this case, are re-shaped. The political meaning of PIM, it is argued in this paper, relates to institutionalised changes in the balance of power among the different interest groups involved in irrigation management that do or do not happen.[2] Or to put it differently: introduction of PIM is not just a 'technical' exercise of management improvement for efficiency and productivity gains, but as a process touches on some rather fundamental features of the social fabric. This

Revised version of a paper presented at the International Training of Trainers Course on Participatory Irrigation Management, 19-25 January, Hyderabad, India. The author would like to thank the organisers of the course for the invitation and incentive for writing this summary paper.

insight provides an important key[3] to understand the failures of many PIM/irrigation reform processes in terms of their stated objectives, as well as of what does happen as a result of policy implementation.

The paper is divided in six sections. In the first section, the inevitability of reform in the irrigation sector[4] is established. This is followed by a discussion of the multitude of reasons why different people want reforms, and thus the diversity of its objectives in section two. Section three discusses in some detail the origins of reform initiatives, taking India as its main example, but also referring to other cases. In the fourth section the notion of 'reform models' is critically examined, as their existence and use has been such a characteristic feature of the history of irrigation reform. In the fifth section the weaknesses of the present, ongoing reform exercises are identified, again with a focus on India. The sixth section gives a very brief summary of the main points of the paper.

I
The Inevitability of Reform[5]

The irrigation sector suffers from a three-fold crisis:

(1) A *financial* crisis: budgets for the sector are generally going down (resulting in less construction of new systems, and lack of funds for operation and maintenance), and governments increasingly see the subsidies to the irrigation sector as a problem because of general budget constraints. Low water fees and low collection rates of these are part of this problem.

(2) A *technical and managerial* crisis: the physical infrastructure of large-scale canal systems is deteriorating and service delivery is poor, resulting in, for example, waterlogging and salinity, low productivity, unequal water distribution and dissatisfied farmers.

(3) An *image* or *legitimacy* crisis: in the public debate irrigation has lost much of its glamour. While it was an undisputed cornerstone of development policy earlier, it is now a much questioned sector. The critique of the negative environmental and social displacement effects of large dams and irrigation systems has raised many doubts about the desirability of this approach to agricultural and economic development. Also the seemingly much better performance of other types of irrigation, notably well irrigation, has cast doubt on the canal irrigation paradigm. Another factor is that the sector is seen as highly corrupt and highly subsidised without delivering the goods it should deliver. Part of the negative image is that the sector is seen as being old-fashioned in its approach to water resources management, incapable and/or unwilling to absorb new insights like integrated water resources management, and very defensive in its interaction with other sectors, and society in general (*cf.* the responses of the Indian water bureaucracy to the activities and report of the World Commission on Dams).

Though the use of 'crisis' terminology is perhaps somewhat alarmist and ideological, there is no doubt that the irrigation sector in many countries is seen rather differently now than, say, 20 years ago. And, the financial, efficiency, productivity and environmental problems of/in the sector are very real, even when their magnitude may vary from place to place. At the same time, there is a history of two to three decades of rather unsuccessful efforts to improve the performance of canal irrigation systems.

Taken together, these considerations and circumstances constitute both the need and inevitability of reform. The basic need for reform lies in the poor performance of the sector, the inevitability in the fact that improved business as usual scenarios have lost their credibility. The logic of all three 'crises' points towards structural changes rather than tinkering with the existing framework. The required reforms of/in the sector revolve around the following three themes:

(1) a redefinition of the sector's financial basis,
(2) a redefinition of the structure of governance and management relations of/in the systems and the agencies that deliver the water, and
(3) a professional reorientation of the engineers and others working within the sector.

The first theme involves questions of the basis and magnitude of resource mobilisation at different levels (financial autonomy/sustainability, water/service fees, subsidies, financial incentives and management and productive performance). It also involves the question whether the economic basis of the sector, that is canal irrigated agriculture, can or should generate (all of) these resources, and *vice versa*, whether canal irrigation or which type of canal irrigation is the best way to generate agricultural and economic development. In the Indian context the impact of liberalisation and globalisation on the agricultural sector will be an important context for developments in the coming years. Qualitative changes in price and market arrangements are being introduced that are likely to have strong implications for the sector, both economically and politically.

The second theme involves the features of a new or adapted institutional framework for governance and management[6] of the canal systems and the agencies and organisations operating within it. Issues like accountability, incentive structures, transaction costs and rights inform the design of new institutions. It can be noted that in the reform debate there are quite a few strong assumptions on how theme 1 relates to theme 2 (like: financially autonomous agencies give better system performance; pricing shapes efficiency of water use; democratic decision making leads to better management, more efficient water use leads to higher agricultural productivity, *et cetera*). Many of these relationships are plausible in theory, but in practice often turn out to be much more ambiguous and complex. However, discussion of this is outside the scope of the paper.

The third theme is the least discussed in the irrigation reform debate so far, but is gaining importance with the increasing popularity of the integrated water resources management concept (IWRM). This in fact implies a rethinking of the notion of sector as such, as well as a reconsideration of the professional scope of the civil engineers, hydrologists and other professionals working within it. The professionals in the sector are mostly water, infrastructure and construction oriented now. However, incorporation of agronomic, managerial, social, environmental and multiple-use considerations into professional expertise and identity is required for IWRM to become a realistic proposition in the irrigation domain.

With this short outline I hope to have sketched the scope of the debate on canal irrigation reform and identified the most important themes within it. I now move to the operational aspects of irrigation reform, beginning with the reasons why different groups of people want reform.

II
Reasons for and Objectives of Reform

The discussion above might suggest that there is consensus regarding the reasons for canal irrigation reform, and the objectives it should try to achieve. Nothing is further from the truth. There is a lot of diversity and difference of opinion on this. What follows is a short sketch of that diversity (for more discussion, see Mollinga, 2000).

The weakest form of canal irrigation reform advocacy, which perhaps cannot be called reform, is the argument that in principle there is nothing wrong with the present setup, neither technically nor institutionally. The problem, from this perspective, lies in poor implementation, lack of funds, non-adherence to rules, lack of enforcement, political interference, and other factors. The solution is doing things as they should be done. In terms of process this perspective usually advocates 'law and order' type measures, and training. The objective is to make the systems work as they were supposed to work.

In a second perspective, reform is advocated for from an economic perspective: in order to maximise the economic benefits of irrigation, reform is necessary. Maximisation of economic benefits can mean different things. It can be argued from the standpoint of the individual farmer (who, with water being the scarce resource is likely to focus on maximisation per unit of land), or from the system/water resource (maximisation per unit of water). It may look at primary agricultural production only, or include the multiplier effects of irrigated agriculture (including employment generation, economic effects through forward and backward linkages). It may be confined to the command area of the system, or take a broader regional view. It may have a focus on growth, or talk about livelihoods.

The third perspective focuses on financial dimensions of the problem and financial incentive structures. One strand emphasises the financial non-sustainability of the present setup, and the incapacity of governments to continue financing canal irrigation. Cost recovery and balancing the budget is a must to make the systems sustainable. Financially autonomous irrigation agencies may be part of the solution. Added to this may be the view that there are no incentives for farmers to use water efficiently, and therefore volumetric water supply and pricing is necessary. Apart from the financial sustainability argument, it may also be thought that financial mechanisms can constitute powerful accountability relations, and thus improve management. Yet another strand focuses on the rent-seeking activities of the government, that is corruption, and the need to eliminate this through 'economic pricing'.

A fourth view, very prominent in the 1970s and 1980s, but less so now, advocates reform from an equity perspective, arguing that the present systems are unjust and need to be reformed because they spread benefits unequally. This may or may not be accompanied by an argument that (more) equal distribution of water and/or benefits also leads to larger overall social and economic benefits. It can be noted that the South Asian debate on canal irrigation reform and equity has not evolved towards issues of rights and entitlements, but has remained stuck in discussions on community participation (for more discussion, see below).

A fifth possible reason for and objective of reform may have to do with decentralisation and democratisation agendas. PIM and the strengthening of water users' associations may be seen as one instance of a general devolution of powers to lower, and more appropriate levels. This may be argued from an instrumental objective (for example, it enhances growth, or, it creates a local political support base), or from a principled perspective (for example, democracy/decentralisation has positive value in itself).

The sixth perspective on canal irrigation reform is that it is highly needed to do something about its problematic environmental effects (from system mismanagement causing waterlogging and salinisation, to the environmental effects of high external input agriculture, to issues of ecological disturbance of coastal water systems, or deforestation through submergence).

Any person's or organisation's position regarding canal irrigation reform is usually a complex combination and hierarchy of different elements discussed above. Also, there is not always a clear difference between reasons, ends and means. What for some is a means to an end, for others is an end in itself. This is perhaps hardly surprising given the complexity of the issues involved, and the fact that the stakes are high. Livelihoods, control, careers, status, institutional survival, political standpoints -- all come into the reform equation.

What can be learnt from this diversity of reasons and objectives is that the notion of 'reform' is very slippery and highly contested. It means very different things to different people. The notion of a 'consensus' on reform (preceding implementation of a policy for example), seems to me to be quite illusory. More relevant may be the question how to negotiate the different viewpoints, objectives, means and interests, to come to an agenda that identifies a sufficient level of agreement (and/or acceptance of disagreement) and acceptable trade-offs to move forward. Or to put it more briefly, instead of consensus, *alignment* of interests is the main issue.

III
Where Does Reform Come From?[7]

The fact that usually governments conceive irrigation reform policies, as stated in the first sentence of this paper, is hardly surprising. Canal irrigation is a government-dominated sector, at least in Asia, and governments have assumed primary responsibility for developing and management of canal irrigation systems. Still the statement needs qualification. Firstly, governments and their administrations are not single things or homogeneous entities. There are many vertical and horizontal divisions, and reform initiatives can come from many compartments of this complex governmental and administrative matrix, and in different combinations and alliances. A recurring pattern is that of an alliance between the political leadership of the state, and a section within the water/irrigation bureaucracy that wants change. In the Philippines there was an alliance between a group within the National Irrigation Administration (NIA) and the president. In Andhra Pradesh, India the alliance is that between the Chief Minister and the secretariat, which consists of non-engineering bureaucratic cadre, but not primarily with the line department (the irrigation department). In Mexico it was an alliance of a section in the irrigation bureaucracy, the president and the World Bank. Such alliances by themselves are not enough, and support networks are needed to sustain them. Those enrolled in the reform initiative may include members of parliament and their political parties, the finance department, international funding agencies like the World Bank, research and development agencies like the Ford Foundation, universities, NGOs, and others.

Secondly, even when policies are and have to be finally consolidated within and by governments, the governments are not necessarily the initiators of policy transformation. This can be illustrated by looking at irrigation reform processes in different states in India. Table 1 provides some of the basic information.[8]

The earliest experiments in India with self-governing irrigation institutions took place in the state of Maharashtra in the 1980s. These took the form of pilot projects for establishing Water Users' Associations (WUAs). These projects were undertaken by NGOs and financially supported by the Ford Foundation. These are significant experiments because they have resulted in sustainable WUAs, to which water is delivered on a wholesale volumetric basis. The experiments further show that with the right effort it is possible to organise farmers for irrigation management across social divisions of caste and class.

Table 1
Canal Irrigation Reform Processes in Different States of India

State	*Maharashtra*	*Gujarat*	*Andhra Pradesh*	*Karnataka*
Role of NGOs	Initiators through pilot projects	Initiators through pilot projects, network in and within government for policy change	Consulted in policy design, supporters of policy implementation	Pilot experiments with WUAs, organising farmers to lobby for policy change
Role of government.	Passive	Some policy changes consolidated in Orders	Main initiator and main implementing agency	Passive till recently
Role of donors and international organisations	Support of NGOs by Ford Foundation	Support of NGOs by Ford Foundation	World Bank loan to fund implementation; Ford Foundation support of NGOs preceding reform process	Minimal support of NGOs, conditionalities attached to project loan by World Bank in earlier phase
Role of water users	Recipients	Recipients	Recipients, with consultation in course of implementation	Claimants (*in statu nascendi)*

The case of Gujarat is very similar: successful experiments and Ford Foundation support, but no volumetric supply. The reference to Ford Foundation support is significant because it was the channel through which the Philippine 'model' became known and was popularised in India.

A difference between the two states is that although the Maharashtra government initially took a favourable attitude, the debate is now fully stuck, and policy change has come to a standstill. In the Gujarat case, the NGOs have been more successful in lobbying the government for policy change. A large number of Government Orders have been issued, which help along the PIM reform initiative, though very slowly.

A similarity again is that in both cases there is an upscaling problem. The NGO-initiated WUAs do not grow beyond a certain number (around 200) while in the order of thousands are needed to cover the full irrigation command area. Typical of the approaches of the NGOs is that they fully focus on *farmer* organisation for water management. They thus focus on the lower levels of the system, with originally no main system or agency reform agenda, and no advocacy or policy formulation role of the water users. The water users are the recipients of policy initiatives, but are not involved in their formulation.

The Andhra Pradesh case is an example of a large-scale (covering all irrigation systems) government-initiated and implemented reform programme. The implementation was preceded by the adoption of an Act (in 1997), which defined the new structure that was to be put in place. The Andhra Pradesh reform process is in some respects a more 'modern' process, incorporating more of the international experience with irrigation reform. It aims at reform of both irrigation management at system level and at agency (irrigation department) level. It devolves real powers to water users; it is building in a financial mechanism that links availability of maintenance funds with water rates collection, it has raised water rates to make the sector more financially sustainable, *et cetera.* An interesting feature also is that, after a few years of implementation, there now is discussion on 'second-generation problems' (like issues of water rights for example), which shows a willingness to learn and adapt. It is difficult to evaluate the achievements of the programme and their sustainability as yet, because the programme is only a few years old. So far the focus has, rather conventionally, been on system repair and rehabilitation (assisted by World Bank funding) and the establishment of WUAs. These are federated into Distributory Committees (one level up) but the Project Committees are yet to be formed. Farmers have been recipients of the policy, and of a lot of training. They were not involved in policy formulation (though field level experience was incorporated in the policy, but mediated through consultants and NGOs). As part of the recognition of second-generation problems, the policy implementation managers are talking of more consultation and negotiation with farmers to help the reform process along.[9]

The fourth example is that of the state of Karnataka, where a reform process is only in the very early stages, and where it is highly uncertain whether it will continue and where it will go. Since 1995 there has been internal process within government circles to formulate policy changes through amendments of the Irrigation Act. This effort was partly triggered by reform conditionalities attached to loans for the Upper Krishna Project by the World Bank, by knowledge about experiences in other states, a very limited number of pilot experiments with WUAs, disappointing results of the Command Area Development Programme (with 200 WUAs existing on paper), by budgetary constraints, and by a few individuals who were motivated to lobby and work for policy change. With a change of government in late 1999 some more momentum emerged, partly inspired by irrigation reform and decentralisation efforts in Andhra Pradesh and Madhya Pradesh, but still the speed of the process is very low and the scope of changes proposed by the government very limited.

This situation led a group of researchers and some others to found an organisation now called Sahayog (and registered as an NGO). It has as its aim to find out whether farmers are motivated to claim reform/policy changes from the government, and if so, to facilitate the organisation of farmers/water users in a platform to discuss desired policy changes and make this known to government. The question to farmers whether they would be interested in such an initiative drew an overwhelmingly positive response.[10] In roughly a year, Federations of Water Users Associations were formed in eight major irrigation systems, and the formation of a state-level apex body is under discussion. The first direct discussion of representatives of the Federations with the Irrigation Minister on the proposed amendments to the Irrigation Act has taken place.[11] Apart from these activities, the group is organising seminars, workshops and writes in newspapers.

The underlying philosophy is that drastic reform initiatives are not to be expected from the government side in the near future, and that reforms therefore need to be demanded by those who will supposedly be the beneficiaries of it, i.e., farmers/water users. The Sahayog group is in the process of finding out, together with farmers, whether and how it is possible to lobby for irrigation reform, and what a farmers' agenda would look like.[12]

What can be learnt from this short discussion of four different reform processes (in very different stages of advancement) is that reform trajectories are very diverse and case specific. There are probably as many types as there are cases. This has implications for the notion of 'models for reform', discussed in the section below.

A second conclusion that may be drawn from this discussion is that, it is remarkable that in none of the irrigation reform processes farmers/water users have been a partner. This is despite the claim that the reform is about *Participatory* Irrigation Management. Participation is reserved for the implementation phase, and for the end-state that needs to be created. This is not only true for India, but for most other countries. One exception to this observation is the intensive public consultation process that was organised in South Africa to formulate a new Water Law. Again, the occurrence of interest group participation in formulation of the Law has to be explained by the specific conditions of the reform process: the highly politicised context of the post-apartheid societal reform process.[13]

It seems to me a likely proposition that reform processes that are imported into areas without the projected beneficiaries having been involved in their formulation, have much lower chances of achieving their stated objectives than those that did have this involvement. And it is not just the involvement *per se*, and not only the higher likelihood of realistic proposals for change, but also, as the South African case shows, the factor that in the consultation (and negotiation and decision-making) process, support for the reform is created, and interest groups rethink and redefine their own positions, enabling workable compromises.[14] For the alignment of interests referred to above, the nature of the process is significant. Indian reform processes have not been characterised by alignment, but rather by confrontation and prescription. Platforms where such alignment could take place have not emerged or been created.

IV
Models for Reform: A Critique

The history of canal irrigation reform is characterised by the succession of a number of 'reform models' that gained international popularity. It was already suggested above that the Philippine model travelled far. The Ford Foundation was instrumental in popularising the reform experience in India, and other places. In South America, the Chilean reform involving privatisation of water rights has been very influential. Many South American countries have modeled their new Water Laws on the Chilean model, sometimes by literally copying it. At present the role of irrigation reform model propagation is mainly fulfilled by the World Bank and INPIM, who have been advertising the Mexican approach to reform for some years (and funded many visits to Mexico by people from all over the world, just like the Ford Foundation did earlier to the Philippines).[15]

The influence of these 'models' on others/other situations seems to have no relation to the results of the approach in the situation where it was conceived and implemented. I mean to say that the failure of the Philippine model in the national systems (the canal irrigation systems) has not reduced its impact elsewhere. Also the serious problems associated with the Chilean approach to privatisation, now even acknowledged by its architects, has not reduced its influence in surrounding countries (nor the zeal of some neo-liberal development experts to advertise it). Qualifications of the success of the Mexican model are starting to come in through field-level research. As a model for others, the case has already done its work however.

I do not want to discuss the merits and demerits of the experiences that have gained model status in past or present, but reflect on the phenomenon that the debate on irrigation reform seems to run from model to model, without seriously looking at the veracity of the 'success' that is claimed. I would like to speculate about the reasons for this. Speculation is necessary as I know of no grounded research on this issue.

The main reason for the popularity and eagerness to think in terms of and search for models might lie in the fact that the notion of a 'model' is attractive to both irrigation professionals (mostly civil engineers) and to policy makers. In engineering the notion of the prototype, or standard design, is very important. Reform models can be understood as prototypes, standard designs or blueprints for a new institutional infrastructure. This way of looking at it strongly resonates with the dominantly technocratic attitude of irrigation engineers. It suggests that the social/institutional world can be engineered in the same controlled manner as the physical world assumedly can.[16] In this way the muddy waters of complex processes of institutional change are avoided, at least in the mind and on paper.

For the policy maker models hold the same attraction of social engineering, but here the fact that models simplify the world, and abstract from the nitty-gritty of specific situations may be of central importance. The act of taking reform approaches out of their context (by lifting them from the Philippines to India for example), is an act of abstraction. The conditions of possibility of the approach become invisible, and thereby the approach becomes a generalised model. The idea that there is a general solution fitting a diversity of local problems, is a very attractive thought to a planner or policy maker, as it makes the task of planning and policy making a lot easier than when locally specific solutions are needed in each and every case. Planners and policy makers prefer to work with standardised solutions.[17]

A third agent that reproduces the working with 'models' is the donor community. The logic of the development industry requires 'success stories' to be able to counter the long list of failed interventions, and to be able to promise success within reasonable time frames. There is obviously nothing wrong with trying to learn from experience elsewhere, and to look for things that might work (*cf.* the 'best practices' approach popular with the World Bank and other development agencies). However, the black-boxing of specific experiences into models seems to serve particular political and ideological causes more than it informs a learning process. I would even argue that it prevents a learning process to take place, exactly because it obscures the conditions of possibility of successes and failures.

The practical implications of the observations and speculation in this section are that we should stop believing in silver bullets, and that there is no way around the complexity of locally specific problems. The specificity of all reform situations calls for strategic analysis and strategic planning, that is, dealing with the political dimension of change and transformation, for which capacity and skills need to be created.

V
TRIP: Four Blind Spots in Canal Irrigation Reform Processes

In this section I summarise what in my view are the major weaknesses, in the sense of missing elements or blind spots in canal irrigation reform processes to date. There are four of them, together forming the acronym TRIP: technology, rights, integrated water resources management and politics. The focus is South Asia.

Technology

The first blind spot is that of technology. In summary, the technology issue has the following four dimensions (see Mollinga, 1998a for more detailed discussion):

(a) The rehabilitation of infrastructure that is part of most PIM/IMT programmes usually boils down, at least in South Asia, to bringing a system back to its original design state. The fact that many systems were designed decades (or even longer) ago for very different farming systems, agrarian economies and socio-political conditions, seems to bother very few people. Institutional reform also requires infrastructural reform, that is rethinking of the design features of the infrastructure, as every design creates demands and constraints for management (institutional decentralisation requires hydraulic decentralisation).

(b) This argument gets amplified considerably when irrigation systems are designed from an integrated water resources management perspective (see below). This creates a whole series of new technical challenges (like for example intermediate storage to increase management flexibility and to decentralise storage, linking of canal water supply with watershed protection and water conservation infrastructure, *et cetera*).
(c) As of now, methodologies for participatory technology development are hardly used in canal irrigation reform processes, that is, farmers/water users play hardly any role in conception and decision-making of/on infrastructural change.
(d) One of the few ways that irrigation reform processes may be made attractive to engineers, is by making clear that such processes provide professional challenges.

Rights

The issue of water rights (and related other rights, like land rights) is a very new issue in the canal irrigation discussion in South Asia (but not so in other parts of the world, like Mexico). The relevance of the rights issue for canal irrigation reform can be understood at the following three levels:

(a) At the level of the WUA and the individual farmers within it. Taking the Andhra Pradesh case as an example, it can be argued that the Act adopted in 1997 effectively consolidates the water rights of those within the command, and excludes all others from access. This reservation of water coming into an area for the irrigators may be questioned. When the systems were designed, the command areas were determined on topographical and soil quality grounds mainly, and the inhabitants of the areas concerned had very little say in it, if they were aware of it at all. Whatever position one may take on who has a right to the water coming into an area through a canal [(landholders, tenants, landless, men, women, farmers (irrigated and rainfed), domestic users, industries, *et cetera*)], and the benefits generated by that water, the point is that such issues are not part of the discussions on canal irrigation reform.

    Apart from the question about who constitutes the community of water users/right holders, there is the question regarding which rights these right holders hold: rights to water or also rights to participation in decision-making, and which obligations come with the rights?
(b) At the intra-system level. Within the community of irrigators, the issue of head-enders *versus* tail-enders has very much to do with the question how enforceable water rights for tail-enders can be created. Depending on the size of the system the head/tail problem occurs at any number of levels within the system (see Mollinga, 1998b for a case study). In India different systems

for rationing of scarce water in canal irrigation systems have been created in different regions (*warabandi* in the north, block system/*sjeh pali* in the west, and localisation in the south), with different degrees of effectiveness, constituting different types of water rights. Practice often diverges from these formal systems, and constitutes actual, negotiated water rights.

(c) At the inter-system or regional level. Along a river where different types of extraction take place at different locations, a head/tail issue also occurs, particularly with increasing water scarcity. To resolve such issues, rights to water need to be defined.

At all these levels the emphasis in India has been on bureaucratic allocation (legal and administrative decisions reserving X, Y and Z quantities of water for different (sub-) systems and sectors). Allocation does not straightforwardly translate into distribution, and allocation mechanisms have given very little protection to tail-enders. More generally, they are not very helpful when disputes arise -- when use patterns change, scarcity increases or for other reasons. These disputes are often about the space and time details of distribution, and the quality of the water involved. For resolution, the *process* of dispute management is also very important. None of these aspects are part of overall quantitative allocations. Workable rights, that is rights that are enforceable and able to deal with real situations, are largely absent in the canal irrigation sector.[18]

Integrated Water Resources Management (IWRM)

Integrated water resources management is the new buzzword in the international and many national debates on water resources management. The concept has a very fundamental logic. Water is part of a cyclical system, organised in basins as far as the on-the-ground part is concerned. Therefore its management should take hydro-geographical units as its starting point, and aim for sustainable resource use to maintain the cycle over time. At this level of abstraction nobody can object to the need for IWRM. At the more concrete level it means that the interaction of irrigation with other water use practices and flows should be taken into consideration. The linkages between irrigation, rainfed agriculture/watershed development, drinking water and sanitation, industrial use of water, and water use for ecological protection need to be problematised. The concept has been put forward very strongly in the World Water Vision exercise concluded in March 2000, but the incorporation in the irrigation sector remains a major challenge. For suggestions on what such an integrated approach to irrigation might look like see for example (Datye, Paranjape and Joy 2000; Paranjpye, 2000).

Politics

The fourth blind spot was already strongly suggested in the preceding sections. It was argued that politics, in the sense of the negotiation of interests/balances of power by different interest groups (and not just in the form of official, state and party politics) is at the heart of irrigation reform policies. The issue of rights discussed above is a highly political issue, as rights define the claims that people have on available (scarce) resources and their relationship with other right holders. Participatory technology development is a political issue because it proposes to change the decision-making structure on infrastructural design. Rethinking the relationship and interactions between different sub-sectors in water resources management is a political issue because it involves a realignment of vested professional, institutional, economic and other interests. And so forth. And at all stages these processes are political – they are conceived in political arenas, their implementation is a political process, and they have political implications. Canal irrigation reform is 'power in motion'.

The practical relevance of giving the political dimension centre-stage, lies in its implications for how reform processes are induced and conducted. This paper started with observing the almost universal exclusion of farmers/water users from the policy formulation process in Indian canal irrigation. In political terms this means that opportunities for alignment of interests regarding reform are missed, and that serious doubt is cast on how genuine the usage of the notion of participation is in the PIM discourse. It is contended here that more systematic analysis of the political dimensions of irrigation reform can and should inform strategic action to strengthen reform programmes.

VI
Conclusion

This paper has tried to broaden the agenda of canal irrigation reform discussion from 'technical' social engineering approaches to institutional change, to a perspective that more critically engages with the received wisdom of the present international debate. It has emphasised diversity in objectives, origins and trajectories of reform processes, and cautioned against the application of 'models'. It has identified four blind spots in present reform processes: technology, rights, IWRM and politics. The paper's major weakness is that it hasn't critically engaged with the financial and economic dimension of the reform debate. That will have to be done in another paper.

NOTES

1. INPIM = International Network for Participatory Irrigation Management; IndiaNPIM = Indian Network on Participatory Irrigation Management.
2. To avoid misunderstanding right at the start, I am not arguing that democratisation or decentralisation of decision-making (as the concrete form of reshaped balances of

power) are, or should be, the main *objective* of PIM exercises, but just that it's political significance is this. On the objectives of PIM, see the discussion below.

3. But by no means the only one, the other major one being the economic dimension of PIM exercises (what does it do to livelihoods, productivity/viability of farming, resource mobilisation by farmer organisations, financial incentive structures, and budgets and incomes of agencies/agency staff). It is worth noting that the economic and the political are different dimensions of one single phenomenon or process, and not two separate 'things'. This is, for example, evident in arguments where financial mechanisms are considered to constitute powerful accountability mechanisms (see below).
4. The focus is on large-scale canal irrigation.
5. For a longer version of the argument presented in this section, see Mollinga (2000)
6. The difference between governance and management can be understood (and refined) by Ostrom's three levels of rules for resource management (operational rules, collective choice rules and constitutional rules). See Ostrom (1990). Another useful division is that between governance and management practices (at different levels) on one hand, and the conditions of possibility of those management practices (in terms of legal structures, socio-economic and socio-political relations, bio-physical and technological conditions, and the like) on the other. See Mollinga (1998b).
7. Based on a draft paper: Peter P. Mollinga *Where does reform come from? Some notes on the origins of irrigation reform processes, with a focus on India,* presented at the Clinic on Triggering and Implementing Management Reform: The Case of the Irrigation Sector, June 5th, 2000, World Bank, Washington, D.C.
8. About each of the states more detailed and complex stories can be told than is done in the table and the text following it. The summary account necessarily simplifies. The aim is not to provide full-blown analyses of the reform processes in the states concerned, but to identify diversity in origin and some other significant features.
9. Ford Foundation support for the present Andhra Pradesh reform exercise has been limited. It did support NGO activities in the 1980s and 1990s similar to those in Maharashtra and Gujarat, and is presently supporting a process documentation research project of an NGO, which looks at policy implementation at field level.
10. The group started by bringing together the presidents/representatives of the 200 WUAs that existed on paper, created through the WUA component of the CAD programme.
11. Activities are now also undertaken by Sahayog in the tank irrigation domain, where it has been possible, after considerable discussion, to include one farmer in a committee that will discuss the technical design features of tank rehabilitation programme planned by the Government of Karnataka (with World Bank assistance).
12. To complete the picture on Ford Foundation involvement, it has very recently decided not to support the Sahayog initiative because it is shifting its priorities from 'single resource' projects to more comprehensive 'natural resources management' projects. This can be interpreted as a sign of the times -- decreasing interest of

international agencies/donors in the irrigation sector, possibly related to fatigue by the slow pace of change in the sector (see discussion above).

13. On the consultation process that preceded the adoption of the new South African water law (and irrigation policy), see Marna de Lange 'Water policy and law review process in South Africa with a focus on the agricultural sector' in Alex Bolding and Peter P. Mollinga (eds.) *The Politics of Irrigation Reform* (forthcoming).
14. Obviously positive outcomes cannot be guaranteed; every participatory process can fail to produce constructive outcomes. When such situations occur, governments will have to consider what kind of mandate (and capacity) they have to force for decision and implementation of policy elements that are only partly shared by those affected by it. In South Africa the political agenda of the majority government and the Constitution was the final arbiter in such cases, as, despite the desire to find win-win situations, these do not always exist, and maybe they rarely exist.
15. The forthcoming book mentioned in footnote 14 contains papers on the Philippines and Mexico models for irrigation reform.
16. That the latter is also questionable is an important point, but outside the scope of this paper.
17. For further implications of these acts of standardisation for the exertion of state power see James C. Scott's work on 'state simplifications'. Another instance of standardisation/simplification is the propensity of governments to produce 'standard bye-laws' for WUAs, often prescribing the envisaged functioning of the WUAs to an amazing level of detail.
18. Outside the scope of this paper is a discussion of why this is so. The fact that governments are the owners of the canal systems, and hold the right to the water flowing through them, is obviously one important factor. However, this does not explain why governments should design rules and procedures that are impractical, to say the least, for the implementation of official allocations.

REFERENCES

Datye, K.R., Suhas Paranjape and K.J. Joy (2000), 'Regenerative agriculture and rural development in India: A new paradigm'. In Peter P. Mollinga (ed.), *Water for Food and Rural Development: Approaches and Initiatives in South Asia.* Sage, New Delhi, pp. 278-95.

Mollinga, Peter P. (1998a), *The Policy Process in Irrigation Reform: Technology, Rural Development and Politics.* Paper presented at the fourth international INPIM seminar in Bali, Indonesia, 14-20 July 1998.

Mollinga, Peter P. (1998b), *On the waterfront: Water distribution, technology and agrarian change in a South Indian canal irrigation system.* Ph.D. thesis. Wageningen Agricultural University.

Mollinga, Peter P. (2000), 'The inevitability of reform: Towards alternative approaches for canal irrigation development in India'. In L.K. Joshi and Rakesh Hooja (eds.),

Participatory Irrigation Management: Paradigm for the 21st Century. Rawat, Jaipur and New Delhi, pp. 67-97.

Ostrom, Elinor (1990), *Governing the Commons: The Evolution of Institutions for Collective Action*. Cambridge University Press, New York.

Paranjpye, Vijay (2000), 'Urban and rural watershed development: examples from India'. In Peter P. Mollinga (ed.), *Water for Food and Rural Development: Approaches and Initiatives in South Asia.* Sage, New Delhi, pp. 296-313.

13

Comparing Gujarat with Andhra Pradesh: Reforms in Irrigation Management and People's Participation

R. PARTHASARATHY

The role of irrigation in the agricultural development of India is well-recognised. During the British rule, India had witnessed an expansion of irrigated area and the successive five-year plans in the post-independence era have not only sought to expand the irrigated area but also to improve the quality of the irrigation systems (Randhawa, 1980). On the quantitative aspect, in the last three-and-half decades, irrigated area has increased by nearly 30 million hectares which is more than the total area under irrigation at the time of independence. The significant aspects of qualitative improvement are the constant endeavour to increase irrigation efficiency, introducing better methods of water control, crop pattern on irrigated lands and concerns on the participatory process in irrigation management. However, the persistent underutilisation of irrigation potential and the continued slow growth of yields despite a rapid expansion in the irrigation facilities suggest that the effectiveness of irrigation is lower than its potential.

Almost all Indian states are considering alternative approaches to improve the effectiveness of irrigation both in the context of modernisation and improvement of physical structures and on the managerial aspects including the organisational structure, people's participation, pricing, etc. The importance of institutions in all phases of water control, namely, construction, maintenance and operation is obvious, yet, there are relatively fewer studies on this subject and even these fail to relate their structure and working to the larger context of development initiatives (Vaidyanathan, 1985).

It is against this background that this paper provides a comparative picture of the reform initiatives in irrigation management being taken up by the government of Gujarat since 1994, and the 1997 policy of promoting Water

Users' Associations (WUAs) in Andhra Pradesh. While there are similarities in the two models on the expected outcomes, there are strong differences in the approach. Section II outlines the imperatives of the reforms and highlights the approaches and the process of implementation. One of the important features of the programmes is the creation of new institutions of the users in both the states. Section III presents a comparative note on structure of the users' institutions in Andhra and Gujarat. Evidently, the crucial issues in putting up new institutions are of viability and sustainability. Section IV analyses these issues. The conclusions and implications of the paper are highlighted in the last section.

II
Imperatives for Reforms in Irrigation Management

Irrigation systems are traditionally managed by the government departments. Most systems whether large or small, have serious problems relating to distribution of water and its use efficiency. Irrigation department, which is responsible for building, operating and maintaining the irrigation structures, has been found to be inadequate to keep the systems in proper shape. Studies suggest that some of systems that were managed traditionally by the water users have performed well both with respect to efficiency and distribution (see for instances, Sengupta, 1991; Datye and Patil, 1987; Maloney and Raju, 1994; Wade, 1987; Pant and Patil, 1983). There have been definite institutional arrangements in which the rewards and sanctions were clearly defined for the systems to work. It was only with the large surface irrigation systems built on the rivers during the British rule that the irrigation and revenue departments assumed total responsibility of building and managing the systems.

Presently, most of the state governments in India are encouraging formation of water users' associations. Some NGOs have also taken up initiatives in this area and have conducted successful experiments.[1] For a long time now, it has been assumed that the farmers in large systems would take up responsibility of construction of field channels and management below the outlet level (Stone, 1984; Chambers, 1988). However, the experience across different systems and across states in India disputes this assumption. Further, almost all governments experience shortage of funds to carry out the necessary repair and maintenance works of the systems (Brewer *et al.*, 1997).

The alternative to demands on government funds is recovering the costs through user charges. So far, however, the government has neither been able to ensure timely payments from the farmers nor it was able to charge an economic price for the water. Also, the necessity of dealing with individual farmers has made water charge collection a costly exercise. One solution that has been tried in various places has been to share the responsibility of the Operation and maintenance (O&M) of the canal system below a certain point with the local farmers. The July 1995 Government of Gujarat Resolution on Participatory

Table 1

Summary Features of the IMT Programmes: Gujarat and Andhra Pradesh

Components	*Gujarat*	*Andhra Pradesh*
Name of policy/law	Government of Gujarat Resolution on Participatory Irrigation Management (PIM)	Andhra Pradesh Farmers Management of Irrigation Systems Act (APFMIS)
Starting date	July 1995	1997
Status	Pilot phase in 13 selected pilot schemes.	State-wide coverage from the start onwards, excluding tanks under the Panchayats and tribal areas
	Over 3,500 WUAs have been formed so far under the plan, to cover 7.5 million hectares (50% of the state's irrigated area) by 2003.	
Involvement of NGOs	Two NGOs played a role in policy making and in 6 pilot schemes. NGOs are also involved in implementation.	One NGO played a prominent role in policy formulation. Under the new Act no clear role for NGOs in implementation.
Number of WUAs formed	Over 3,500 WUAs presently (An estimated 45 WUAs have been formed prior to the PIM).	10,292 WUAs have been formed in all the 22 districts of Andhra Pradesh covering major, medium and minor irrigation projects in the state
Federating WUAs	Higher-tier federation is not possible in the 13 pilot schemes, as long as the tertiary schemes along the same main system are not handed-over to users	Three-tier management structure
		At the primary level is the WUA; at the secondary level, a distributory committee (DC); and at the apex level, there is a proposed Project Committee (PC).

Irrigation Management (PIM)[2] and the Andhra Pradesh Farmers' Management of Irrigation Systems Act, 1997 (Act) are a step in this direction (for a summary of features, see Table 1).

While the PIM in Gujarat envisages a complete turnover of O&M to a *Piyat Mandali* (Water Users' Association), the Andhra Act of 1997 empowers the farmers to manage the irrigation systems themselves. In both the cases, the canals remain government property and major repairs continue to be the responsibility of the irrigation department, however, the day-to-day functioning of the system passes on to the users' associations.

The first goal of the PIM programme in Gujarat is to 'turn over' 50 percent of the command area of its extant irrigation schemes to WUAs (called *Piyat Mandalis*) by the year 2003. As a precursor, thirteen pilot projects were chosen and the implementation of the policy began in 1994. Expectedly, the experience of these pilot projects and the lessons learned from them are to provide inputs for future expansion to cover the existing command area of 1.5 million hectares, and another 1.8 million hectares of land to be covered under the impending Narmada command.

III
Approaches Towards PIM

In an era when the development initiatives from below are being advocated, the Andhra Pradesh programme has clearly followed a top-down approach. The state has legislated an Act that has made the formation of WUAs compulsory to run the irrigation systems. However, there have been extensive consultations at the user level, and in a year's time the Andhra government has mustered political will, bureaucratic commitment and peoples' confidence. For instance, the draft Act was widely circulated among the farmers in the entire state and feedback was obtained. Consultants at the regional level had the responsibility to carry out serious consultation work and provide feedback on the ground realities. Although the approach has been top-down, a realistic ground level assessment seems to have been carried out.

Under the programme, 12 major, 74 medium and 12,240 minor irrigation works are to be covered. So far, the Act does not cover the tanks under the panchayats and tribal areas. All farmers in the command area were compulsorily made members. Other users, including the hitherto illegal users are members without voting rights.

The PIM programme in Gujarat has been in operation for over five years now. In some of the PIM projects, non-governmental organisations (NGOs) are involved, while in others the government departments take care of social and technical aspects. The notable feature of PIM scheme is its emphasis on the social engineering aspect. The inspiration for the programme, specially the emphasis on social engineering came from the Philippines experiment (for details, see

Parthasarathy and Iyengar, 1998). WUAS are perhaps the most noteworthy feature of social engineering and they are formed as a result of the efforts by the irrigation department or the NGOs. These agencies besides motivating the farmers also aim to build their capacity to manage the WUAs.

Evidently, formation and functioning of any cooperative set-up is influenced by various parameters. In the case of water users, farmers' participation in Gujarat appears to be a function of the regional differences, presence of NGOs, social characteristics of the participants, cultivation practices, and their past experiences in water supplies by the irrigation department. These parameters besides being important in determining the time taken to form a WUA were also found to crucially affect the effectiveness of the implementation process. Unlike the case of Philippines, concerns for identifying such variations in detail while deciding on the steps to achieve certain degree of people's participation appeared to be weak, except in some of the pilot schemes (see, Parthasarathy, 2000).

As a result, in the PIM programme, there seems to be a set procedure of process. In the village meeting (in some cases only with the village leaders), forming of a WUA is emphasised. *Inter alia*, promises to repair the canals and redressing of long pending complaints follow. Once the token amount of Rs. 11 or Rs. 51 is collected from a majority of the farmers whose lands fall under the command area as share capital, the process of election of office-bearers began. Soon thereafter, application for registration of the WUA was forwarded to the concerned government department. Around this time a joint survey of the command area was conducted to identify the extent of repair work to be carried out and an estimate of costs was finalised. The participant farmers in the PIM programme are expected to contribute 10 percent of the estimated cost of repair and rehabilitation works. This component of farmers' contribution is expected to inculcate a sense of ownership among them.

In some cases it has been observed that farmers are not even aware of the need to contribute the share capital. Further, there are variations across the areas in the implementation of the policy to collect the ten percent from farmers as their contribution. While in some cases such variations are a result of a conscious decision on the part of the implementing agencies, in others there appears to be a gap in the understanding of the PIM scheme by the implementers. However, these variations also point out to the flexibility in the programme to accommodate variations. By and large, this flexibility is a result of the approach, i.e., building up the programme from below. In contrast, the Andhra model is distinct in its approach.

The Process of Implementation in Andhra Pradesh

If the Philippines model formed the basis of the PIM in Gujarat, it is IRDAS' work on a pilot basis in the Sriramsagar (SRS) project that led to a large-scale adoption of the WUA concept in Andhra Pradesh. IRDAS is a Hyderabad-based NGO which took up the pioneering initiative in participatory irrigation management in the state. IRDAS and SONAR are the two NGOs that took part in the two-year pilot phase of the SRS project in 1995. Each of these two NGOs worked on 20,000 acres of command area. Initially, the NGOs organised the farmers for one season and then for the other two seasons. Thus, the interventions have been gradual. During the pilot phase, while IRDAS was concentrating in D51, SONAR worked on D64 systems in the SRS project with a culturable command area (CCA) of 7,777 and 13,785 acres respectively.

After about 11 years of commissioning of the SRS project, in 1985 the agriculture department had introduced 'Warabandhi' system. Chak level plans and the cadastral plans were drawn and warabandhi arrangements made. The experiment failed because farmers were not involved and there was no coordination between the engineering and the agriculture department. In 1991, IRDAS introduced Pipeline Committee experiment in selected areas. Some of the outlets were selected. Farmers having lands in the command area of these outlets were organised into a committee and it was envisaged that they will supervise the distribution, operations and ensure equity. Farmers were also to attend to small faults and repair works (for details on the pipe cooperatives, see Singh *et al.*, 1994).

The capacity utilisation in the outlets selected for the pipeline committee project went up to 80 percent without any substantial investment. The success was mainly because of the active participation by the farmers and served as an eye opener for the irrigation department. The World Bank, which had been supporting other irrigation projects, observed the working of the participatory irrigation management programme and offered further support. At IRDAS, the thinking was that these experiments should be taken up at a large scale and the government too believed in a 'big bang' approach so that a comprehensive impact is made.

Interactions between IRDAS and officials from irrigation and other departments of the government of Andhra Pradesh led to a realisation that if PIM is to become a reality, a new legislation has to be brought out. The idea took shape in 1996. A study of the existing laws and its provisions was undertaken. A new draft legislation was prepared and was widely circulated among farmers for discussion. Seven consultants were appointed for different regions to guide the consultation process. The first reaction of the farmers was that whatever the department was unable to do, it wanted to pass on to them. However, when the total implications of the programme became clear, most farmers realised the scope for their own empowerment. It is reported that the consultation took about a year and the legislation was passed in April 1997.[3]

Typically the consultations began in the second half of 1997 when Chief Engineers or the engineers-in-charge of the projects visited some of the villages and conducted meetings. They introduced the government's new idea of empowering farmers to manage the water systems themselves by providing financial support for repairs and rehabilitation of the structures. Visits by higher level officials followed by those of the assistant engineers. Subsequently, the farmers were convinced that government was serious in its intention to supply water to their fields and hence they formed the WUAs in their respective territories and undertook the rehabilitation of the system.

So far, a total of 10,292 WUAs have been formed in all the 22 districts of Andhra Pradesh covering major, medium and minor irrigation projects in the state. Forming associations and completing the process of electing the presidents of WUAs in July 1997 are indeed significant achievements for any government. The test, however, is the sustenance of these institutions and ensuring adequate supply of water for the WUAs. Apparently, the farmers seem to assume adequate supply of water at their outlet. Although the Act places the responsibility of evolving ways and means to make economic use of water, in practice the farmers' primary concern is supplying water by turn to all parts of the command.

The Implementation Process in Gujarat

In Gujarat, however, no such systematic effort in promoting the concept of PIM has been evident. A High Level Committee was formed and one of the sub-committees it appointed was tasked with the selection of the pilot areas. The sub-committee selected thirteen project areas of which at least three had NGOs already working on them. After selecting the pilot projects, senior officials visited the area and explained the new policy. Like Andhra farmers, in Gujarat too their initial reaction was one of suspecting the government's motives for a sudden change in the functioning.

This is where the motivation work assumed importance. In some places, motivation work shaded into coaxing (for an analysis, see Parthasarathy, 1998). Barring some WUAs, imparting of skills to the office-bearers on book-keeping and on the technical aspects of repair and rehabilitation work were superficial. Perhaps, this is one of the reasons why in later stages one observes that a few persons control the WUA management and by and large people's participation in a popular sense is absent.

WUAs Management

Under the PIM programme in Gujarat, the command area farmers elect committee members and they in turn elect the office-bearers consisting of a chairman, a secretary and a treasurer. Though constituent villages are represented, the elected members invariably belong to the dominant caste or group. This phenomenon is especially true of the chairmen of WUAs, who almost always belong to the dominant caste. Further, in a committee set-up, the members were expected to jointly shoulder the responsibilities. Some of the implementing agencies admit that, "after so many years of work with the community an age old social reality has not been comprehensively dealt with". Essentially it is a failure to "evolve process where common grounds can be identified, proper incentives created so that divergent stakeholders (are able to) identify common grounds and find it worth confronting the divisive forces" (EMS, 1998). Similar is the case of women representation in the committee. Though the by-laws of the WUAs state that one-third of the members should be women, in practice however, there are WUAs where not even one woman member is in the committee.

Though the executive committee members take active part in the meetings, almost all the work related to the WUA's management are being carried out by the president. There are evidence to suggest that though the (ordinary) members of the WUA have a good understanding of the PIM programme, there is an absence of training on the importance of the cooperative behaviour and on the various functions of WUAs. Evidently, the inputs of the implementing agencies are a continuous process and that there is a difference between educating the participants about the programme, motivating them to contribute their share of the estimated costs and their actual participation. It is in this context the direct election procedure for the president of the WUA in the AP model is interesting.

According to the 1997 WUA Act, the president has to be elected from among the general body consisting of members with voting rights. The command area under a WUA is divided into territories depending upon the hydraulic gradients. The farmers in a defined territory form the Territorial Constituency (TC). A member representing each territory is elected by the territory farmers. The WUA president along with the TC members run the WUA's administration. There is no special provision for including women members either in the TC or in the WUA committee. However, women, by virtue of being landowners can become members of the WUA and can also contest elections.[4] The WUA office-bearers are to hold office for three years. However, if a majority of the WUA members are not satisfied with the performance of their president, there is a provision in the Act to recall him/her. This is a unique and an important provision (Table 2).

In large irrigation projects, a three-tier management structure has been planned. At the primary level, there is a WUA and at the secondary level, a Distributory Committee (DC) is formed. All WUA presidents are members of the DC and amongst them a president is elected. At the apex level, there is to be a Project Committee (PC) that has plans to include DC presidents as well. The PC, when formed, will be responsible for attending to the problems relating to repairs and rehabilitation (R&R) and operation of the main delivery systems, while the DCs are responsible for the O&M of the distributory level structures. And, at the WUAs level, the farmers are responsible to manage the field level

Table 2
Summary Organisational Design of WUAs in Gujarat and Andhra Pradesh

Components	*Gujarat*	*Andhra Pradesh*
Membership criteria in WUA	Voluntary	By law (APFMIS Act 1997)
	Land owners in the WUA area.	Land owners and tenants in the WUA area. Other water users to be co-opted in to the WUA
Members' voting rights	Land owners in the WUA. One person one vote	Land owners in the WUA. One person has one vote for the WUA President and one vote to elect the representative of his/her territory who will serve as Territorial Constituency member on the WUA committee.
Procedure for electing WUA management committee	Varies between WUA, but members of the WUA usually elect the president, secretary, and treasurer. Territorial representation is common but optional.	The District Collector is in charge of the election of President and TC members.
		The elected Managing Committee will holds office for five years
		Provision exists to recall the President or a Chairman or a TC member of the WUA.
		The first elections were held in 1997. New WUAs received a grant of Rs. 50,000 if the president was selected by consensus and Rs. 30,000 if there was a contest.
Procedure for electing higher-tier bodies	Not Applicable	The WUA presidents are members of the DC and amongst them a president is elected. At the apex level, there is supposed to be a Project Committee (PC) which has DC presidents as members.
Quota for women in committees	Rules of the WUA state that one-third of the managing committee members will be women, but this rule is very rarely implemented	No explicit provision. 94 of the new WUAs (one percent of total) have a women president.

distribution. In Gujarat, however, so far federating the WUAs has not been attempted.

Membership and Jurisdiction of the Users' Association

The real stakeholders in the WUAs are the land holders. A positive aspect in the Andhra Pradesh's Act is that the tenants are treated as members. The Act states that where the owner and the tenant are land holders in respect of the same land, the tenant is to be recognised as a member. The other water users outside the area delineated by the District Collector may be co-opted in a water users' area provided he is a cultivator (owner or tenant). However, the right to vote is vested only with the land holders falling in the delineated water users' area. Other households do not have any stake in the WUA.[5]

Similar to the Andhra Pradesh model, the stakeholders in the Gujarat WUAs typically live in different villages and the command area also covers lands in more than one village. Unlike Andhra, in the case of Gujarat only the landowners in the command area are entitled to become members of the WUA, yet, different types of land lease practices are prevalent. Under these circumstances, almost all the WUAs find it difficult to manage a large area under its jurisdiction. For example, in one of the PIM project area, the command area falls in five villages and the farmers reside in 11 villages. The problem is further compounded when farmers lease out their land and the WUA has to locate the tenants to collect water charges. In this scenario, perhaps the Andhra model of making tenants eligible to become members of the WUAs has merits. To circumvent this problem, most of the members across WUAs are of the opinion that the basis of formation should be a village and not the minor. This in fact questions the viability of these new institutions.

The questions of viability and sustainability of this class of interventions vitally hinge on the financial provisions made, political commitment, training and skill formation modules and on the monitoring and feedback mechanisms besides the jurisdiction of the users' associations. Both the programmes have noticeable features and also elements that cause concern.

Financial Provisions

One of the most important elements which has improved the credibility of the Andhra programme among the farmers is the financial resources earmarked for repairs and maintenance of the distribution systems. Whether it is an irrigation tank or an outlet of a medium or major irrigation project, funds for repairs have been disbursed to all the WUAs. Earlier, Rs. 40 per acre was allocated for O&M, out of which Rs. 35 were spent on salaries and only five rupees per acre were left for repairs and maintenance of the structures. Under the new scheme, Rs. 5 has been raised to Rs. 100 per acre. A total of Rs. 106 crores has been allocated during

1997-98 for repairs and rehabilitation to be eventually disbursed to WUAs and the federating bodies like the DCs.

While the WUAs in the major irrigation projects are allocated Rs. 100 per acre for operation and maintenance, in the case of tank irrigation projects Rs. 90 has been fixed (this is in addition to Rs. 50,000 for those WUAs which have an uncontested president and Rs. 30,000 for the WUAs which had a contest). This lumpsum grant is shared in the following way: 50 percent to the WUA, 20 percent to DC, and 20 percent for the PC. The remaining 10 percent is for the village panchayats to undertake development programmes. Further, the funds are released in three installments of 40, 40 and 20 percent[6]. In the medium and major projects, the grants are meant for repairs of the distribution channels above the minors. Hence, out of Rs. 100 per acre, a proportional allocation has to be made for the repairs of distributories and main canal network.

In Gujarat, there are two types of grants for the WUA to cover the O&M expenditure, yet none of them, unlike the Andhra model, attempts to integrate the existing village level organisations like the panchayats. First is related to the performance of the WUA. In that, the WUA can retain 30 percent of the water charges collected for O&M expenditure and another 20 percent for administrative expenditure. The second grant consists of a contribution of Rs. 250 per hectare from the state government, Rs. 200 from the Central government and Rs. 50 from the members of the WUA. In so far as the first type of grant is based on a proportion of the total water charges collected, there is an incentive for the WUA to raise the water charges. So far, only one of the 13 WUAs has attempted to raise water charges. Even in this case, initially, the water charges were fixed at four times the government rate. Eventually, due to resistance from its members the difference was narrowed down to one-and-half times the government rate (Parthasarathy, 1998a).

It is indeed difficult to say which of the two models will ensure financial sustainabilty of the WUAs. However, Gujarat's experience points out that the financial viability of the WUAs is crucial to its credibility, sustenance and performance.

Politics and WUAs

Unlike Gujarat, where politics through a provision in the by-law of the WUA is consciously kept out of the purview, the post of president of the WUA in Andhra has already become a coveted one. It was learnt that the contestants had spent money up to Rs. 50,000 in the elections. It is possible that such posts enhance their visibility and help further their political ambition. Interestingly, in many places farmers reported that party politics did not play any role. Use of money in the election may also be due to the financial stakes involved in managing the WUAs. The funds handled by a WUA annually are likely to be substantial and hence a vested interest may develop in acquiring control over them. Probably, the contractors appointed by the department could be replaced by the proxies of the WUA presidents and influential members.

Interestingly however, it was reported in many places that those persons who had been voluntarily helping the village level functionaries -- *Laskars*[7] and *Neeradis*[8] -- and had a thorough knowledge of the system and its problems, were elected as the president of the WUA.

Indeed it is important to note that the WUAs are by design different from the panchayati raj institutions and cooperatives. Thus, one finds no mention of political affiliations coming in the way of holding office in the WUAs. In fact, some of the higher officials in the irrigation department stress the important role of politics in a democracy and the useful role it plays in shaping political opinion towards this programme. Perhaps, it is this aspect that may ensure its continuity, whichever political party comes to power later.

Programmes Toward Training and Skill Formation

In both the states, training of farmers and engineers has been identified as the key factor for putting the new system in place. There are differences in approach. In Andhra Pradesh, a comprehensive training plan was chalked out to train engineers and farmers in the new system. The government has also spent a few crores on training and production of material. To popularise the programme, poster campaigns and slide shows were also organised. In Gujarat, however, popularising the PIM programme is still a low-key affair. The task of preparing posters, visuals and handouts has been entrusted to an NGO and the distribution of these is limited to select pilot project areas. Perhaps it is one reason why there is perceptible difference in the awareness of the programme and its salient features among farmers in the two states. Evidently, the awareness among Andhra farmers seems to be higher than among Gujarat farmers.

As far as training is concerned, in Gujarat the emphasis seems to be more on the exposure visits of the farmers to projects where PIM is working well. Tailor-made training programmes for farmers are few and confined to some of the project areas. In Andhra however, after the formation of the WUAs, the presidents were trained by the officials of the irrigation department during October-December 1997. The training was intensive and the subjects covered were water distribution and management, water budgeting, accounts and social audit. What is important is that there is concurrent documentation of the happenings in the field in Gujarat whereby the missing elements in training programmes or the felt needs of the farmers are periodically highlighted. It is this component that seems to be the weakest in Andhra Pradesh programme.

Table 3
Some Characteristics of Study WUAs

States/details	*Gujarat*				*Andhra Pradesh*		
WUA	Thalota	Laxmipura	Tranol	Digas	Ellabotharam	Peddapala-Kaluru	Jantaluru
Canal scheme	Dharoi	Dantiwada	Mahi Kadana	Ukai – Kakrapar	Sriramsagar	Nagarjuna-sagar	Cuddapah-Karnool and Thungabhadra
Region	Mahesana	Patan	Anand	Bharuch	Telangana	Coastal	Rayalseema
Water supply reservoir	Scarce	Scarce	Perennial	Perennial	Scarce	Neither scarce nor perennial	Scarce
Main irrigation season	Rabi	Rabi	Rabi	Rabi	Kharif	Kharif	Kharif
Two main irrigated crops	Wheat Mustard	Mustard Wheat	Tobacco Wheat	Sugarcane Wheat	Paddy Maize	Cotton Chilies	Groundnut Paddy
Command area (ha)	224	246	356	921	464	2600	1369
No. of villages/territories	1	2	1	6	4	12	12
Number of WUA members	210	174	168	169	500	2325	1200
States/ details	*Gujarat*				*Andhra Pradesh*		
Implementing Agency	*NGO*	*NGO*	*Irr. Dept.*	*Irr. Dept.*	*Irr. Dept.*	*Irr.Dept.*	*Irr. Dept.*
Particularities	Almost all households have access to tubewell water. In all, 5 tubewell 'companies' operate in the village. People have contributed to R&R work	Only one watering in 1998-1999; none in 1999-2000. People's contribution in R&R work.	Seepage of kutchha water courses. Logging, and inaccessibility of tail ends. People's contribution to R&R work.	Water excess and logging; lack of drainage.	Female President. No water supplied in 98-99 due to repair work. R&R work done under different programmes like Jhanmabhoomi. Voluntary labour for R&R work reported.	Large no. of villages and big command area. Mini dams in up-stream of canal for power generation affects water flow. WUA helped in improving water supply.	Lack of water access in tail ends. Illegal appropriation of water by jail authorities in the up stream of canal. Large no. of villages. Traditional water distribution practices observed.

Table 4
Distribution of Number of Sample Households by Farm Size

Category	*Gujarat*					*Andhra Pradesh*			
	Thalota	*Laxmipura*	*Tranol*	*Digas*	*Total*	*Ellabotharam*	*Peddapalakalur*	*Jantaluru*	*Total*
< 0.5 ha	30	28	30	2	90	30	28	31	89
0.51-1 ha	30	30	30	7	97	30	31	29	90
1.01-2.5ha	15	15	15	15	60	15	15	15	45
2.51 + ha	15	15	15	15	60	16	14	15	45
Tenants	23	3	8	-	34	30	30	30	90
Male-headed	107	86	94	39	326	113	108	105	326
Female-headed	6	5	4	-	15	8	10	15	33
Total	113	91	98	39	341	121	118	120	359

IV
People's Participation and WUAs: A Comparative Analysis

In a recent IWMI-GIDR study, seven Water Users' Associations (WUAs) were selected that represent different agro-ecological regions in both states (the characteristics of the WUAs are given in Table 3). Within each region, the selection was random (Parthasarathy *et al.*, 2000). In the case of Gujarat, the study built upon an earlier work carried out by the Gujarat Institute of Development Research in which the WUAs were randomly selected as well (Parthasarathy, 1994; Parthasarathy and Iyengar, 1998). Some characteristics of the sample households are given in Table 4 (for details see, Parthasarathy *et al.*, 2000).

The study revealed that 22 percent of the sample households in Gujarat are not aware of the WUA (Table 5). Awareness is highest in Thalota. The rather high incidence of awareness in Thalota may be partly due to Irrigation Management Transfer (IMT) implementation by an NGO. It is reported (Parthasarathy and Iyengar, 1998) that this NGO had concentrated considerable efforts in this WUA but the same NGO was less successful in Laxmipura, which could explain the incidence of unawareness of about 27 percent. This suggests that implementation of IMT by an NGO is not a sufficient condition for well informing stakeholders.

Table 5

Percentage Distribution of Households Reporting Unawareness About the WUA

State	*Gujarat*					*Andhra Pradesh*			
WUA	*Thalota*	*Laxmipura*	*Tranol*	*Digas*	*Average*	*Ellabotharam*	*Peddapalakalur*	*Jantaluru*	*Average*
% of HH unaware	7	27	33	23	22	40	48	68	52

In Andhra Pradesh 52 percent of the households were not aware of the WUA. Unawareness about the WUA was more prominent among the two categories of farmers with less than one hectare of operational holding and tenants, than other categories of farmers. In Andhra Pradesh, a majority of them belonged to scheduled castes and scheduled tribes. The variation between the WUAs is significant in Andhra Pradesh as well. In Jantaluru, two-thirds of its members have reported to be unaware of the WUA and awareness about the WUA was found to be 40 percent among Ellabotharam farmers who are known as politically active and have a female president. This general high level of unawareness indicates a large gap between the paper declaration of the APFMIS Act in which all irrigators are members of WUA and the reality. However, generalising to the whole state, it would be quite an achievement to inform half of the state's irrigators of initiatives to reform water management, even though the prevailing hierarchies among farmers continue to be reflected in their respective levels of awareness.

Participation of Members in WUA Activities

Overall rates of participation in the various activities of the WUA in institution building, R&R, and water distribution and fee collection are low, and roughly as low in Gujarat as in Andhra Pradesh (Table 6). So, in spite of the higher rates of awareness in Gujarat, rates of active participation were generally even lower than in Andhra Pradesh. The only exception is the attendance of annual meeting, which is done by 22 percent of the sample households in Gujarat. The pattern of a relatively high level of participation among the relatively few who are aware in Andhra Pradesh is especially noted for the participation of scheduled castes in annual meetings. Although one-third of all farmers who are unaware of the WUA belong to scheduled castes, among the scheduled caste farmers who are aware, a relatively high proportion of 30 percent also have taken active part in annual meetings.

Table 6
Percentage of Households Participating in WUA Activities

Activity	*Gujarat* *N = 341 (100%)*	*Andhra Pradesh* *N = 359 (100%)*
Institution building		
Motivation	4	9
Election of committee members	N.A.	7
Attendance in annual meetings	22	13
Infrastructure R&R		
Joint survey	12	N.A.
Supervision R&R	6	6
Labour input R&R	7	17
Water distribution and fee collection		
Demand forms/collection water charges and contributions	3	4
Supervision water distribution	1	4

There is a clear bias according to farm size in the attendance of annual meetings. In Gujarat the proportion of farmers attending the annual meetings were found to steadily increases with size of farm, however, no tenant reported to attend meetings. A similar pattern was observed in Andhra Pradesh, with the exception that tenants were found to participate as well and as frequently as the smallest farmers. Labour provision for R&R was, after attending annual meetings, the second most often cited activity in which the respondents participate in Andhra Pradesh. A strong involvement in labour provision is found for women in Andhra Pradesh. Out of the 54 female-headed households and households with women landowners, 25 percent have contributed voluntary labour. However, hardly anyone of these women had participated in the activities of the WUA. On the other hand, women's contribution to R&R work as labourers was significant (for details, see Parthasarathy, *et al.*, 2000).

To conclude, the PIM and APFMIS programmes find a limited response among farmers at local level. In Gujarat, years of intensive intervention by an NGO to a water-scarce WUA, as in Thalota, had impact. The APFMIS approach for multi-tiered WUAs with strong public support by the highest government officials and state-wide elections or appointments of WUA committees has left half of the irrigators still unaware of their new rights and obligations as members of Water Users' Associations. However, the farmers who are aware of the APFMIS programme come to a comparable degree of involvement within three years. Especially in Andhra Pradesh, labour provision by poor farmers and tenants, even though the latter are no official members, is somewhat stronger than that of their fellow-farmers. This is not surprising given the hierarchical agrarian structure, but may still reflect their interest in the R&R component of the new programmes and improvement in the canal network, as canal water is cheaper than water from other sources of irrigation. Neither of the programmes has been able to avoid reinforcement of a higher unawareness and lower participation in annual meetings among poorer farmers than among better-off farmers in both Gujarat and Andhra Pradesh. The lesser interest in WUA activities like the annual meetings among the largest farmers in Andhra Pradesh needs further research.

V
Conclusions and Implications

The basic philosophy behind the IMT seems to be to transform the irrigators from being beneficiaries to partners in planning and implementing irrigation development programmes. Though the IMT programmes of Gujarat and Andhra Pradesh were also based on this premise, there are variations in the components of the process.

Partly, the efficacy of IMT in Andhra Pradesh was because of the fact that it was nested in a much wider set of agricultural and economic liberalisation policies. It was not only accompanied by new laws that supported the transfer of authority and responsibility to the users, but also had high political commitment. Orientation and training of leaders of WUAs and governmental staff also emerge as one of the essential ingredients of IMT. On the other hand, the Gujarat model of the bottom-up approach is found to be more sensitive to local conditions. However, the slowness of implementation of the programme also permits the entrenchment of vested interests and a less equitable outcome.

An important aspect is whether the new systems create more equitable water rights. A right in this case is a recognised claim to irrigation water of a specified quantity and at specified times. In both Gujarat and Andhra Pradesh, the new system appears primarily to confirm the rights of landholders. Since equity is one of the objectives of IMT, opportunities have been created to bring about a change in attitudes towards participation in WUA functions, though nominally.

The key to successful operation is expected to be related to the kind and degree of participation exercised by water users. There is evidence to suggest that participation varies according to local socio-cultural, economic and political factors. Top-down versus bottom-up approaches are also found to create different contexts for participation and the initial arrangements heavily influence the establishment of norms of participation as well as long-term impacts. From the analysis it is clear that women's and tenants' participation can be stimulated and it is found to be dependent upon political (as in the case of Ellabotharam) or NGO (in the case of Thalota) support from outside the WUA.

Though in some cases it is clear that IMT has brought about new consciousness of empowerment to communities and individuals, especially, those marginalised like SC households in Andhra Pradesh, the traditional system seems to have been bypassed in the process. Even so, the newly 'empowered' do not seem to assert their rights and interests on crucial matter like improved access to water or greater participation in the WUA matters.

However, it is important to note that whenever water management has improved under IMT, it seems to benefit all farmers, including the poor. There is little evidence to suggest that the poor suffer or benefit more than others. In this respect, the potential for a strong say in water management by the poor is apparently high in Andhra Pradesh, provided the efforts towards educating these sections about their rights are sustained.

Yet another important facet of access to water by the various sections of the society is the role of alternative sources. A large majority of the farmers seem to depend on sources other than canal to meet their water requirement. This is not surprising as the prevalence of wells in the command area is known. The flip side of this argument however is, with lining of canals (under IMT) it is possible that less ground water recharge to the wells will take place and hence it is highly desirable that an integrated management of water in the command area by the WUAs is encouraged.

In the current development debate of involving users for improving the efficiency of systems, indeed the two models are in either end of the spectrum. Evidently, there can be no one approach to the development of infrastructures, be it rural or urban based; also there is no method to involve or promote people's participation. Clearly, the Andhra model is based on the tenet that the state can and in fact should promote popular awareness towards sustainable development and that participation of the users can be forced through an Act. On the other hand, the Gujarat case follows the now trendy route to generate people's participation and then trying to rather meekly force the concept of sense of ownership among the users. This, it is hoped, will lead to better utilisation and equitable distribution of the available resources. Interestingly, the Andhra model starts with this premise of utilising the available resource and then traces the steps backwards. The similarity however, is the faith both the models place on the user farmers ability and willingness to rally together in the government's efforts. The crucial difference of course is, while Andhra policy has been approved by the legislature, the Gujarat PIM policy still remains an attempt by the bureaucracy supported by a few NGOs and enthusiasts.

As a direct consequence of the differences in the approach, there is also a difference in the pace and coverage of the reforms aimed. There are obvious advantages and disadvantages in both these approaches. If cost effectiveness and speedy implementation are considered as relevant parameters, then the Gujarat experiment apparently is a more expensive route and prolongs the risk of knowing the outcomes. This issue arises especially in the context of the experiences of NGOs in creating islands of excellence and the Philippines experience of tapering interest in such programmes both of the government's and of the participants. In fact, the Philippines experience being followed in Gujarat should also be periodically reviewed to understand the latest developments there and bring about corrections on stages where the programme has either faltered or belied the expectations. However, this issue is equally valid for the alternative model of Andhra Pradesh where the sustenance of the initiatives of a particular party depends on the commitment of successive governments.

Summing up in the words of Dantwala (1947): "Institutions are not built in a day. Their streamlined exterior hardly gives any idea of the toil and sweat invested in the spade-work. The evolution of institutions is not mere history, it is an object lesson for the builders of tomorrow. Trial and error, not omniscience, go to the building of an institution. The journey is as thrilling as the journey's end. Perhaps, there is no end, but a continual process of growth and adjustment. This process of adjustment with concrete situations makes a fascinating study. Institutions cannot lag behind the exigencies of a situation, nor can they shoot ahead of times without causing serious maladjustment. Posterity may smile at the obvious crudeness of yesterday's institutions. Possibly it cannot always view them in historical perspective. A few years hence the apparently perfect institutions of today may be as anachronistic in the context of tomorrow as those of yesterday are in the context of today."

NOTES

1. For example, *Pani Panchayats* in Maharashtra and water user cooperatives in Tamilnadu, Gujarat and Karnataka.
2. The PIM policy is an offshoot of the consultation workshop held under the auspices of government of Gujarat and the Ford Foundation in 1993 at WALMI, Anand.
3. In fact, during the consultation process, besides the provisions of the proposed Act, even land delineation was discussed with the farmers.
4. In one of systems command area, about thirty percent of landowners were women and hence they were member of WUAs. A woman among them had been elected as a TC member.
5. The water bodies in a village, especially tanks, are used by the entire village community. Further, there are variety of purposes other than irrigation for which tank water is used. Thus, a tank is a common access resource. Perhaps, the constitution of the stakeholders for the village-level water body could be broad-based, encompassing all type of users.
6. Accordingly, some of the WUAs had received funds in early 1998, and in the first week of June 1998, the repair and rehabilitation work was almost completed and the WUAs were ready with water distribution plans. Some of the WUAs, like the Ghanpur WUA have also collected Rs. 20 from every member as their contribution. This however is a voluntary effort.
7. Laskar is a village level functionary of the irrigation department who is responsible for operating the canals and inspecting the water distribution. His counterpart in Gujarat is known as *chokidar* or *pagi*. He is paid by the irrigation department.
8. Neeradi is a village level functionary appointed by the revenue department. His main function is to collect irrigation charges from the farmers which is collected as a part of the land revenue.

REFERENCES

Andhra Pradesh Farmers' Management of Irrigation Systems Act (1997), *Act and Rules*, (Act 11 of 1997), Department of Irrigation and Command Area Development, Government of Andhra Pradesh.

Brewer J., S. Kolavalli, A.H. Kalro, G. Naik, S. Ramanarayan, K.V. Raju and R. Sakthivadivel (1997), "Irrigation Management Transfer In India: Policies, Processes and Performance", (*unpublished report*), Indian Institute of Management, Ahmedabad and International Irrigation Management Institute, Colombo, December.

Chambers, R. (1988) *Managing Canal Irrigation: Practical Analysis from South Asia*, Oxford and IBH Publishing, New Delhi.

Dantwala, M.L. (1947), *A Hundred Years of Indian Cotton*, The East India Cotton Association Limited, Bombay.

Datye, K.R. and R.K. Patil (1987), *Farmer Managed Irrigation Systems: Indian Experiences*, Centre for Applied Systems Analysis in Development, Bombay.

(EMS) E.M. Shashidharan, Executive Director, Development Support Centre, Ahmedabad, Private Correspondence, 1998.

Maloney, C. and K.V. Raju (1994), "The Role of Farmers' Organisations in Environmentally Beneficial Water Management", in M.V.K. Sivamohan and C.A. Scott (eds.), *India: Irrigation Management Partnerships*, Booklinks Corporation, Hyderabad.

Pant, N. and R.K. Patil (1983), *Farmers' Organisation and Irrigation Management*, Ashish Publishing House, New Delhi.

Parthasarathy, R. (1994), "Policies, People's Participation and the Importance of Documenting the Process of Implementing Irrigation Projects: An Overview of the Experiences of India and the Philippines", A background paper presented at the Workshop on Process Documentation Research, Gujarat Institute of Development Research, Ahmedabad, August 11, 1994.

Parthasarathy, R. (1998), "Water Users Association in Irrigation Management Programme: Participation Strategies in a Game Theoretic Framework", in Vishwa Ballabh (ed.), *Proceeding of the International Workshop on Shared Resource Management in South Asia: The Next Step*, Institute of Rural Management, Anand, February 17-19, 1998.

Parthasarathy, R. (2000), "Participatory Irrigation Management Programme in Gujarat: Institutional and Financial Issues", *Economic and Political Weekly*, Vol. XXXV, Numbers 35 and 36, August 26-September 2.

Parthasarathy, R. Barbara van Koppen, and Harish Joshi (2000), "Access to Water of Land-Poor Men and Women under Irrigation Management Transfer: A Comparison of Gujarat and Andhra Pradesh, India: A Summary of Findings", Paper Presented in the Workshop on Poverty, Gender and Water in South Asia, organised by International Water Management Institute, Colombo and Gujarat Institute of Development Research, Ahmedabad, August 10-11, Ahmedabad.

Parthasarathy, R. and Sudarshan Iyengar (1998), "Participatory Water Resources Development in Western India: Influencing Policy and Practice through Process Documentation Research", in David Mosse, John Farrington and Allen Rew (eds.), *Development as Process: Concepts and Methods for Working with Complexity*, Routledge Research/ODI Development Policy Studies, Routledge, London.

Randhawa, M. S. (1980), *A History of Agriculture in India*, Vol. I, Indian Council of Agricultural Research, New Delhi.

Sengupta, N. (1991), *Managing Common Property: Irrigation in India and the Philippines*, Sage Publications, New Delhi.

Singh, K.K., Syed Turab Ul-Hassan, C. Sithapathi Rao and G. Kemuel (1994), "Promoting Farmers' Participation: Action Research in Sreeramsagar Project", in Sivamohan, M.V.K. and C. Scott (eds.), *India: Irrigation Management Partnerships*, Booklinks Corporation, Hyderabad.

Stone, I. (1984), *Canal Irrigation in British India: Perspectives on Technological Change in a Peasant Economy*, Cambridge University Press.

Vaidyanathan, A. (1985), "Water Control Institution and Agriculture: A Comparative Perspective", *Indian Economic Review*.

14

Andhra Participatory Irrigation Management Experience Demonstrates the Need for NGO Role: A Prescriptive Note

S.N. LELE, R.K. PATIL and S.B. SANE

Reforms in any field -- social, cultural, religious -- are generally initiated or triggered by either individual politicians, active workers or activists, or individuals who may be from any institution or government organisations, who are able to think beyond their normal tasks. The bureaucracy as an institution is generally trained and instructed to carry out orders given by the government and implement programmes as planned. The approach of the bureaucracy is to maintain status quo, maintain law and order and report the activities undertaken faithfully whenever asked for.

Irrigation Sector Reforms

The reforms in irrigation sector through PIM were initiated in India in 1985 on some requests/pressures from:

- donors,
- some individuals in the government, and
- institutions like Ford Foundation, Administrative Staff College, Hyderabad, etc.

The CADA Wing, Ministry of Water Resources, issued instructions to the states/CAD administrators to initiate action research programme/pilot projects in organising farmers at minor level and entrusting Operation & Maintenance (O&M) responsibilities to them so as to provide timely, reliable and sustainable irrigation services and guaranteed supply of water to all the farmers in the command. These instructions were followed only by two CADAs out of 50 CAD authorities in India. One WUA was initiated in Maharashtra (Bhima Project) and a few of them initiated by one Chief Engineer in Tamil Nadu in Lower Bhavani Project.

Role of NGOs in Participatory Irrigation Management

However, the NGOs from Maharashtra, Karnataka and Andhra Pradesh, namely, CASAD, ISARD and IRDAS, had taken more initiative and approached the concerned state governments to undertake Action Research Programme to evolve methodology and approach for motivating and organising the farmers and thereafter established that:

- The farmers if approached/motivated properly can manage irrigation waters efficiently and judiciously.
- They have skills and willingness to participate and undertake O&M responsibilities, if they see benefits like reliable/sustainable service and they get some rights in allocating/using water.
- They can contribute resources, if properly motivated for self reliance, sustainability and rights over getting water.

As these voluntary efforts were going on, the AP government took a decision to universalise PIM throughout the state via legislative action. The legislation was enacted in 1997 and the Act was passed unanimously.

AP Model of Participatory Irrigation Management

The AP model of PIM is politically driven and the conceptual framework was conceived and prepared by the bureaucracy. This is called 'big bang' or top-down approach. The AP experience shows that this is perhaps a better method for creating awareness among the users/suppliers and also in implementing the programme in all areas quickly through statutory provisions. PIM as it is now functioning in AP has following characteristics:

- 10,000 of WUAs have been formed throughout the state and brought into existence through elections conducted by the government machinery.
- The irrigation department is now responsive to the WUAs/ Distributory Committees (DCs).
- Due to physical improvements in the canal system (desilting and doweling) made possible through World Bank assistance, the water supply in tail reaches has improved, area under irrigation has increased due the regularity in water supply.

Essential Functions of WUAs

For effective, efficient and sustainable services, it is expected that, the farmers should take over certain functions and responsibilities such as:

- Distribution of water equitably to all the farmers.
- Training/educating the farmers in better water management, economic and judicious use of water, crop diversification for enabling more farmers in utilising available water.

- Decide water fees and other service charges, collect water fees from the farmers to meet the O&M costs of the WUAs, and systems and administration charges of WUAs and DCs.

Present Status of Turning over O&M to WUAs

At the moment, though the above activities/responsibility are planned to be handed over to the WUAs/DCs as per the Act, they have not, so far, been entrusted to them by the government. The discussions with the farmers and WUA office-bearers indicate that the WUAs are not very keen to take over recovery of water fees, nor do they think of any other means for increasing financial sources to meet the O&M costs. On the contrary, they expect that the government should:

(a) retain all the field staff (laskars, patrols, junior engineers) for water distribution/allocation within the jurisdiction of the WUAs, and pay them directly, but place this staff under the control of the WUAs,
(b) provide financial assistance for maintenance and repairs of the physical system, and
(c) contribute management cost of the WUAs/DCs. The government on the other hand has extended the financial assistance arbitrarily on ad hoc basis for: (i) initial or mobilisation grant of Rs. 50,000 and Rs. 30,000 as per unopposed/contested elections; (ii) maintenance of system at Rs.100 per ha for the first year and Rs. 45 per ha in successive years; and (iii) Rs. 2,500 per WUA towards management costs of WUAs. Further, the WUAs are demanding pecuniary benefits such as fishing rights on reservoirs and canals and plantation of trees on the lands within their jurisdiction, but are not willing to take on the responsibility of recovering the water fees from farmers.

Shortcomings in the Present Programme

Neither the irrigation department nor the WUAs/DCs are so far thinking, planning or considering the following:

- Volumetric supply of water for introducing optimal, economical and judicious use of available water.
- Introducing crop diversification for increasing the irrigation area especially in Rabi season and increase the value of irrigated crops.
- Educating the farmers to take precautions such as not crossing the canals/field channels except at bridge points, not grazing the animals, washing/bath in canal, so as to prevent damages and thereby reduce/minimise maintenance/repair costs.
- Contribute either in cash/kind in meeting cost of management/repairs.

- Exploring possibilities of raising revenue/income of the WUAs to make them self-reliant.
- Coordination or interfacing with existing organisations such as gram panchayats, and credit societies.
- Conjunctive use of surface and groundwater for increasing the irrigated areas or increasing frequency of irrigation.
- Charging for use of water from wells in their jurisdiction (as this water is recharged from canal water).

Non-implementation of New Act

There are some provisions in the new Act on revenue/income generation of the WUAs, but the farmers and the Irrigation department officials are not able to act on these provisions in the absence of specific instructions, precedence and projections. Similarly, regarding the crop diversification, detailed discussions need to be carried out with WUAs/farmers after studying the soils, water availability, frequency of water supply, ground water availability, transport and marketing facilities. The irrigation and water use efficiency can be increased only after testing the canals, finding out capacities, seepage/transit losses, crop water relationship, methods of on-farm water management. As these conditions will change from project to project, WUA to WUA, and also depending upon farmers attitude/aptitude, some sort of piloting or undertaking Action Research Programme to identify various possibilities and areas of success will be necessary.

Training by WALAMTARI

The training imparted by the WALAMTARI is more focussed on provisions in the new PIM Act, agronomic aspects like crops, inputs and cropping practices, etc. But, it lacks in business and financial management, group working, motivating the farmers for raising their incomes, making WUAs self-reliant, increasing irrigation efficiency, improvements in scheduling of irrigation water, allocation and measurement of water diversification of crops, equity in water allocation and raising income/revenue of the WUAs.

Need for Hands-on Training

The classroom training at present imparted by the WALAMTARI would not be enough for this purpose. Repeated training/demonstration and discussion with farmers groups along with hands-on training will be essential to strengthen the WUAs and guide them in future.

Training of Trainers

Large-scale, continuous as well as 'hands-on training' to lakhs of farmers would obviously need large number of trainers. This can be done by the O&M

irrigation officers of the rank of Executive Engineers and down below up to junior engineers. These trainers would further need training in respect of methods of training, training material, standard lesson plan that can be organised by the WALAMTARI.

Also WALAMTARI faculty themselves will need some training from outside experts in motivation of farmers, and other social aspects of irrigation management, business and financial management.

Performance of WUAs

Briefly, the ground level data as available from various studies indicate:

(a) Operation of the water distribution system is still in the hands of bureaucracy. No doubt, the office-bearers are consulted as to the timings of water releases, but basically the irrigation department takes the decisions. Though such occurrences are inevitable during the transition stage, no sustained efforts are visible in the last two to three years to actively involve the office- bearers in the process of preparing a realistic operation plan.

(b) Presently, maintenance (and rehabilitation/restoration) of the physical system is done by the department through the funds made available by the World Bank. Here too some consultation with the office-bearers is seen, but WUAs have no liability. None of the WUAs are collecting membership fees or contribution from members towards maintenance expenditure.

(c) Though in the principle it is accepted that water would be supplied on volumetric basis, no steps have been taken to install measurement devices and work out volumetric water charging system.

(d) WUAs have no powers to fix rates that are to be charged to the farmers nor are they responsible for assessment and collection. Presently, assessment and collection of water charges are the responsibility of revenue department. No dialogue has been started with that department to shed off these responsibilities. There are no proposals to empower the WUAs, though this aspect is mentioned in the official proclamations. Without this power, WUAs would be very weak bodies.

(e) Though a regulatory authority in the form of commissioner is contemplated in the legislation to overview the functioning of the WUAs (audit accounting, annual reports, etc.) there is no machinery on the ground to supervise the working. Some WUAs do print and circulate the accounts and audit among the members; a perusal of these reports shows that apart from the grants made by government, there is no source of income. One gets the impression that these WUAs are the creation of statute. They are not self-dependent entities.

This picture or present state of affairs of WUAs needs to be changed. Such change cannot be brought about by the bureaucracy. The essence of participation, viz., in decision-making, implementation and cost sharing is lost in the present set-up.

Criteria for Evaluation of WUAs

The programme has been operating for the last three years and a good deal of literature (critiques, comments, rapid evaluations, etc.,) is now available. The performance needs to be judged against some norms and if the performance is not nearer the mark, reasons thereof have to be investigated. Select criteria for judging the success or otherwise of WUAs are:

(1) Water allocation, distribution and use -- part of operating the system -- would be handled by the WUAs, particularly by the managing committee. Equity and social justice should be the guiding principles.
(2) Maintenance would be the sole responsibility of WUAs.
(3) Waters are to be supplied on volumetric basis and charged at volumetric rates.
(4) Fixation of water fees, assessment and collection of water fees would be in the domain of WUAs.
(5) Audited accounts and annual reports need to be submitted and approved by the general assembly of WUAs.

These are the minimum functions that should be performed by the WUAs to qualify as an effective instrument of PIM. While nobody expects the WUAs formed three years ago to fulfill these criteria in full, one has to satisfy oneself that the progress is made on the expected lines. Viewed in this way, some of the happenings in the area should cause grave concern as to the long-term success of PIM in the state. Such evaluation by government bodies will be self-defeating. In this case also, NGOs have to play a major role as they can present non-partisan and realistic picture and suggest various measures for bringing about desired changes.

Need for Deployment of NGOs

The Government of Andhra Pradesh, i.e., the policy makers and high-level officers from water resource department feel that as the PIM Act covers all different aspects of farmers' involvement and turnover of the irrigation schemes to the WUAs/DC, there is no scope nor a need for deploying NGOs in this area. However, the progress made by the water resource department in AP and the available studies of PIM programme indicate that, on the contrary, large-scale efforts are needed to scale up the WUAs qualitatively and for their future progress. This task cannot be fulfilled by the bureaucracy alone.

Further mobilising resources, self-reliance, undertaking of other services by the WUAs to assist farmers in reducing cost of production or increasing coordination with other organisations for mutual benefits, conducting business and financial management, increasing irrigation efficiency, crop diversification, etc., needs close study of the situation and frequent and repeated dialogue with farmer/WUAs/DCs, and more persuasion, which cannot be handled by the irrigation/ agriculture department in the existing situation. Here, deploying a few NGOs in various commands to evolve solutions to second/third generation problems and evolving a few model WUAs/DCs can be done in a short period by the NGOs. The NGOs have or can mobilise suitable and experienced experts quickly, can change/modify the approach according to the situation, have better access ranging from farmers to highest levels in government, other agencies and have better social background to work with farmers. Hence, the NGOs can be utilised as pace-setters to take the PIM to new heights.

15

Irrigation Resources and Implementation of PIM with Special Reference to Andhra Pradesh

S. TURABUL HASSAN

Introduction and History Prior to Independence

In India, the idea that water resources are God's gift and as such the common property of all members of the community is taken as a religious doctrine. No single water user has the right of excess use or misuse, nor can he deny others their legitimate share. That water is considered a resource common to all, to be used for agricultural purposes by all who have access to its course of flow, was the principle in the past and remains so to this day. In many communities in India deep reverence and respect for flowing water has been an important aspect of culture. Harnessing the flow of water for the benefit of the community is considered an act of great merit and any misuse – its contamination and pollution – a sin. The country's national anthem with its mention of the major rivers indicates the importance of water.

Development of irrigation from surface flows has always been the result of communal or state enterprise, whereas irrigation from wells is mostly the result of private enterprise. Wells for drinking water excavated by philanthropists or government are for common use and their upkeep is the responsibility of the community.

Small reservoirs behind earthen dams are called tanks in India. In the villages of Andhra Pradesh, there are three types of tanks: those providing drinking water for people and animals, those belonging to temples, and tanks for irrigation. Tank irrigation is an old-established practice in most of the semi-arid tropical parts of India. Here, the monsoon is erratic and rainwater is to be had only a few months in the year. Irrigation tanks serve to store and regulate the flow of water for agricultural purposes. In AP, tank irrigation has been prevalent since pre-Vedic times. The *Ramayana* mentions two tanks on the Tungabhadra, named Pancha Prarathataka (Lake of the Five Nymphs) and Madakarni respectively (Yazdani, 1960). Further, there are references to tank irrigation in records dating back to many centuries before the commencement of the Christian era. Many of the tanks found in the erstwhile Hyderabad state (parts of which are now in AP) have been in existence for several generations and are referred to in inscriptions of the 8th and 9th century AD (Harris, 1954).

A rock inscription dated 1030 AD praises the many tanks built by the local ruler, Kota Gonka. An inscription of 1071 AD also mentions how new land was brought under cultivation by digging a tank, Settikere (Department of Information & Publications, Hyderabad).

In different periods tanks were donated to temples as acts of piety and lands were earmarked for their upkeep. A number of inscriptions dating around the 11th and 12th century describe tank construction activity in Warangal taken up by small chieftains and big landlords.

Telangana has been termed a land of tanks where tank irrigation was developed extensively right from ancient times. The districts of Warangal and Karimnagar offer some instances from the medieval period. The Pakal, Ramappa, Laknavaram, and Shanigaram lakes were excavated by the 12th and 13th century kings. The Kesari-Tatakam tank was commissioned by Prola, the first of the Kakatiyas. Beta II constructed two tanks, Setti Kerayam and Kesari Samudram. Prola II excavated two tanks. Elsewhere, in what is now Andhra Pradesh, Pratapa-Charitra states that the ruler Ganapatideva had tanks dug at Nellore, Gangapuram, Ellore, Ganapapuram, Ekasilapuri (Appadurai, 1936).

There is a system of tanks at Kotagiri referred to in inscriptions of 1100 AD and they bring to light the technique of constructing tanks in a series of different levels of a watershed. In 1108 AD, a merchant named Dasi Setti renovated and increased the size of a tank at Bhanvir. In 1201 AD, after a famine, two persons built tanks in memory of their mother (Appadurai, 1936). The Somavaram inscription dated 1213 AD states that Rachela Bete Reddy constructed two tanks and donated it to the people of the area. In what was a widespread practice, kings, chieftains and well-to-do locals constructed tanks for the use of the community. They also maintained and regulated the same in their lifetimes, and specified in their wills that after their death the tanks be transferred to a temple trust or to the local panchayat. These two organisations then took up the task of organising distribution of water, controlling the cultivation process and, if required, raising money for maintenance purposes.

In the 16th century, the Vijayanagar kings constructed 15 anicuts across the main Tungabhadra river, 12 of which are still operative.

Tank construction in the past played a key role in the ritual-based system of entitlements wherein land resources were controlled by the temple priest. By constructing a tank the local chief generated resources for gifts to the temple which, in turn, enhanced brahminical ritual power and provided material support that left this class ample time to perform not only rituals but also to guide locals on many matters.

To maintain the water balance and help farmers during the dry season, self-discipline was inculcated. Rules and regulations with regard to use of irrigation water and penalties for excess use or wastage of water were laid down as early as in the Vedic times. The preservation of most of the centuries-old irrigation systems attached to temples to this day bears testimony to the efficacy of the guidance provided by the Vedic hymns.

In the past, panchayats also used to play an important role in disciplining users, but now the onus has largely been transferred to governmental authority. Nowadays, the native wisdom of the farmer and the teachings of tradition are hardly taken into account. This has adversely affected tank and private irrigation channels to such an extent that irrigated area has started dwindling, affecting the agricultural productivity of the state.

It will not be out of place to mention Kautaliya's statement that a good state does not depend on rain alone for cultivation. The development of water resources is critical for crop growth in some areas and is essential for higher productivity in others. And, as such, surface and ground water both are to be exploited for irrigation in all parts of the country. This statement is valid even today.

The British were highly impressed by the extent of tank irrigation they found in the country. In 1853, R. Bird Smith wrote: "The extent to which tank irrigation has been carried throughout in all the irrigation regions of Madras Presidency is truly extraordinary. As per imperfect record, the number of tanks in the fourteen districts was 43,000 and 10,000 needed repairs in the crop season." (Smith, 1856). Sir Authur Cotton, a pioneer British engineer, who took up several irrigation works in South India (especially an anicut across the Godavari at Dowleshwaram in 1852 which was considered one of the great engineering enterprises of the 19th century), wrote in 1874.

> "There are multitudes of old native works in various parts of India . . . . These are noble works, and show both boldness and engineering talent. They have stood for hundred of years . . . . it was from them (the native Indians) we learnt how to secure a foundation in loose sand of unmeasured depth. In fact, what we learnt from them made the difference between financial success and failure, for the Madras river irrigations executed by our engineers have been from the first the greatest financial success of any engineering works in the world, solely because we learnt from them . . . . with this lesson about foundations, we build bridges, weirs, aqueducts, and every kind of hydraulic work . . . . we are thus deeply indebted to the native engineers".

Irrigation has been the key force behind agricultural growth in AP. The state is blessed with many major rivers, the most important being the Godavari, Krishna, Vamsadhara and Pennar. In AP, rainfall occurs only for a few months in the year and in varying amounts in most areas. Storing, regulating and diverting rivers and other flows have been tried since time immemorial. There are many water-harvesting systems, chiefly anicuts and tanks, that were built by rajas and their chiefs several hundred years ago. They help farmers exploit groundwater for agriculture. This tradition was emulated by later rulers, chiefly the Qutub Shahis and the Asaf Jahis. There are many water-harvesting structures based on regulating and diverting -- the Anantharaj Sagar in Cuddapah district, the Cumbum tank in Nandyal in Kurnool district, and the Pakhal, Ramappa, and Laknavaram lakes in Warangal district, all bear testimony to this tradition.

In the entire period of British rule no new works were completed and only improvement and modernisation of the old irrigation system in the deltas was taken up with an eye on profitability for the East India Company/British government. The British were aware of the utilisation of the Tungabhadra, a tributary of the Krishna, by the Chola and the Vijayanagar dynasties (some of the anicuts and canal systems are in good condition even today). They did nothing to further utilise the excess Tungabhadra water and surveyed no area except for the Kurnool-Cuddapah canal which was excavated by the Madras Irrigation and Canal Company in 1860 (the doom of which was already predicted). No maintenance work was taken up in this area. In fact, the system was badly affected by experiments with different sub-canals to enlarge the area under irrigation. This continued till about the middle of the century.

The British East India Company administration undertook repairs on anicuts across the Godavari, Krishna, and other major rivers. Spurred on by the profits that these repair ventures generated, they started floating debentures to attract British private capital by offering government guarantees of five percent interest on the new ventures. This was basically to excavate an irrigation-cum-navigation canal. Their first big venture, the Kurnool-Cuddapah canal, ended in disaster. Consequently, the British government's policy was to take up maintenance of only those schemes which promised an adequate return on the capital invested. The only major project taken up by them by way of insuring against drought and benefiting the farmers of the erstwhile Madras presidency was the Tungabhadra project, that too because the ceded districts of Bellary, Anantapur, Cuddapah and Kurnool were famine-affected areas and the British government had to spend recurrently, once in five years, Rs. 3 crores on famine relief in the period 1876 to 1932.

This state of affairs entailing expenditure of huge amounts called for remedial action and so the British government along with Madras presidency officials conceived various proposals for the utilisation of the waters of the Tungabhadra. But due to one reason or the other no measures were initiated until 1944. It will not be out of place to mention that the taking up of the Tungabhadra project was made possible only after an agreement reached by Mysore, Madras, and Hyderabad on water sharing. A joint venture of the government of Madras and the Hyderabad government was sanctioned in December. Work began on 28 February 1945.

The other important irrigation projects taken up in present-day AP before independence were the Nagavalli barrage in Vishakapatnam district and the Nizamsagar in Nizamabad. The total surface irrigation potential created prior to the First Five-Year Plan was 30.47 lakh hectares. The state has large water resources whose supply is, however, variable both within and over the years.

Intervention Strategies after Independence

The development of irrigation in India after independence was subject to a number of forces, the most important being the effect of partition in the shape of large-scale migration of rich and efficient farmers, loss of developed irrigated area to Pakistan and the resultant food scarcity and in some cases even famine conditions. To mitigate the same, enormous funds were allotted to increase food production by constructing major and medium irrigation projects under the government sector and to provide all facilities to those prepared to invest and take the risk of trying to convert dry land to irrigated land. Progressive farmers and big landlords, along with refugees with experience of irrigated farming, took advantage of these projects and the lot of the poor farmers saw no improvement.

Given the anxiety to see the country self-sufficient in food and fibre in the shortest period, no serious action was taken up against farmers even if they misused the irrigation network and deprived poor farmers or tail-enders of their due share of water. Consequently, while the country did march rapidly towards self-sufficiency in food grain production, little heed was paid to the fact that the irrigation potential was not being properly utilised and long-run problems of waterlogging and salinity were being created.

This policy was vigorously followed till 1970, and rapid utilisation of irrigation potential went hand-in-hand with consolidation of the existing system biased towards rich progressive farmers and disregard of the rights of the poor farmers whose lands lay in the tail end of the command area. However, whatever the state of the poor farmer, irrigation potential has increased from 22.6 mha in 1951 to about 89.56 mha at the end of the Eighth Plan with unutilised potential amounting to 8.80 mha.

Slip-ups in utilisation of potential began to be noticed by the Fourth Plan itself and the central government began to make suggestions to the states from time to time to ensure that benefits from land and water resources were fully availed of. To ensure that large dams built at high cost were utilised to the full, the concept of command area development programmes was introduced in December 1974. Priorities were spelt out – instead of new projects, emphasis was laid on completion of ongoing projects and extension of canal networks to cover the entire command area. Farmers were encouraged to take up on-farm development (OFD) work considering the chak as one unit which requires cooperation of all farmers under the pipe.

Interest in farmers' participation under large irrigation systems in India stems from the difficulties faced in the execution of command area development programmes chiefly with regard to OFD works and their maintenance. In spite of huge investment by state governments, OFD works taken up without involvement of farmers deteriorated within months of completion. Farmers' involvement as a group was sought as a means to prevent breaches in the irrigation network and to impress upon them the need to take up new OFD works as an absolute prerequisite for proper agronomic practices and increase in productivity.

Some States formed pipe committees and at a later date gave statutory recognition to these committees for operating below the pipe outlet. Some encouraged NGOs to form farmers' organisations (FOs), while others motivated farmers to register themselves under the Societies Act or form their own cooperative organisations to manage the irrigation network from the minor itself.

On government-owned irrigation systems, FOs are created to perform many important functions with regard to irrigation management. These include procurement, allocation and distribution of water, regular donation of funds and labour for organised maintenance, enforcement of discipline, settlement of disputes and sound overall supervision and control of the irrigation system. Some find it helpful to broad-base activities to include procurement of farm inputs, marketing of produce and a variety of social and income-generating projects. An FO can be deemed sustainable if it performs its various functions independently with a fair degree of success over a period of time.

In most states, FOs have been formed to reduce the gap between the potential created and utilised but have failed to achieve the goals set. While this can be attributed to lack of administrative will, farmers and politicians are also found wanting in this regard.

The main purpose of FOs is to benefit farmers by improving water delivery and system maintenance. Accordingly, timely access to water in dependable quantity holds the key to good performance. Since the irrigation system in large and medium projects is owned and operated by state irrigation department (ID) a FO has to comply with the operational plan of the ID. Decisions about water releases and canal stoppages are taken without consulting farmers, as a result of which many FOs collapse soon after they are formed.

Since FOs are designed as social organisations, an external agent is needed to help establish them. This could be an NGO or a government department such as ID, CAD, or AD or a resourceful individual, say, a local leader or activist. Being new social entities, the members have to be apprised of the objectives, functions and resource requirements of an FO and what they must do to make it strong. Typically, the agent (organiser) first explains the need for an FO, what water users stand to gain and what they have to do, individually and as a group, to be able to profit from self-management. Joint inspection of the irrigation system along with the officers of the irrigation department to identify system deficiencies and the scope for improvements, and the resources required for the same are undertaken. Several group meetings and a few training programmes are organised. In the process, a close working relationship is sought to be nurtured between the outsider and the members of the FO. Wherever formal recognition through registration is given by the government, the FO often succeeds in taking over system management.

Most FOs, however, depend on funds earmarked by the government or obtained from multinational and national organisations without realising that external assistance is not limitless. A time soon comes when active association is phased out. The FO is left to function on its own within the administrative framework of the state's irrigation department. While it is hoped that the experience and competence gained in self-management will help them carry on independently, in reality this expectation is rarely fulfilled. FOs are known to collapse within a year or two of assistance being withdrawn.

An adequately functioning FO needs strong leadership and group cooperation for the management of its affairs. Under unfavourable conditions of water supply or when system defects and unsatisfactory management distort water delivery in canal minors and outlets, the leadership loses its credibility in the eyes of the farmers. Water users are not keen to support group management. Potentially good leaders too do not come forward. The very purpose of group action is negated.

These experiments ended by 1985 when it became clear that FOs were not up to the task of providing assured water to all its members, chiefly when the irrigation network deteriorated or the providers (ID officials) failed to release to the minors or outlet the quantum of water to which they were entitled. The presence of government officials and their active support might have kept the FOs alive but once they were transferred out of the project it was difficult for the FOs to convince farmers of the usefulness of the organisation. Some irrigation societies working effectively for more than a decade were suddenly rendered defunct by change of leadership and the hostile attitude of the ID. In the early 1980s, participatory irrigation management (PIM) was not an important issue. The main concern of the central government was maintenance of irrigation works and involvement of farmers' organisations only as an extension of the government's role. Thus, its survival depended upon the attitude of the government officials.

New Policy of Government of India

To promote utilisation of the irrigation potential already created under the major and medium projects and to restore all tanks which were the main source of water for flow irrigation as well as ground recharge, technical and other input supports were provided by the government. These steps increased irrigation potential from 22.6 mha in 1951 to about 94.73 mha (provisional) at the end of 1999-2000. But it was noticed that the gap between potential created and utilised is widening due to defective irrigation supplies as a consequence of the degradation of the irrigation system resulting in tail-end problems, waterlogging and salinity.

To arrest this trend, the time lag between potential created and potential utilised should be shortened so that the benefits of building large dams at high costs are not lost. In the Eighth Plan, therefore, the government took note of the financial and social constraints of states eager to help small and marginal farmers. But no positive action was suggested. Only at the end of 1997 did the central government place high priority in providing early and comprehensive support to 'reforming states' that were implementing programmes for poverty alleviation, giving priority to social and environmental concerns and promoting private sector involvement in developmental activities. The 'country assistance strategy' supported in particular the increased focus in the rural areas (where the majority of the poor live), on policy and institutional reforms aimed at giving the weaker sections a say in all programmes that can foster sustained agriculture growth (for which water is a critical input).

Even three years after the above two policy directives, if we look into the working of irrigation projects we find that sheer management lapses have failed to ensure equitable supply of water to small and marginal farmers thereby negating the egalitarian thrust of policy directives. Currently, most government-managed surface irrigation schemes are in disrepair, with many canals heavily silted, their lining punctured or missing and outlets damaged.

The main reasons for such poor maintenance, according to ID officials, are inadequate funds for O&M and the "interference" of users' groups in water schedules and upkeep of the irrigation network. For their part, farmers point to inadequate emphasis on operation and neglect of maintenance works, and say that ID staff are more interested in taking up new works, which they find more remunerative. They also feel the ID staff are mainly preoccupied in ensuring that the water reaches the fields of big farmers and political heavyweights.

Studies taken up by our organisation and others have indicated that the main reason for tail-end problems and the widening gap between potential created and utilised is due to poor maintenance of the irrigation system and misuse by farmers in the upper reaches. In addition to this, in some projects almost 20 percent to 25 percent of land is subjected to waterlogging and salinity.

New Trends in Irrigation Management

Most policy-makers in the states since 1997 feel that a water-scarce situation demands fundamental institutional changes. More specifically, one needs to examine the irrigation bureaucracy and its rules of the game that have degraded, overexploited, or underutilised water resources. One needs to know how this well-entrenched powerful institution has precipitated this crisis and adversely affected the quality and quantity of water supply to the command area.

Our studies in Andhra Pradesh, Bihar, Haryana, and Orissa indicate that the main factors affecting the efficiency of the irrigation system are:

1. Inability of ID and other government officials to work out crop-specific water requirements based on rain, soil moisture and irrigation available from their canal system and from groundwater; and
2. Failure of departments to take steps against anti-social elements and water-grabbers so as to ensure flow in the entire system.

With population growing alarmingly and land remaining constant, wastage and improper use of water spells catastrophe. The need for scientific utilisation is imperative. Careful management should bring about more yield per drop of water so as to maintain a system capable of supplying water equitably in the entire command.

Hitherto, the emphasis in irrigation has been on technical and physical aspects while managerial and organisational questions have not been given much emphasis. It is high time the structure of departments (irrigation, agriculture, and cooperation) was revamped to suit farmers' needs. There has been very little inter-departmental consultation in planning the scheduling of irrigation. Operational procedures require to be reviewed in the light of agronomical practices which have changed with the introduction of high yielding varieties (HYVs) and farmers' gaining the right to choose their own cropping pattern.

A lot of research has gone into assessing crop water. Water releases need to be worked out to match crop requirement at different stages of growth. This requires rapport between farmers, agricultural experts and irrigation officials and calls for frequent meetings at the field level where each understands the others' problems and a plan to allocate water resources among the users' group is arrived at. If a management process that enables integrated and environmentally sustainable planning is devised and if there are financially sustainable farmers' organisations at the project, distributory and farm gates, it will not be difficult to develop an efficient irrigation system.

Checking anti-social elements and water-grabbers requires increased focus on policy, on institutional reforms and on programmes that will foster sustainable agricultural growth in the basin and the non-command areas of the project (uplands). This can be achieved by utilisation of the existing tanks and exploitation of groundwater, which will simultaneously help solve the problem of waterlogging and continuous decline in the water table. With comprehensive management of water resources on a multi-sectoral and river-basin basis, efficiency in the use of irrigation water can be enhanced to 50 percent from the present figure of 35 to 40 percent.

As regards policies now being followed by the central government and progressive irrigation states, the brightest part is the emphasis on the participation of civil societies (users' groups) in all water projects, i.e., major, medium, and minor. The strategy for translating water vision into action reaffirms participatory and institutional mechanisms to involve in decision-making all sections and sectors of society irrespective of caste, creed, sex, landholding, age, and literacy.

The stratified nature of rural societies, the unequal power structure and the wide variation in landholding with a large number of small and marginal farmers, and the dependence syndrome nurtured over the decades all act against effective and responsible people's participation and collective action. To overcome this, the irrigation system needs to be reoriented to provide for effective participation of users' groups. Government organisations and irrigation departments suffer from a number of infirmities which are inherent in the very nature of bureaucratic functioning and go against the concept of self-regulation for responsible utilisation of water without state control. They hesitate even to accept in principle the collective regulation by farmers' organisations above the last point of their delivery system.

They are of the opinion that major irrigation is complex in nature and has essentially three components: (i) the irrigation system itself with its extensive network of distributories and drains; (ii) the large number of farmers in the command area who form heterogeneous groups with different socio-economic backgrounds; and (iii) the irrigation department functionaries who are expected to maintain and operate the system with almost no funds for system maintenance.

PIM requires that all these components are effectively brought together so that dependable and reliable irrigation supplies are made to users on an equitable basis (depending upon the extent of landholding). To governmental bodies this appears an unachievable task for they feel that the irrigation scenario has vastly altered from what it used to be in the '60s to '80s of the previous century. The sector, in almost all states, suffers from:

(a) Financial crisis with budgets going down, low water fees, and low collection rates, etc.;
(b) Technical and managerial crisis in the light of the deteriorating physical system and poor delivery of water; and
(c) Image crisis, with major dams and canal -- irrigation being questioned on account of their adverse environmental impact.

If against this background farmers' organisations/water users' associations and government functionaries are to adopt PIM, they need to be backed by a law nurtured, supported, and groomed for sometime so that it can overcome the initial hiccups and become mature enough to carry on not only with the assigned mandate provided by law but also to innovate on their own, to improve the all-round efficiency not only of the system but of the sector as a whole which is now reported to vary from 35 to 40 percent to at least 50 percent or more.

This was pointed out by the Planning Commission at the end of the Sixth Plan itself, through its circular stating that "though considerable progress has been registered, India's development record does not match its potential, and that most of the deficiencies are rooted in the lack of involvement of the people". Thus, the need for involving people in order to accelerate the pace and enhance the judicious use of irrigation was emphasised.

It is in this background that government of AP went into framing a new Act forming Water Users' Associations (WUAs) at appropriate levels with the objective of devolving power and responsibility to beneficiaries. An awareness programme was launched and a chain of meetings were held to appraise beneficiaries and leaders of the problems ID was facing and would face and why government felt WUAs where more appropriate for tackling those problems. Farmers were encouraged to boldly narrate problems they were facing in getting their share of water and what in their view was the right solution. ID was meant to ensure that problems currently encountered would not occur once the new system of joint management started working and that ultimately the system would be transferred to beneficiary organisations. Only after general acceptance by farmers and their leaders became evident, did the state promulgate the new Act.

Andhra Pradesh Farmers' Management of Irrigation Systems Act, 1997

Implementation

The Act was passed by the state legislature in March 1997 without any opposition and was approved on 8 April 1997 by the Governor of Andhra Pradesh. The Act has 43 sections and enables the constitution of WUAs and specifies their functions. Delineation of the area of WUAs and their territorial constituencies is covered under sections 3 and 4. The Act provides for a three-tier system consisting of WUAs at the minor level, Distributory Committees (DCs) at distributory level and Project Committee (PC) at the project level. At the state level, as per section 9 of the Act, an apex policy-making body is envisaged. There is a unique provision under section 10 for recall of the president or a member of the WUA by the farmers. One-third of the members can serve a notice and the recall motion can be carried through by a simple majority in the general body meeting. Section 13 of the Act provides for readjustment of the area of operation, should it appear at a later time that some areas might be better irrigated if these were under adjacent WUAs. Sections 16 and 17 deal with the functions, rights, and responsibilities of the WUAs. Section 18 deals with the functions of DCs and section 19 with the PC. Keeping in view the importance of raising of resources, section 20 provides for fixing of fees by the farmers with the approval of the general body. Section 22 deals with the provisions for the WUA to acquire income from grazing lands, sale of fishing rights and renting out of properties attached to the Association. Section 26 pertains to disputes. The DCs were to decide disputes between WUAs and members and the PC those between DCs and WUAs. A dispute between a DC and a PC will be referred to the apex committee. Section 28 deals with the records a WUA has to maintain, while section 29 provides for a yearly financial audit of the Association. Section 31 of the Act provides for a 'competent authority', and a government functionary to assist the WUA in the discharge of its functions. To enable the competent authority to be effective, the government has delegated powers of technical sanctions to the authority. An Assistant Engineer may give technical sanctions for works up to Rs. 50,000.

Elections to the WUAs and DCs were held in June and November 1997, and 10,292 WUAs were formed. Of these, around 80 percent were on minor irrigation schemes/tanks with over 40 ha (100 acres) command. The major irrigation projects covered about 80 percent of the irrigated area in the state and around 1,700 WUAs were formed in the command of these projects, depending upon intensity and extension of irrigation in various districts. The largest number of WUAs (800) were in Karimnagar and the smallest (140) in Ranga Reddy district. Steps for the formation of these organisations can be summarised as:

(1) Generating political support;
(2) Creating favorable environment through public debate;
(3) Developing the legal framework;
(4) Forming organisations -- WUAs and DCs;
(5) Laying down with clarity the roles of farmers' organisations and of the irrigation department;
(6) Building up the skills and capacity of WUAs and departmental functionaries;
(7) Ensuring transparency and accountability in the working of WUAs; and
(8) Regular monitoring of WUAs and mid-course evaluation.

Three months after the formation of WUAs, C. Sithapathi Rao took a quick survey of 6,500 WUAs based on which he reported that about 69 percent of the presidents of WUAs were in the age-group of 25 to 50 years. Twenty-three percent were big farmers. About 80 percent were literate and of these 11 percent were graduates/postgraduates. He found even those who were illiterate had leadership qualities. In regard to the farmers' perceptions and expectations from the WUAs, Rao stated that the farmers wanted transparency in the distribution of water so that everybody knew who was getting how much water. They expected improved decision-making, management, supervision, maintenance, and quality control. Farmers' organisations were also expected to put an end to the corrupt management. In the very first year of WUAs' formation, two state level conferences of the presidents of WUAs had been organised by the ID to get feedback on their performance, firstly during crop seasons and secondly in the maintenance season. The Chief Minister participated in both conferences. The feedback indicated that maintenance works were efficiently done, irrigation disputes effectively resolved and distribution of water was satisfactory. It was also found that the farmers wanted more training to make their participation more effective. Maintenance of accounts, mobilisation of resources, development of operational plans and their implementation required training. The farmers also suggested that they be entrusted with the authority to collect water charges. Social audit was also mentioned as one of functions of WUAs which would help in:

(1) stimulating farmers' thinking about the performance of WUAs and the system;
(2) equity in distribution of water at the head and tail reaches;
(3) increase in productivity;
(4) crop diversification, if water was available;
(5) improving the quality of work; and
(6) spreading maintenance works in the command.

In January 1999, a five-day long national conference on PIM was held in Hyderabad. Soon after the conference, four groups consisting of officials of central government from 16 states, scholars, representatives of national and international NGOs, the World Bank, and the Ford Foundation were formed. Two groups visited the Sreeramasagar project, the third went to the Nagarjunasagar left bank and the fourth to the Krishna delta. They formed themselves in smaller groups and conducted field visits and had the opportunity to interact with farmers, local administrators, and officials of the irrigation department after consulting among themselves and with the chairman and other officials and farmers of WUAs, the groups met separately and reported on their field visits.

Their common observation was that there was general enthusiasm about implementing PIM and keenness to improve the irrigated area and crop yield. They also felt that the irrigation network should be transferred to beneficiary organisations, if not from the dam at least below the main canal, and that the government functionaries and office-bearers of the WUAs and DCs should be made accountable to the general body of the WUAs and DCs. They further pointed out that effective and sustainable WUAs called for greater clarity in the concept of PIM and implementation of the operational plan.

The impressions of the four groups and their recommendations were as follows:

- Legal and administrative support had built up the confidence of WUAs;
- Although as per the APFMIS Act, 1997 functions of water allocation and distribution were with the DC, an MoU between the WUAs and the agency may be necessary to allot water rights and protect the interests of WUAs;
- Continuous guidance in financial accounting was needed;
- Regular dialogue on policy issues was required to take care of changing water demands with the passage of time;
- Effective functioning of WUA required institutionalisation of procedures of resource mobilisation, levying of irrigation fees and their collection as per the Act;
- Training should be imparted to cover agricultural production, marketing, financial management, administrative skills, self-appraisal and on-farm development under the CAD programme;
- The role of women in the WUAs was currently limited and no special mention made about it in the Act;
- The present emphasis on works should be reoriented towards water distribution and regulation, and obtaining the allocated quantities at the beginning of the WUAs' area;
- Extensive maintenance work done in irrigation and drainage system by WUAs with government funding helped water to flow even to the tail-end areas, enabling early plantation and prevention of submergence of low-lying lands. This had resulted in extra yield of paddy; and
- The other strengths included political and legal support, efficient water management, existence of local leadership among WUAs, bureaucratic and technocratic support, availability of funds, and enthusiasm and confidence in participants. There were not bogus members in the WUAs.

Andhra Pradesh's concern for effective and not merely symbolic participation of users' groups is reflected in the acts and rules framed (APFMIS Act of 1997). This has already been discussed in the earlier part of this paper. In March 2001 we took up fieldwork to ascertain how far the WUA groups and competent authorities had succeeded in their role as water-users and decision-makers. A host of issues were addressed. Had they been able to create water awareness and monitor and control wasteful use of water, and to suggest and implement measures to improve the effectiveness of water supplies? How far had government officials in the role of facilitators helped WUAs? Had measures to improve the effectiveness of the system been suggested and technical and accountancy assistance provided when required? Had the Managing Committees (MCs) apprised individual farmers of the different methods with which they could generate enough funds so as to afford the full cost price of water supplied by government or private organisations?

How were the decisions taken by the MCs and general body of the WUAs? What methods were being adopted to ensure transparency and accountability? Was a social audit being carried out?

Participation, when genuine, finds its own agenda. Appropriate roles and linkages are evident, as are the democratic values of franchise (number of stakeholders with a voice in decision-making), and scope (extent or range of items on which these groups/individuals have a say in decision-making).

Responses to a questionnaire covering all the above issues directly or indirectly reflect the present scenarios; this was gleaned from interviews with 60 presidents, 104 TC members, and 23 CAOs, from books and ledgers maintained by the WUAs, other primary sources, and field visits in the command area. The picture that emerged is presented below:

There are different field-level situations, namely, areas irrigated by canals from projects, those irrigated by medium irrigation sources (chiefly tanks) and minor irrigation sources (mostly wells and ponds). In these different situations, sustainable development required improved maintenance of the irrigation system for timely and more equitable supplies.

The stratified nature of rural societies, the unequal power structure and the wide variation in landholding with a large number of small and marginal farmers, the dependence syndrome nurtured over decades, all act against effective and responsible people's participation and collective action. To address these lacunae, Andhra Pradesh Farmers' Management of Irrigation Systems (APFMIS) Act for the first time provided for collective self-regulation and paved the way for responsible utilisation of water without state control, and collective regulation below a particular level of the irrigation system.

To find out the field-level situation, our study selected from the three districts, i.e., Krishna, Kurnool, and Karimnagar, 18 WUAs in major, 12 WUAs in medium, and 30 in minor irrigation projects. In all these WUAs, the area and the command structure under the Act was fixed. Based on the area, the number of TCs varies from three to 10 in the major and medium irrigation projects. In minor projects it is at the most two (In 60 WUAs selected, there were 116 TCs, of which 36 belonged to 18 major WUAs, 24 to medium, and 56 to minor irrigation). The strength of the managing committee is laid down under the Act. It varies from five to 11. In 60 WUAs the total strength of MCs is to be 300 and in no WUA should the MC be of less than five members, i.e., at least four MC members and one president. The existing strength is as follows:

Presidents	60
MC members	277

The reason for the shortfall in MC members was that there were no nominations to 17 seats and six MC members died after election.

Eighteen major WUAs have 111 MC members along with their respective chairmen. Twelve medium WUAs have 51 MC members along with their 12 presidents, and 30 minor WUAs have 115 MC members with their respective presidents.

As for location of identified WUAs, in the case of major WUAs six are from the head-reach, five from the middle-reach, and seven from tail-end. In medium irrigation WUAs, four are from the head-reach, six from the middle-reach, and two from the tail-end. With respect to the minor irrigation system the question of location does not arise, since as per policy the entire command under the minor is entrusted to one WUA.

The area of operation of WUAs under major irrigation systems falls between one to five mandals, covering in three cases more than 15 villages per WUA. That is, the command of four WUAs falls within a mandal, that of 13 between two and three mandals, and the area of only one WUA extends over four mandals. In the case of medium irrigation the command area of five WUAs is in one mandal, and that of remaining seven in two to three mandals. The majority of WUAs, i.e., about 47 fall under five villages. The area of eight WUAs extends between five to 10 villages, that of two WUAs is between 10 to 15, and the area of three WUAs extends to more than 15 villages. The command of most of the minor irrigation system WUAs is within one village.

Command areas of the sample WUAs vary under major, medium, and minor systems. The maximum area under major WUAs is 15,970 acres and the minimum 1,151 acres. Under medium WUAs, the maximum area is 4,200 acres and the minimum 788 acres. Whereas in the minor irrigation it is 866 acres and 118 acres respectively.

The total command area of 18 major WUAs is 94,678 acres, that of 12 medium WUAs 25,448 acres and the cumulative total of all the 30 minor WUAs only 9,815 acres. District-wise localised area of identified WUAs is 74,968 acres of wet in Krishna, 10,486 acres under wet and 23,792 acres under irrigated dry in Kurnool, and 10,449 wet and 10,231 acres under irrigated dry in Karimnagar.

Within their area of operation, groups of WUAs, after forming DCs, were to be given powers that today are vested in the irrigation department so that the institution-based DC level associations could be strengthened to represent various interest groups and make the irrigation department accountable to the farmers.

That this transfer of power had not taken place was reflected in the very fact that in two WUAs under the medium projects no water was let out in the years 1997-98 and 1999-2000, and in quite a number of distributories designed discharges were not registered. But no effective remedial measures were taken up. The reason given out by the ID was inadequate flow or no water in the two medium projects. No compensation for this non-supply was paid which clearly indicates that the farmers' rights to water is not given due recognition in the Act. This has to be remedied by the framing of suitable rules, failing which the accountability and transparency required for effective PIM will not be established.

As per the Act, irrespective of caste, gender, and landholding, all the beneficiaries under the command have one vote to elect president and for MC members, an individual member, enjoys as many votes as number of MC members of WUA, but will have to cast only one vote for one person in a territorial constituency. Significantly, in the sample set of 60 WUAs no woman was elected or nominated as president and only four were chosen out of 277 MC members. It is also worth noting that 34 WUAs were constituted through election and 26 were unanimously formed. It was mainly the major irrigation WUAs that were contested -- 15 out of 18. The predominance of unanimously formed WUAs can be accounted for by the government offer or Rs. 50,000 for bodies so formed as compared to Rs. 30,000 for elected WUAs.

In two WUAs in the minor irrigation system there was no nomination of MC members. The president formed an ad hoc management committee. In 17 constituencies no proposals were received for MC membership. The reluctance could be attributed to other avocations or unwillingness to work without remuneration.

In traditional Indian village, caste still plays an important role despite 50 years of democratic rule. Though education, awareness, and motivation, are important prerequisites for the formation and functioning of WUAs, our study indicates that more importance was attached to caste and size of landholding. Of the 60 presidents, 47 belonged to other castes (higher castes) and only 12 to the backward classes. There was only one scheduled caste president, that too in the minor irrigation systems. Likewise, in the case in the composition of MC members, 137 MC members were from other castes followed closely by 105 from the backward classes. Only 35 belonged to the scheduled castes. Scheduled caste MC members composition in major, medium and minor projects was 10, 6 & 19 respectively.

The landholding pattern of the presidents indicated the dominance of big farmers with 28 of them possessing landholdings of more 10 acres, followed by 9 middle-level farmers with holding varying between 5-10 acres, and 12 small farmers with holding in 2.5-5 acre range. There were 11 marginal farmers owning less than one acre. Of these 10 belonged to minor irrigation systems.

With regard to age composition, only one president was over 70 years old, eight were below 35 years, and the remaining 51 were middle-aged.

As presidents and MC members, farmers must be able to take up responsibilities of system (operation & maintenance), and water distribution. This calls for qualities of leadership and commitment. The present lot of presidents is very effective in dealing with irrigation authorities but fails to provide leadership in physical maintenance of the system.

It is hoped that with more awareness members of the WUA will be amenable to the idea of voting for chairman and members on the basis of their leadership qualities, rather than on consideration of caste, size of landholding, and age, as at present.

The Act lays down that irrigation department will make available all maps and hydraulic records to WUAs within their area, so that they may understand the hydrological system and be aware of the number of beneficiaries. But the available records with the president of the society or with its other members indicates that out of 60 WUAs, ayacut maps of only 38 were provided with register of landholding, and property registers of only 18 WUAs were made available

Allocation of water depends on composition of the landholdings, and the setting up of hydraulic boundaries in the absence of ayacut maps and property registers, is bound to be problematic. Laying down or revising field channels to avoid tail-end problems is rendered difficult. Attempts made by the WUAs to obtain the relevant records failed. It is now imperative that the government steps in to see to it that all records needed are made available at the earliest. Other than these records which are to be updated, WUAs are expected to maintain cash-books, pass books, minutes books, cash maintenance books, ledger books and the like. Out of the 60 WUAs studied, 57 were found maintaining the cash-books and pass books, and 59 the minute books. So it is evident that with proper training and motivation it is not difficult for the WUAs to maintain all books and registers as per the Act and rules.

The president is the custodian of all the records and is responsible for their maintenance. In 44 WUAs, records and accounts were being maintained by their respective presidents, in 5 cases they were being maintained by the MC members, and only in 11 cases were services of outsiders utilised. Records of four WUAs were in the custody of the competent authority, one WUA's record was with an MC member and rest were in offices of the respective WUAs.

The financial transactions of the WUAs indicate that these 60 WUAs received Rs. 118.31 lakhs as incentive grant at the time of their formation and later Rs. 7.78 lakhs as special grant. With the introduction of PIM, government transferred to the WUAs a sum of Rs. 58.78 lakhs to take up O&M. For rehabilitation work by the WUAs in their area of operation, government provided Rs. 29.17 lakhs from the Andhra Pradesh State Hazard Mitigation (APSHM). In addition, WUAs raised Rs. 7.78 lakhs from their internal resources. Of these financial resources amounting to Rs. 221.87 lakhs, Rs. 18.90 lakhs were spent towards repairs to shutters, sluices, gauges, irrigation channels, UTs, gear boxes, culverts, channels, tank bunds, and ayacut roads; Rs. 63.19 lakhs for rehabilitation of water distribution works which included items like lining of irrigation canals, rehabilitation of water distribution canals, construction of structures on canals, revetment of canal banking, construction repairs to guide wal m(.6Uaddition to silt removal and in medium irrigation sources chiefly desilting of tanks and major distributories.

The success of an organisation can be judged by general body meetings allowing open discussion. It was noticed that in all general bodies the chairmen of the WUAs reported comprehensively on the working of the WUAs, accounting for the various decisions taken by the MC members as well as the spot decisions made at the time of the 'walkthrough' besides presenting audited accounts.

The very fact that more and more general body meetings were convened year after year indicates that the WUAs are not only keen on making their working transparent to their members but also on seeking out their views at frequent intervals. The meetings organised in the period 1997-2000 indicate that in the first year of the working of the 60 WUAs, 54 general body meetings were held while in subsequent years 47and 93 meetings were conducted in addition to the statutory ones to discuss steps to be taken with regard to various aspects of management of the irrigation system. Increase in meetings was noticed in major and minor systems, whereas in medium irrigation out of 12 WUAs in the first year only seven conducted statutory meetings. But in the second year of their working, these two rose to 25. In the third year, 20 meetings were held. This clearly indicates that WUAs under all the three systems conducted the statutory general body meetings from the second year onwards.

Effective management of water resources entails that attention is paid to four major aspects: water regulation, routine maintenance, major repairs, and social issues. It was found that general body meetings held between 1997-99 did take up the task. In 46 such meetings, problems and solutions in efficient water regulation were addressed, members were informed of maintenance works to be taken up in 117 cases. In 52 meetings, repairs taken up, costs involved, and benefits accrued were discussed, while 44 meetings were called exclusively to place social audit reports before their members indicating how social issues hindering water delivery were tackled.

In addition to the above mentioned topics, in the 104 routine general body meetings organised under the statute, steps taken for control of water through regulation, maintenance of the irrigation network and field channels, and methods to be adopted to prevent of soil erosion and seepage in the canal system, and water logging due to uncontrolled releases were discussed. Specialists were invited to provide latest knowledge with regard to the most productive and economic use of water.

In all the 60 WUAs, MC meetings were convened at least thrice a year at different locations to discuss water regulation, distribution, and resource management. In 197 meetings, work programmes to be taken were discussed, and in 270 MC meetings administrative approvals were obtained for all the works decided on a priority basis such as bund repairs, drainage works, and maintenance of the main drainage system. To meet the extra amount needed for the work over and above government grants, 10 MC meetings were called resulting in the collection of about Rs. 1,00,000.

To assess sustainability of WUAs and members' interest in the same, direct multiple choice questions were asked at different time intervals. Answers to the same by MC members, ordinary members, and presidents reflect a cross-section of views regarding fund-raising, social audit, formation of sub-committees, etc.

With regard to the possibility of raising internal resources, 38 initially were not in favour, 20 were non-committed, and only two responded favourably. But by the final round many reported a positive response. Similarly, 54 WUAs favoured the idea of social audit by its members. Forty-nine WUAs were keen that the general body select the committee members for social audit, and six were for the formation of the sub-committee by the competent authority. Most of them were of the opinion that it will be more beneficial if social audit reports reveal in more detail social problems and how they are tackled and by whom. They were also of the view that in general body meetings along with social audit reports of their own WUAs, one or two good working WUAs' social audit be placed to help them in finding out how in different situation problems occurred and were tackled.

As regards formation of sub-committees, almost all WUAs' members were for it. But only 12 out of 18 major WUAs formed separate sub-committees for water regulation, works execution, finances and resources, and monitoring and & evaluation. Whereas out of 12 medium WUAs, six sub-committees were formed for water regulation, five for work execution, five for financial resources and six for monitoring and evaluation. As for 30 minor irrigation WUAs, six sub-committees were formed for water regulation, eight for work execution, six for finance and resources, and seven for monitoring and evaluation. It is therefore, clear that there has not been appreciable progress in the formation of sub-committees. The reason appears to be that the role of these sub-committees is not effectively project before the general body. Moreover, in some cases WUAs feel that the MC can tackle most of the issues earmarked for sub-committees.

To ensure optimal handling of water, it is imperative that the state works out and adopts a strategy to mobilise funds and, among other things, bring about appropriate coordination between engineering department officials on the one hand and the beneficiary farmers on the other. Under the Act, the state created competent authority and depending on the area and type of work to be undertaken by the WUA or DC, nominated Assistant Engineer (AE)/Assistant Executive Engineer (AEE)/Junior Engineer (JE)/Deputy Executive Engineer (DEE)/Executive Engineer (EE) for different levels of work.

Improving irrigation management is an indispensable part of the water-augmenting scheme. This cannot be left entirely to departmental officers or their field staff, however effective they may be. Their best endeavors would remain insufficient unless farmers were actually involved. Keeping this in view the joint role of WUAs and the competent authority was envisaged. In most of the WUAs this close coordination was noticed in the following activities:

> "Undertaking walk-through survey, conducting meetings, preparing estimates, maintaining records, executing works, ensuring quality control, recording measurements, making bill payments, preparing operation plans and making area assessments."

For PIM to succeed it is necessary that departmental personnel do not deem community participation as an infringement on their work. PIM needs to be carried out in a carefully planned manner which would allow adequate time for training to be imparted so as to motivate farmers and field functionaries to modify their existing practices. As regards improving irrigation management, certain steps were taken like improving the distribution system for providing water flow. Measurement devices were fixed at the head of the distributories and moves to have the same done at the head of the WUAs was considered necessary.

To see to it that water theft was controlled if not eliminated, and to plug all loopholes responsible for losses in the system, groups were organised to patrol the minors, walkthroughs were organised for identification of works to be taken up, and irrigation department functionaries along with WALAMTARI and NGOs took steps to impart training to MC members with regard to maintenance of the irrigation system and water-related records. Farmers in general were requested to adopt new practices of application of water for wet as well as irrigated dry crops.

Based on our field assessment and perusal of records, we noted that major changes were brought about by implementation of all the above-mentioned activities, thanks to effective PIM. The major improvements noticed are as follows:

Increase in irrigated area of 24 WUAs, increased supply of water to 43 WUAs, decline in water-related disputes in 38 WUAs, better departmental coordination in 52 WUAs, increase in the water charges recovery in 13 WUAs, and improved crop productivity in 44 WUAs. Forty-three WUAs received water for the first time in the tail-end and involvement of farmers in O&M has improved in 35 WUAs. Training of water users on PIM was under way in 36 WUAs, and raising of internal resources was noticed in 36 WUAs. Awareness of social problems and the desire to resolve the same due to presentation of social audit was noticed in 51 WUAs. Groundwater level increased in three WUAs. But the most important change was that for the first time there was equity in water supply in three WUAs, and water was made available to all members of all the 60 WUAs.

To check how far principles of PIM were absorbed and implemented by the irrigation functionaries, 23 competent authority officials (CAOs) were contacted, ten looking after major irrigation projects, five from medium, and eight from minor irrigation systems. Their multiple responses were gathered so as to find out their perceptions about PIM as envisaged in the Act, and to assess ID officials' involvement in it.

According to them, between 1997-99, they organised 269 meetings to motivate farmers and to explain the complexities of the irrigation system. In the 18 major WUAs, 108 meetings were conducted, 59 in medium, and 30 in minor. Of this, the specific purpose of 46 meetings was to discuss problems based on the present status of water regulation, and obtain views of both users and suppliers for improving the same. The action taken on the basis of these meetings resulted in improved water supply in 26 WUAs. Eight ID officials stated that supply remained as it was earlier and no marked change was noticed in the upkeep of the system also. Ten officials mentioned they were not aware whether any changes were effected by the members of the WUA. This clearly indicates that some competent authorities were not taking PIM seriously. Similarly, lack of interest was revealed when no replies were forthcoming when they were asked about steps taken for maintenance of water-related records. It emerged that records were maintained by only 26 of the 60 WUAs, with the rest blaming the department for not only not transferring the irrigation schedule records to be maintained but also not fixing the measuring devices or providing alternate methods of collecting information with regard to inflows.

As for water distribution practices, it was reported that 15 WUAs were still continuing with flood irrigation. Only 24 WUAs adopted the practice of supplying water as per the demand and satisfied their members fully. Eleven WUAs were found to meet demand largely through mutual consent of the farmers.

The overall perception of the competent authorities was that in their areas, water allocation was as per the availability in the system and not as designed. If water was insufficient they informed the WUAs of the same. They also mentioned that in 50 WUAs though allocation of water was far short of the designed amount, in 47 WUAs water was allocated in proportion to the area under the command. They noticed that 46 WUAs succeeded in convincing their members that if they took up the maintenance of irrigation system along with cleaning of field channels water could be had up to the tail-end. Most of the government officials were of the opinion that water distribution in 45 WUAs was better than it was before the implementation of PIM. Information as regards water availability, water distribution, time of release, and schedule of release was now being made available to all the members by their respective WUA managing committee, and well before the release the irrigation network was made to confirm to the original design with WUAs clearing debris and weeds and strengthening the canals bunds.

As reported by the competent authorities (CAs), there was excellent transparency in the functioning of 50 WUAs. This they attributed to the presence of competent authority and MC members in the general body meetings of the WUAs. They also reported that in 15 general body meetings, senior project officials, territorial constituency members, and farmers also took active part in discussing various issues raised. The presence of CAs in 83 meetings indicated the interest taken by them in trying to understand the working of the WUA.

In addition to these meetings, discussions were organised by forming 98 groups with an average participation of 12 to 13 members in each group, the total number of participants being 1,076. The suggestions made to improve the performance of WUAs listed below reflected the shift from passive use of the irrigation system to active participation in all sectors leading to efficient and sustainable irrigation:

(a) MC interaction with general body to be improved;
(b) interaction of members of the MC with farmers should be improved;
(c) auction right (fishing & grass) should be given to WUAs;
(d) more funds to be sanctioned;
(e) formation of ayacutdars;
(f) drains to be maintained;
(g) weed removal to be performed every year;
(h) repairs to the canal;
(i) TC members should be active;
(j) mandal level committees on minors should be formed;
(k) regular monitoring;
(l) control irrigation of non-ayacut area if not possible to stop it completely;
(m) raising funds through membership drive to form corpus fund;
(n) WUAs to be given right of water cess collection;
(o) water cess should be collected from non-ayacutdars; and
(p) improved agriculture department cooperation.

What is significant is that in all WUAs group members along with other farmers took active part in walkthroughs not only identifying works to be taken up, but later following up their progress and reporting lapses in quality, if any. This active involvement was responsible for 82 works being completed in time. Moreover the quality of work in 85 cases out of 88 was reported as good. It was also noticed that in the execution of the works additional funds when required were always to be had. For 25 works cash was provided, and in 28 works material and labour contributed by the WUAs. The competent authorities were found cooperating actively with WUAs and individual farmers in water regulation, cleaning, and maintenance of the irrigation network. The most effective change was noticed with regard to walkthroughs which were taken up by 58 WUAs, the two exceptions being two WUAs under minor irrigation. In almost all walkthroughs, MC members, CAs, and a majority of the farmers actively participated to identify 308 works to be taken up, and prioritised 240 works. Due to close coordination and better understanding between MCs and its CAs, out of 240 prioritised works, administrative and technical sanction were obtained in 204 cases and work was taken up without loss of time.

Information with regard to improved water supply was gathered from 98 groups involving all the presidents, one or two members of the MC, irrigation officers, and neighboring WUAs. Interviewed at different intervals, their replies indicate that though water supply was not quite perfect, availability had improved and that PIM was working on the right lines towards achievement of equity. They appreciated the official keenness to provide designed discharge at schedule fixed, although due to many mishaps they failed to do so in the entire system. But anyway, seven out of 18 major WUAs and four out of 12 medium WUAs, almost achieved the same. When 23 CAs were contacted, they painted even a brighter picture reporting that water supply was improved in nine major, nine medium, and eight minor WUAs.

Over these three years, PIM has now emerged as an administrative and technical system substituting the traditional, administrative machinery which was more bureaucratic, more costly, and less responsive to the real problems at the field level. But PIM still has some way to go, such as formation of project level committees (PLCs) which could give beneficiaries a say in policy matters like allocation of the water, its release to different sub-areas of the project, collection of water cess by the WUAs and conjunctive use of water by taking the basin as a hydraulic unit. Also, collection of water charges remains with the revenue department, though some improvement in the same was noticed.

Working of special schemes and addressing problems of the tail-enders are still in early stages.

Despite the shortcomings and the non-implementation of some of the suggestions made by the groups, the three districts from where these WUAs were randomly selected indicate a positive change in the area irrigated and the relationship between water users and government authorities. In spite of poor rainfall resulting in inadequate inflows in the tanks, there is a continuous increase in irrigated area. While in 1997-98, 10,045 acres of the localised area of 1,29,941 acres (i.e., 77.33 percent) was irrigated, in 1998-99 it was 1,01,342 acres (77.99 percent), and in 1999-2000 1,26,629 acres (97.54 percent).

The cropping pattern has also undergone a change with the introduction of pisciculture where sufficient water for wet or irrigated dry crops is not to be had.

As regards the working of CAs, we noticed that most of them were attached to the same WUAs for the past three years, but only few were actually monitoring the working of the WUAs. When suggestions were called for improving their performance, four of them were of the view that farmers should cooperate with CAs as this would go a long way in improving the efficiency, and 26 CAs were of the opinion that MC members' cooperation in system maintenance needed to be strengthened. Seven CAs were of the view that if the WUAs took up area assessment not only would the gap between the designed command and actual irrigated area become clear, the unauthorised area now being irrigated would also be revealed.

A majority of the CAs felt that the government needed to share up funding; WUAs should also be enabled to generate funds on their own through right to auctioning grass, and collection of cess from fishing rights being given to them. They were also of the view that WUAs be allowed to fix and collect water cess. Some pointed out that holding MCs at shorter intervals and taking up more frequent walkthroughs would greatly improve the performance of the irrigation system. Twenty out of the 23 CAs considered appointing professionals for maintaining the records highly essential for sustainable maintenance of the system.

Restructuring Irrigation Administration in India

The results of our study show that in all states of India, it is possible to motivate farmers to adopt PIM and that irrigation officials are not necessarily bent on undermining PIM and its turnover. Some are attracted (chiefly in Tamil Nadu, Karnataka, Bihar, and Haryana) by new ideas and would like to try them out. But they find the existing rules and regulations of the department a constraint. They feel within the framework of the existing administrative rules and procedures, FOs have a little chance of surviving as an independent self-managing social organisation. This is substantially proved by FOs like Mohini (Gujarat) failing after successfully working for 10 years. The next step forward for them therefore does not lie in how to organise FOs, what manpower to have or what training to impart, but how to overhaul the administrative system by changing the Act itself so that the state irrigation department and farmers can be bound in a productive functional relationship.

Making a success of PIM and turnover calls for comprehensive overhauling of the old irrigation acts that have served as models to support a culture of exclusiveness no longer in tune with democratic values and the growing dominance of the public domain in post-independence India. State irrigation departments will have to be restructured in order to give them new direction. The functions and powers of officials will have to be redefined with reference to new functional (territorial) units wherein farmers will have greater responsibilities as regards maintenance and supervision and greater say in the administration of irrigation. The irrigation officer will have to be made more directly accountable to water users but must also have the authority to enforce discipline and to bring law-breakers to book. Hopefully, a "culture of partnership" will replace present attitudes wherein each party sees the others as not doing what is expected. This means that state irrigation acts and rules under them will have to be rewritten. The initiative clearly lies with top policy-makers in the states and the government of India.

Irrigation laws have to be changed to provide for the involvement of the users and their participation in management. The following changes in irrigation laws seem to be essential:

1. Irrigation Acts should be enabling laws, so that PIM initiatives become possible. Many PIM models have been tried out but few have been adopted for general implementation. In the laws enacted for command area development, most states provided for village-level committees consisting of water users and villages. Thousands of committees were formed but few have survived. They were given many responsibilities, but not the authority or the resources to enforce decisions. Most significantly, they could not hold irrigation officials accountable for non-delivery of water at the outlet point.
2. Within their area of operation, water users' associations (WUA) or farmers' organisations (FOs) have to be given powers that today are vested in state irrigation departments so that the institutional base of village-level associations can be strengthened. In particular, WUAs/FOs require an independent resource base and an enabling organisational structure which represents various interest groups and makes the organisations accountable to farmers.
3. The farmers' right to water (that is, the agreed quantum) has to be recognised in law and also the rules framed under it. Without this, the overriding principle of accountability and transparency cannot be established. Both are needed to make PIM strong and functional.
4. The old irrigation Acts were based on the concept of a social contract that is not appropriate in the present times. Instead of legitimising a top-down hierarchical relationship, the new social contract has to legitimise relationship of co-equals or partnership between the irrigation bureaucracy and the water users.

Addressing Survival Factors in Sustainability

Sustainability in government-owned irrigation systems depends on two sets of factors: start-up factors, that are involved in the establishment and functioning of FOs in the initial years, and survival factors that contribute to sustainability over the long run. Vital among the start-up factors are the emergence of strong leadership, group management, efficient communication and information sharing within the management group and between the management and water users. Important survival factors are self-renewing leadership and group management, independence, legal legitimacy; authority/independence resource generation and control over funds, periodic main system upgradation by ID, etc.

A Note of Caution

PIM and turnover should not be implemented gradually, starting from outlet commands to minors, to distributories and then on to the irrigation project as a whole. This would ensure failure. There are two reasons: First, each level is interconnected. The relations between succeeding and alternate levels will have to be established, problems identified and resolved and acceptable work procedures developed. These cannot be accomplished in a step-wise fashion. Revolution rather than evolution mark the space of decisive changes in the social realm. The entire irrigation system should be brought under the participatory regime if turnover is to work successfully.

Secondly, PIM in India would directly challenge the value system of the administration where hierarchy, deference to superiors and lack of transparency in public dealings are taken for granted. Unless these values are challenged, PIM and turnover cannot take root. The only way to attack them directly in the public realm is by bringing about wholesale changes in one go rather than adopt a step-wise "let us see" approach to change.

Not all irrigation projects will be equally hospitable to PIM. States under Warabandi where water allowance is low, relative to the size of the command, might find it difficult to implement. States following the demand system or where irrigation supplies are generally continuous should find it less arduous. Implementation strategies will have to be worked out taking into account administrative practices, soils, crops, groundwater exploitation and the cultural background of the communities involved.

REFERENCES

An Approach to Organizational and Procedural Change in Irrigation Sector, GOI, New Delhi, 1992.

Anarchy, Paternalise or Collective Responsibility under the Canal, EPWD, December 1978, Henry Hart.

Andhra Pradesh Economic Restructuring Project, Irrigation Component, Project Implementation Plan, Government of AP, 1998.

Guidelines for Preparation of Project Report for Surface Water Minor Irrigation Work, Irrigation Department, Government of AP, 1986.

Improving Irrigation Water Management with Farmers' Organization and Participants, Norman Uphoff, 1985.

India: Irrigation Management Partnership, Sivamohan, M.V.K. & C.A. Scott edited by Book Links, Hyderabad, 1994.

Irrigation and Its Management in India: Need for a Paradigm Shift, National Conference on PIM, January 20-22, 1997, L.K. Joshi (ed.), MOWR, GOI Publication.

Irrigation System in Decay, Singh K.K., Turabul Hassan, & C. Sithapathi Rao, 1997.

Multi-disciplinary Management of Irrigation Schemes & Command Area Development Approach to India, HCM Rajasthan State Institute of Public Administration, Jaipur, Paper No.2, Hooja Rakesh, 1995.

National Seminar on Water Management: The Key to Development Agriculture, (ed.) J.C. Kakgar.

National Water Resource Policy, Ministry of Water Resources, GOI, 1987.

On the Water Front, Mollinga, Peter P., 1998, Wageningen University, Netherlands.

Paper on Optimum Utilization of Water, Its Control and Remedies, S. Turabul Hassan, Workshop Paper, Gandhi Institute of Studies, Varanasi, July 1981.

Programme on Policy and Management of CAD, 30th September to 5th October, 1985, ASCI, Hyderabad (reading material & edited by Dr. K.K. Singh).

Report of Irrigation Commission, Ministry of Irrigation & Power, GOI, 1972.

Report of the Commission for Irrigation Utilization, Government of AP, 1982.

Report of the National Commission on Agriculture, Ministry of Agriculture & Irrigation, New Delhi, 1976.

Status of PIM in India: A Documentary Critique, ISS Manuscript Report (report prepared by Deepak Gupta & L.R. Srivastava) November, 1999.

The Andhra Pradesh Farmers' Management of Irrigation Systems Act of 1997.

The History of Cauvery Mettur Project, Mr. C.C. Barber, 1936.

The History of Lower Bhavani Project, published by Irrigation Department of Tamil Nadu, 1965.

The History of Periyar Project, Mr. A.T. Machinze, 1896.

Vested Interest and Administration -- Paper of Seminar, Kandapal presented in various workshops, 1981.

Vijayanagar Channels: A Traditional Irrigation System in Transition, M.V.K. Sivamohan, article published in "Farmers in the Management of Irrigation System", edited by K.K. Singh, published by Sterling Publications, 1991.

Vision 2020, Hyderabad, Government of AP, 1999

Water Users' Associations in AP, paper presented by Raymond J. Peter in 1999 at the National Conference of Participatory Irrigation Management, NIRD, Hyderabad.

APPENDICES

APPENDIX I

Andhra Pradesh Farmers' Management of Irrigation Systems Act, 1997

The following Act of the Andhra Pradesh Legislative Assembly received the assent of the Governor on the 7th April, 1997 and the said assent is hereby first published on the 9th April, 1997 in the Andhra Pradesh Gazette for general information:

Act No.11 of 1997.

An act to provide for farmers' participation in the management of irrigation systems and for matters connected therewith or incidental thereto.

Whereas the state of Andhra Pradesh in essentially an agricultural State depending on an efficient and equitable supply and distribution of water which is a national wealth, ensuring optimum utilisation by farmers for improvement of agricultural production is the immediate need;

And whereas scientific and systematic development and maintenance of irrigation infrastructure is considered best possible through farmers' organisations;

And whereas such farmers'organisations have to be given an effective role in the management and maintenance of the irrigation system for effective and reliable supply and distribution of water.

Be it enacted by the Legislative Assembly of the State of Andhra Pradesh in the Forty-eighth Year of the Republic of India as follows:

Chapter I
Preliminary

Short Title, Extent and Commencement

1. (1) This Act may be called the Andhra Pradesh Farmers' Management of Irrigation Systems Act, 1997.

(2) It extends to the whole of the state of Andhra pradesh.
(3) It shall come into force on such date as the government may, by notification in the Andhra Pradesh Gazette, appoint and they may appoint different dates for different areas and for different provisions.

Definitions

2. (1) In this Act, unless the context otherwise requires:
(a) 'area of operation' in relation to farmers' organisation means a contiguous block of land in the command area of an irrigation system as may be notified for the purposes of this Act.
(b) 'ayacut road' means a road within the area of operation of a farmers' organisation for the purpose of irrigation and agriculture but does not include a road vested in gram panchayat, mandal parishad, zilla parishad, municipality, municipal corporation or roads and buildings department of the government.
(c) 'command area' means an area irrigated or capable of being irrigated either by gravitational flow or by lift irrigation or by any other method from a government or the corporation source and includes every such area whether it is called 'ayacut' or by any other name under any law for the time being in force.
(d) 'competent authority' means the authority appointed under section 21.

Act 12 of 1997

(e) 'corporation' means the Andhra Pradesh Water Resource Development Corporation constituted under the Andhra Pradesh Water Resources Development Corporation Act, 1997;
(f) 'distributory system' means and includes:
(i) all main canals, branch canals, distributories and minor canals constructed for the supply and distribution of water for irrigation;
(ii) all works, structures and appliances connected with the distribution of water for irrigation; and
(iii) all field channels and other related channels and structures under a pipe outlet.
(g) 'district collector' means the collector of the district in which the irrigation system is situated and includes any officer specially notified by the government to perform all or any of the functions of the district collector under this Act;
(h) 'drainage system' in relation to an irrigation system includes:
(i) channels either natural or artificial, for the discharge of waste or surplus water and all works connected therewith or ancillary thereto;

(ii) escape channels from an irrigation or distribution system and other works connected therewith, but does not include works for removal of sewage;

(iii) all collecting drains and main drains to drain off surplus water from field drains; and

(iv) all field drains and related structures under pipe outlets.

(i) 'farmers organisation' wherever it occurs, shall mean and include:

(i) water users' association at the primary level consisting of all the water users, as constituted under section3;

(ii) distributory committee at the secondary level, as constituted under section5;and

(iii) project committee at the project level, as constituted under section7;

(j) 'field channel' includes a channel existing or to be constructed by the government or by the land holders or by any agency to receive and distribute water from a pipe outlet.

(k) 'field drain' includes a channel excavated and maintained by the other agency, to discharge waste or surplus water from the land holding under a pipe outlet; and includes drains, escape channels and other similar works existing or to be constructed.

(l) 'financial year' means a year commencing from the 1st April of the relevant year to the 31st March of the ensuing year.

(m) 'financing agency' means any commercial bank, or any co-operative society or any other bank or organisation established or incorporated under any law, for the time being in force, which lends money for the development of the area of operation of the farmers' organisation.

(n) 'government' means the state government of Andhra Pradesh

(o) 'hydraulic basis' means the basis for identifying a viable irrigated area served by one or more hydraulic structures such as head works, distributories, minors, pipe outlets and the like.

(p) 'irrigation system' means such major, medium and minor irrigation system for harnessing water for irrigation and other allied uses from government or corporation source and includes reservoirs, open head channels, diversion systems, anicuts, lift irrigation schemes, tanks, wells and the like.

Explanation

(1) 'Major Irrigation System' means irrigation system under major irrigation project having irrigable command area of more than 10,000 hectares.

(2) 'Medium Irrigation System' means irrigation system under major irrigation project having irrigable command area of more than 2,000 hectares and upto 10,000 hectares.

(3) 'Minor Irrigation System' means irrigation system under minor irrigation project having irrigable command area upto 2,000 hectares.

Act 26 of 1971

(q) 'land holder' means an owner and/or a tenant recorded as such in the record of rights under the Andhra Pradesh Record of Rights in Land Act 1971 in respect of land in the notified ayacut area of an irrigation system.

(r) 'maintenance' means execution of such works on the irrigation system as are necessary to ensure that the physical system designed to the standards operates for proper distribution of water to the land holders in the area of operation.

(s) 'notification' means a notification published in the Andhra Pradesh Gazette, and the expression 'notified' shall be construed accordingly.

(t) 'operational plan' means a schedule of irrigation deliveries with details of the mode and duration of supplies drawn up for regulation of irrigation in the command area of an irrigation system.

(u) 'prescribed' means prescribed by the government by rules made under this Act.

(v) 'warabandi' means a system of distribution of water allocation to water users by turn, according to an approved schedule indicating the day, duration and the time of supply.

(w) 'water allocation' in relation to an irrigation system means distribution of water determined form time to time by a farmers' organisation in its area of operation.

(x) 'water user' means and includes any individual or body corporate or a society using water for agriculture, domestic, power, non-domestic, commercial, industrial or any other purpose from a government or the corporation source of irrigation.

Act 15 of 1984

(2) The words and expression used in this Act, but not defined, shall have the same meaning assigned to them in the Andhra Pradesh Irrigation Utilisation and Command Area Development Act, 1984

Chapter 11
Farmers' Organisation

Delineation of Water Users' Area and Constitution of an Association

3. (1) The district collector may, by notification and in accordance with the rules made under this Act, in this behalf, delineate every command area

under each of the irrigation systems on a hydraulic basis which may be administratively viable; and declare it to be a water users' area for the purpose of this Act.

Provided that in respect of the command area under the minor and lift irrigation systems, the entire command area may, as far as possible, form a single water users' area.

(2) Every water users' area shall be divided into territorial constituencies, which shall not be less than four but not more than ten, as may be prescribed.

(3) There shall be a water users' association called by its local distinct name for every water users' area delineated under sub-section (1).

(4) Every water users' association shall consist of the following members, namely,

(i) all the water users who are land holders in a water users area; provided that where both the owner and the tenant are land holders in respect of the same land, the tenant;

(ii) all other water users co-opted in a water users' area;

(iii) members specified in clause (i) and (ii) shall constitute the general body for a water users association;

(iv) a person eligible to become a member of more than one territorial constituency of a water users' association under clause (i) shall be entitled to be member of only one territorial constituency and he shall exercise his option thereof as prescribed;

(v) members specified in clause (1) shall alone have the right to vote.

Election of President and Members of the Managing Committee of Water Users' Association

4. (1) There shall be a managing committee for every water users' association.

(2) The district collector shall make arrangements for the election of President of the managing committee of the water users' association by direct election by the method of secret ballot in the manner prescribed.

(3) The district collector shall also cause arrangements for the election of a managing committee consisting of one member from each of the territorial constituencies of a water users' area, by the method of secret ballot in the manner prescribed.

(4) If at an election held under sub-sections (2) and (3), the president or the territorial constituencies of water users' association are not elected, fresh elections shall be held in the manner prescribed.

Provided that the government for the reasons to be recorded in writing may, from time to time, postpone elections.

(5) The president and the members of the managing committee shall, if not recalled earlier, be in office for a period of three years, from the date of the first meeting.

(6) The managing committee shall exercise the powers and perform the functions of the water users' association.

Delineation of Distributory Area and Constitution of the Distributory Committee

5. (1) The government may, by notification and in accordance with the rules made in this behalf, delineate every command area of the irrigation system, comprising of one or more water users' associations, and declare it to be a distributory area for the purpose of this Act.

(2) There shall be a distributory committee called by its local distinct name for every distributory area declared as such under sub-section(1).

(3) All the presidents of the water users' associations in the distributory area, so long as they hold such office, shall constitute the general body of the committee.

Election of President and Constitution of Managing Committee

6. (1) There shall be a managing committee for every distributory committee.

(2) The district collector shall cause arrangements, in the manner prescribed, for the election by the method of secret ballot of the president and members of the managing committee who shall not be more than five from among the members of the general body of the distributory committee. Provided that the government may, for the reasons to the recorded in writing, may from time to time postpone elections.

(3) If, at an election held under sub-section (2), the president and the members of the managing committee are not elected, fresh elections shall be held in the prescribed manner.

(4) The term of office of the president and the members of the managing committee shall, if not recalled earlier, be coterminous with the term of the general body specified in sub-section (3) of section 5.

(5) The managing committee shall exercise the powers and perform the functions of the distributory committee.

Delineation of Project Area and Constitution of Project Committee

7. (1) The government may by notification and in accordance with the rules made under this Act in this behalf, delineate every command area or

part thereof, of an irrigation system and declare it to be a project area for the purposed of this Act.

(2) There shall be a project committee called by its distinct name for every project area declared under subsection (1).

(3) All the presidents of the distributory committees in the project area, so long as they hold such office, shall constitute the general body for the project committee.

Election of Chairman and Constitution of the Managing Committee

8. (1) There shall be a managing committee for every project committee.

(2) The district collector shall cause arrangements in the manner prescribed for election by the method of secret ballot of chairman and managing committee consisting of not more than nine members from among the members of the general body of the project committee. Provided that the government for the reasons to be recorded in writing may, from time to time, postpone elections.

(3) If, at an election held under sub-section (2), the chairman and the members of the managing committee are not elected, fresh elections shall be held in the prescribed manner.

(4) The term of office of the chairman and the members of the managing committee shall, if not recalled earlier, be coterminous with the term of general body specified in sub-section (3) of section 7.

(5) The managing committee shall exercise the powers and perform the functions of the project committee.

Apex Committee

9. (1) The government may by notification, constitute an apex committee with such number of members as may be considered necessary.

(2) The committee, constituted under sub-section (1) may exercise such powers and functions as may be necessary to:

(a) lay down the policies for implementation of the provisions of this Act; and

(b) give such directions to any farmers' organisation' as may be considered necessary, in exercising their powers and performing their functions in accordance with the provisions of this Act.

Procedure for Recall

10. (1) A motion for recall of a chairman or president or members of a management committee, as the case may be, of a farmers' organisation may be initiated by giving a written notice as may be prescribed, signed

by not less than one-third of the total number of members of the farmers' organisation, who are entitled to vote.

Provided that no notice of motion under this section shall be made within one year of the date of assumption of office by the person against whom the motion is sought to be moved.

(2) if the motion is carried with the support of majority of the members present and voting at a meeting of the general body specially convened for the purpose, the district collector or the government as the case may be shall by order remove him from office and the resulting vacancy shall be filled in the same manner as a casual vacancy.

Constitution of Sub-committees in Farmers' Organisation

11. The managing committee of a farmers' organisation may constitute sub-committees to carry out all or any of the functions vested in each organisation under this Act.

Farmers' Organisation to be a Body Corporate

12. Every farmers' organisation shall be a body corporate with a distinct name having perpetual succession and a common seal and subject to the provisions of this Act vested with the capacity of entering into contracts and of doing all things necessary, proper or expedient for the purposes for which it is constituted and it shall sue or be sued in its corporate name represented by the chairman or the president, as the case may be.

Changes in Farmers' Organisation

13. The government may in the interest of a farmers' organisation in the command area by notification and in accordance with the rules made in this behalf:
    (a) form a new farmers' organisation by separating the area from any farmers' organisation;
    (b) increase the area of any farmers' organisation;
    (c) diminish the area of any farmers' organisation;
    (d) alter the boundaries of any farmers' organisation; or
    (e) cancel a notification issued under this Act for rectifying any mistake, provided that no such separation, increase, diminution, alteration and cancellation shall be effected unless a reasonable opportunity is given to the organisation likely to be affected.

Disqualifications of Candidates or Members

14. (1) No village servant and no officer or servant of the government or of a local authority or an employee of any institution receiving aid from the

funds of the government shall be qualified for being chosen as or for being a chairman, or president or a member of a managing committee.

Explanation

For the purpose of this section the expression 'village servant' means in relation to

(i) the Andhra Area, any person who holds any of the village offices of neeruganti, neeradi, vetti kawalkar, toti, talayari, tandalagar, sathsindhi or any such village office by whatever designation it may be locally known;

(ii) the Telangana Area, any person who holds any of the village offices of neeradi, kawalkar, sathsindhi or any such village office by whatever designation it may be locally known.

(2) No person who has been convicted by a criminal court for any offence involving moral turpitude committed under any law for the time being in force shall be qualified for being chosen in or for being a chairman or President or a member of a managing committee.

(3) A person shall be disqualified for being chosen as a chairman or a president or a member of the managing committee if on the date fixed for scrutiny of nominations for election, or on the date of nomination he is:

- (a) of unsound mind and stands so declared by a competent court;
- (b) an applicant to be adjudicated as an insolvent or an undischarged insolvent; or
- (c) a defaulter of land revenue or water tax or charges payable either to the government or to the farmers' organisation.
- (d) Interested in a subsisting contract made with, or any work being done for, the gram panchayat, mandal parishad, zilla parishad or any state or central government or the farmers' organisation.

  Provided that a person shall not be deemed to have any interest in such contract or work by reason only of his having share or interest in:

  - (i) a company as a mere share-holder but not as a director; or
  - (ii) any lease, sale or purchase of immovable property or any agreement for the same; or
  - (iii) any agreement for the loan of money or any security for the payment of money only; or
  - (iv) any newspaper in which any advertisement relating to the affairs of the farmers' organisation in inserted.

Explanation

For the removal of doubts it is hereby declared that where a contract is fully performed it shall not be deemed to be subsisting merely on the ground that the gram panchayat, mandal parishad, zilla parishad, the farmers' organisation, the state or central government has not performed its part of the contractual obligations.

(4) A chairman or a president or a member of managing committee shall also become disqualified to continue in office if he:

(a) is convicted in a criminal case involving moral turpitude; or

(b) absents for three consecutive meetings without reasonable cause. Provided that such disqualification under item (b) shall not apply in the case of women who are in an advanced state of pregnancy and for a period of three months after delivery, and he shall cease to hold the office forthwith.

(5) A person having more than two children shall be disqualified for election or for continuing as a chairman or a president or a member of the managing committee.

Provided that the birth within one year from the date of commencement of this Act, hereinafter in this section referred to as the date of such commencement of an additional child shall not be taken into consideration for the purposes of this section.

Provided further that a person having more then two children (excluding the child of any born within one year from the date of such commencement) shall not be disqualified under this section for so long as the number of children he had on the date of such commencement does not increase.

Filling up of Vacancies

15. (1) (a) A vacancy arising either due to disqualification under sub-section (4) of section 14 or due to death or resignation or by any reason, such vacancy shall be filled up by nomination in the following manner:

(b) a vacancy in the distributory committee shall be filled, by nomination by the managing committee of the project committee in the manner prescribed; and

(c) a vacancy in the project committee shall be filled, by nomination either by the apex committee or by the government, as the case many be, in the manner prescribed

(2) the district collector shall take necessary steps to conduct elections to fill up any vacancy caused within a period of one month from the date of occurrence of such vacancy.

(3) The term of office of member of a president of a chairman of the farmers' organisation, elected under sub-section (2), shall expire at the time at which it would have expired, if he had been elected at the ordinary election.

Chapter III
Objects and Functions of the Farmers' Organisation

Objects

16. The objects of the farmers' organisation shall be to promote and secure distribution of water among its users, adequate maintenance of the irrigation system, efficient and economical utilisation of water to optimise agricultural production, to ensure ecological balance by involving the farmers, inculcating a sense of ownership of the irrigation system in accordance with the water budget and the operational plan.

Functions of Water Users' Association

17 The water users' association shall perform the following functions, namely:

(a) to prepare and implement a warabandi schedule for each irrigation season, consistent with the operational plan, based upon the entitlement, area, soil and cropping pattern as approved by the distributory committee, or as the case may be, the project committee;

(b) to prepare a plan for the maintenance of irrigation system in the area of its operation at the end of each crop season and carry out the maintenance works of both distributory system and minor and field drains in its area of operation with the funds of the association from time to time;

(c) to regulate the use of water among the various pipe outlets under its area of operation according to the warabandi schedule of the system;

(d) to promote economy in the use of water allocated;

(e) to assist the revenue department in the preparation of demand and collection of water rates;

(f) to maintain a register of landholders as published by the revenue department;

(g) to prepare and maintain a register of co-opted members;

(h) to prepare and maintain an inventory of the irrigation system within the area of operation;

(i) to monitor flow of water for irrigation;

(j) to resolve the disputes, if any, between the members and water users in its area of operation;

(k) to raise resources;

(l) to maintain accounts;

(m) to cause annual audit of its accounts;

(n) to a assist in the conduct of elections to the managing committee;

(o) to maintain other records as may be prescribed;

(p) to abide by the decisions of the distributory and project committees;
(q) to conduct general body meetings, as may be prescribed;
(r) to encourage avenue plantation on canal bunds and tank bunds by leasing such bunds; and
(s) to conduct regular water budgeting and also to conduct periodical social audit, as may be prescribed.

Functions of Distributory Committee

18. The distributory committee shall perform the following functions, namely:
(a) to prepare an operational plan based on its entitlement, area, soil, cropping pattern at the beginning of each irrigation season, consistent with the operational plan prepared by the project committee;
(b) to prepare a plan for the maintenance of both distributories and medium drains within its area of operation at the end of each crop seasons and execute the maintenance works with the funds of the committee from time to time;
(c) to regulate the sue of water among the various water users' associations under its area of operation;
(d) to resolve disputes, if any, between the water users' associations in its area of operation;
(e) to maintain a register of water users' associations in its area of operation;
(f) maintain an inventory of the irrigation system in the area of its operation, including drains;
(g) to promote economy in the use of water allocated;
(h) to maintain account;
(i) to cause annual audit;
(j) to maintain other records as may be prescribed;
(k) to monitor the flow of water for irrigation;
(l) to conduct general body meetings as may be prescribed;
(m) to abide by the decisions of the project committee;
(n) to cause regular water budgeting and also the periodical social audit as may be prescribed;
(o) to assist in the conduct of elections to the managing committee; and
(p) to encourage avenue plantation in its area of operation;

Functions of Project Committee

19. The project committee shall perform the following functions; namely:
(a) to approve an operational plan based on its entitlement, area, soil, cropping pattern as prepared by the competent authority in respect of the entire project area at the beginning of each irrigation season;

(b) to approve a plan for the maintenance of irrigation system including the major drains within its area of operation at the end of each crop season and execute the maintenance works with the funds of the committee from time of time;
(c) to maintain a list of the distributory committees and water users' association in its area of operation;
(d) to maintain an inventory of the distributory and drainage systems in its area of operation;
(e) to resolve disputes, if any, between the distributory committees;
(f) to promote economy in the use of water;
(g) to maintain accounts;
(h) to cause annual audit of its account;
(i) to maintain other records as may be prescribed;
(j) to conduct general body meetings as may be prescribed;
(k) to cause regular water budgeting and also the periodical social audit as may be prescribed; and
(l) to encourage avenue plantations in its area of operation.

Power to Levy and Collect Fee

20. A farmers' organisation may, for carrying out the purposes of this act, achieving the object of the organisation and performing its function, levy and collect such fees as may be prescribed from time to time.

Appointment of Competent Authority and his Functions

21. (1) The government may by notification appoint such officer from the irrigation and command area development department, or any other department or corporation including irrigation development corporation, as they consider necessary, to be the competent authority to every farmers' organisation for the purpose of this Act.

(2) The competent authority appointed under sub-section (1) shall be responsible to the respective farmers' organisation in the implementation and execution of all decisions taken by the farmers' organisation.

Chapter IV
Resources

Resources of Farmers' Organisation

21. The funds of the farmers' organisation shall comprise of the following, namely:

(i) grants received from the government as a share of farmers' organisation;
(ii) such other funds as may be granted by the state and central government for the development of the area of operation;
(iii) resources raised from any financing agency for undertaking any economic development activities in its area of operation;
(iv) income from the properties and assets attached to the irrigation system within its area of operation:
(v) fees collected by the farmers' organisation for the services rendered in better management of the irrigation system; and
(vi) amounts received from any other source.

Chapter V
Offences and Penalties

Offences and Penalties

22. Whoever without any lawful authority does any of the following acts, namely:
(a) damages, alters, enlarges, or obstructs any irrigation system;
(b) interferes with, increases, or diminishes the water supply in, or the flow of water from, through, over or under any irrigation system;
(c) being responsible for the maintenance of the irrigation system, neglects to take proper precautions for the prevention of wastage of the water thereof or interferes with the authorised distribution of water there from or uses water in an unauthorised manner, or in such manner as to cause damage to the adjacent landholdings
(d) corrupts or fouls, water of any irrigation system so as to render it less fit for the purposes for which it is ordinarily used;
(e) obstructs or removes any level marks or water gauge or any other mark or sign fixed by the authority of a public servant; and
(f) opens, shuts, or obstructs or attempts to open, shut or obstruct any sluice or outlet or any other similar contrivance in any irrigation system,

shall, on conviction, be punished with imprisonment which may extend to two years or with fine which may extend to five thousand rupees or with both.

Punishment Under Other Laws Not Barred

23. Nothing in this Act shall prevent any person from being prosecuted and punished under any other law for the time being in force for any act or omission made punishable by or under this Act.

Provided that no person shall be prosecuted and punished for the same offence more than once.

Composition of Offences

25. (1) A farmers' organisation may accept from any person who committed or in respect of whom a reasonable belief can be inferred that he has committed an offence punishable under this Act or the rules made thereunder, a sum of money not exceeding rupees one thousand by way of composition for such offence.

(2) On payment of such sum of money, the said person, if in custody, shall be discharged and no further proceeding shall be discharged and no further proceedings shall be taken against him in regard to the offence, so compounded.

Chapter VI
Settlement of Disputes

Settlement of Disputes

26. (1) Any dispute or difference touching the constitution, management, powers or functions of a farmers' organisation arising between members, shall be determined by the managing committee of the farmers' organisation.

(2) Any such dispute or difference arising between a member and the managing committee of a water users' association or between two or more water users' associations shall be determined by the managing committee of the distributory committee.

(3) Any such dispute of difference arising between a member and the managing committee of a distributory committee or between two or more distributory committees shall be determined by the managing committee of the project committee.

(4) Any such dispute or difference arising between a member and the managing committee of a project shall be determined by the apex committee, whose decision shall be final.

(5) Every dispute or difference under this section shall be disposed off within fifteen days from the date of reference of the dispute or difference.

Appeals

27. (1) A party to a dispute or difference aggrieved by any decision made or order passed by the managing committee of a water users' association

may appeal to the managing committee of the distributory committee, whose decision thereon shall be final.

(2) Any party to a dispute or difference aggrieved by any decision made or order passed by the managing committee of a distributory committee may appeal to a project committee, whose decision thereon shall be final.

(3) Any party to a dispute of difference aggrieved by any decision made or order passed by the managing committee of a project committee may appeal to the apex committee, whose decision thereon shall be final,

(4) Any appeal under sub-section (1) or sub-section(2) or sub-section (3) shall be preferred with fifteen days of communication of the decision of the order to the person aggrieved.

(5) Every appeal under this section shall be disposed off within fifteen days from the date of filing of the appeal.

Chapter VII
Miscellaneous

Records

28. (1) Every farmers' organisation shall keep at its office the following accounts, records and documents, namely:

- (a) an up-to-date copy of this Act;
- (b) a map of the area of operation of the farmers' organisation along with map of the structures and distributory networks prepared in consultation with the irrigation department;
- (c) a statement of the assets and liabilities;
- (d) minutes book;
- (e) books of account showing receipt and payments;
- (f) books of account of all purchases and sales of goods by the farmers' organisation;
- (g) register of measurement books, level field books, work orders and the like;
- (h) copies of audit reports and enquiry reports;
- (i) all such other accounts, records and documents as may be prescribed from time to time.

(2) The books of accounts and other records shall be open for information to the members of the farmers' organisation.

Audit

29. Every farmers' organisation shall get its accounts audited in the manner prescribed.

Recovery of Dues

30. The meetings of the farmers' organisation and the managing committees thereof, at such intervals, the procedure, the presidency and the Quorum thereof, and the cessation of membership thereof shall be, as may be prescribed.

Meetings

31. The meetings of the farmers' organisations and the managing committees thereof, at such intervals, the procedure, the presidency and the quorum thereof and the cessation of membership thereof shall be as may be prescribed.

Resignation

32. (1) A member of managing committee of a farmers' organisation may resign his office by a letter sent by registered post or tendered in person to the chairman or president of the managing committee concerned.
(2) The president of the managing committee of a water users' association may resign his office by a letter sent by registered post or tendered in person to the president of the distributory committee concerned.
(3) The president of the managing committee of a distributory committee may resign his office by a letter sent by registered post or tendered in person to the chairman of the project committee concerned.
(4) The chairman of the managing committee of a project committee may resign his office by a letter sent by registered post or tendered in person to the chairman of the apex committee.
(5) Such resignation as above mentioned shall take effect from the date of its acceptance or on the expiry of 30 days from the date of its receipt whichever is earlier.

Appointment of a Commissioner

33. (1) The government may by notification appoint a commissioner to exercise general control and superintendence over the competent authorities and the district collector in performance of their functions under this Act or the rules made thereunder.
(2) The powers to be exercised and the functions to be performed by the commissioner shall be such as may be prescribed.

Transitional Arrangements

34. The government may by notification appoint an officer or officers to exercise the powers and perform the functions of a farmers' organisation and

the managing committee thereof till such time such farmers' organisation is duly constituted or reconstituted and such managing committee assumes office under the provisions of this Act.

Authentication of Orders and Documents of the Farmers' Organisation

35. All permissions, orders, decisions, notices and other documents of the farmers' organisation shall be invalid by reason only of the existence of any vacancy in, or defect in the constitutions of, the said committee.
Acts not to be invalidated by informality or vacancy, etc.
36. No act or proceedings of the managing committee of a farmers' organisation shall be invalid by reason only of the existence of any vacancy in, or defect in the constitution of, the said committee.

Deposit and Administration of the Funds

37. (1) The farmers' organisation shall keep their funds in a nationalised bank or a co-operative bank namely a Primary Agricultural Cooperative Society or the District Co-operative Central Bank or the Andhra Pradesh State Co-operative Central Bank.
(2) The funds shall be applied towards meeting of the expenses incurred by the managing committee of the concerned farmers' organisation in the administration of this Act and for no other purpose.

Sinking Fund

38. (1) The managing committee of the farmers' organisation shall maintain a sinking fund for the repayment of moneys borrowed and shall pay every year into the sinking fund for the repayment of moneys borrowed and shall pay every year into the sinking fund such sum as may be sufficient for repayment within the period fixed of all moneys so borrowed.
(2) The sinking fund or any part thereof shall be applied in or towards, the discharge of the loan for with such fund was created, and until such loan is wholly discharged, it shall not be applied for any other purpose.

Budget

39. The managing committee of a farmers' organisation shall prepare in such form in every financial year a budget in respect of the financial year next, showing the estimated receipts and expenditure of the committee and shall place before the general body of the farmers' organisation for its approval as may be prescribed.

Protection of Acts Done in Good Faith

40. No suit, prosecution of other legal proceedings shall be instituted against any person for anything which is, in good faith done or intended to be done under this Act or under the rules made thereunder.

Power to Remove Difficulties

41. (1) If any difficulty arises in giving effect to the provisions of this Act or this Act or as to the first constitution or reconstitution of any farmers' organisation after the commencement of this Act, the government, as the occasion may require, by order published in the Andhra Pradesh Gazette, do anything which appears to them necessary for removing the difficulty.

(2) All orders made under sub-section (1) shall as soon as may be, after they are made, be placed on the table of the legislative assembly of the state and shall be subject to such modifications by way of amendments or repeal as the legislative assembly may make either in the same session or in the next session.

Savings

42. (1) Nothing contained in this Act shall affect the rights or properties vested in a gram panchayat, mandal parishad, zilla parishad, municipality or municipal corporation under any law for the time being in force.

(2) Nothing contained in this act shall apply to the minor water bodies in the Scheduled Areas in the state of Andhra Pradesh.

Power to Make Rules

43. (1) The state government may, by notification in the Official Gazette, make rules to carryout the purposes of this Act.

(2) Every rule made under this Act shall immediately after it is made, be laid before the legislative assembly of the state, if it is in session and if it is not in session, in the session immediately following for a total period of fourteen days which maybe comprised in one session or in two successive sessions, and if, before the expiration of the session in which it is so laid or the session immediately following, the legislative assembly agrees in making any modification in the rule or in the annulment of the rule, the rule shall from the date on which the modification or annulment is notified, have effect only in such modified form or shall stand annulled as the case may be, so however,

that any such modification or annulment shall be without prejudice to the validity of anything previously done under that rule.

G. Bhavani Prasad
Secretary to Government,
Legislative Affairs and Justice,
Law Department

APPENDIX II

The Karnataka Irrigation and Certain Other Law (Amendment) Act, 2000

Note from Editors: Though this Act is in the form of amendments to existing Acts, the proposed functions of Water Users Societies/Federations and the approach to PIM encouraged become clear from the Act.

An Act further to amend the Karnataka Irrigation Act, 1965 and the Karnataka Irrigation (Levy of Betterment Contribution and Water Rate) Act, 1957.

Whereas it is expedient further to amend the Karnataka Irrigation Act, 1965 (Karnataka Act 16 of 1965) and the Karnataka Irrigation (Levy of Betternment Contribution and Water Rate) Act, 1957 (Karnataka Act 28 of 1957) for the purposes hereinafter appearing.

Be it enacted by the Karnataka State Legislature in the fifty first year of the Republic of India, as follows:

1. Short Title and Commencement

(1) This Act may be called the Karnataka Irrigation and Certain Other Law (Amendment) Act, 2000.
(2) It shall be deemed to have come into force with effect from the fourteenth day of June, 2000.

2. Amendment of Karnataka Act 16 of 1965

In the Karnataka Irrigation Act, 1965 (Karnataka Act 16 of 1965).
(1) In section 2.
    (i) after clause (b), the following clause shall be inserted, namely:

(bb) "Command Area Development Authority" means the Command Area Development Authority established under the Karnataka Command Area Development Act, 1980 (Karnataka Act 6 of 1980)."

(ii) after clause (o), the following clauses shall be inserted, namely:

(p) "Water Users Apex Level Federation" means a Water Users Apex Society registered under the Karnataka Co-operative Societies Act, 1959 (Karnataka Act 11 o 1959) which shall consist of members as specified in its bye-laws:

(q) "Water Users Distributory Level Federation" means a Federation of Water Users Societies at Distributory Level registered under the Karnataka Co-operative Societies Act, 1959 (Karnataka Act 11 of 1959) the members of which consist of not less than fifty one percent of the Water Users Societies within an area in relation to a distributory of a major or medium irrigation project, as may be notified by the State Government:

(r) "Water Users Project Level Federation"means.

(i) in the case of a major irrigation project, a Federation of Water Users Distributory Level Federations at the project level registered under the Karnataka Co-operative Societies Act, 1959 (Karnataka Act 11 of 1959) the members of which consist of not less than fifty one percent of Water Users Distributory Level Federation within an area in relation to the major irrigation project as may be notified by the State Government:

(ii) in the case of a medium irrigation project, a Federation of Water Users Societies at the project level registered under the Karnataka Co-operative Societies Act, 1959 (Karnataka Act 11 of 1959) the members of which consist of not less than fifty one percent of Water Users Societies within an area in relation to the medium irrigation project as may be notified by the State Government.

(iii) in the case of a minor irrigation or lift irrigation works, a Federation of Water Users Societies at relation to such minor irrigation or lift irrigation works registered under the Karnataka Co-operative Societies Act, 1959 (Karnataka Act 11 of 1959) the members of which consist of not less than fifty one percent of Water Users Societies within such area as may be notified by the State Government.

(s) "Water Users Society" means a Water Users Co-operative Society registered under the Karnataka Co-operative Societies Act, 1959 (Karnataka Act 11 of 1959),

(a) members of which consist of not less than sixty percent of the Water Users for cultivation of land; or

(b) majority of members of which consists of Water Users for cultivation of land and holding not less than fifty one percent of land under an irrigation work within the area of operation of the society.

Explanation

For the purpose of this clause "area of operation" means an area comprising a contiguous block of land within an irrigation work as may be notified by the Irrigation Officer.

(2) in section 4, after sub-section (1), the following proviso shall be inserted, namely:
"Provided that the State Government may entrust to a Water Users Society control, maintenance and monitoring of any irrigation work either wholly or in part and thereupon such Water Users Society shall be responsible for the control, maintenance and monitoring of such irrigation work".

(3) in section 10, after the words "the State Government" the words "or the Water Users Society in respect of any irrigation work entrusted to it, as the case may be" shall be inserted.

(4) in section 14,
   (i) for the words, "appears to the State Government", the words "appears to the Water Users Society in case of any irrigation work entrusted to it and in other cases to the State Government" shall be substituted;
   (ii) for the words "the State Government may" the words "the State Government or as the case may be, Water Users Society" shall be substituted;
   (iii) after the words "authorised by the State Government" the words "or as the case may be, Water Users Society" shall be inserted.

(5) in section 24, after sub-section (1), the following proviso shall be inserted, namely:
"Provided that in respect of an irrigation work entrusted to a Water Users Society such expenses shall be deemed to be dues by such user to a Co-operative Society and shall also be recoverable in accordance with the provisions of the Karnataka Co-operative Societies Act, 1959."

(6) in section 27, the following proviso shall be inserted at the end, namely:
"Provided that where an irrigation work is entrusted to a Water Users Society, the Irrigation Officer, instead of consulting the committee so appointed by the State Government, shall,
   (i) in the case of a Water Users Society whose area of operation extends to one or more minor irrigation or lift irrigation works, consult such Water Users Society; and
   (ii) in other cases, consult the Water Users Project Level Federation."

(7) in section 28,
   (i) in sub-section (3), after the words "water rate" in the two places where they occur, the words "or water charges, as the case may be" shall be inserted.
   (ii) in sub-section (5),
      (a) after the words "water rate" the words "or as the case may be, water charges" shall be inserted;
      (b) after the words "thirty times the rate" the words "or the water charges, as the case may be" shall be inserted;

(8) in section 29, after clause (f), the following shall be inserted, namely:
   (g) "whenever and so long as water is used for sowing planting or growing crops in contravention of notification issued under sub-section (2) of section 32;
   (h) whenever and so long as stoppage of supply of water is necessitated due to any cause beyond the control of Irrigation Officer;
      (i) whenever and so long as a person does not pay the arrears of water rate or as the case may be water charges."

(9) in section 32, in sub-section (4), in clause (b), after the words "water rate" in the two places where they occur, the words "or water charges as the case may be" shall be inserted;

(10) in section 44, in sub-section (1),
   (i) for the words "Taluk Development Board" the words "Zila Panchayat or a Taluk Panchayat" shall be substituted.
   (ii) after the second proviso, the following proviso shall be inserted, namely:
   "Provided also that no maintenance cess shall be levied and collected under this sub-section in respect of any land in the area benefited by any irrigation work maintained by a Water Users Society".

(11) in section 45, for the words "Taluk Development Board" in the two places where they occur, the words "Zilla Panchayat or Taluk Panchayat" shall be substituted.

(12) after section 60, the following section shall be inserted, namely:

60A. Compounding of Offences

(1) "Subject to such restrictions and conditions and in such manner as may be prescribed, Irrigation Officer may either before or after initiation of any proceeding for an offence punishable under this Act or the rules made thereunder accept a sum of money not exceeding five thousand rupees by way of composition of an offence which such person has committed or is suspected to have committed.

(2) On payment of such money, such person if, in custody, shall be discharged, property, if any, seized shall be released and no further proceeding shall be taken against such person or property."

(3) after Chapter IX and section 62, the following chapter and sections shall be inserted, namely:

Chapter IXA
Functions of Water Users Society, etc.

62A. Functions of Water Users Society

Water Users Society shall perform the following functions, namely:

(1) to develop irrigation infrastructure by availing institutional finance.

(2) to procure water in bulk on volumetric basis from the Irrigaion Department or Krishna Jala Bhagya Nigam or Karnataka Neeravari Nigam and distribute it to the land holders in accordance with the principles laid down by the General Body for equitable distribution of water.

(3) to operate and maintain canals situated within its jurisdiction.

(4) to levy and collect water charges and service charge from the land holders.

(5) to educate and train land holders in the efficient and economical use of water and adoption of new technology as well as to implement necessary programmes.

(6) to prepare water budget and financial budget for each irrigation season.

(7) to resolve disputes that may arise among land holders.

(8) to mobilise financial resources for undertaking activities and efficient management of the society.

(9) to implement or execute on-farm development works.

(10) to assist the Irrigation Department and Command Area Development Authority in implementing irrigation and drainage works.

(11) to levy water charges and service charges on non-members at rates approved by the General Body of the Society.

(12) to send annually to the prescribed authority on or before the prescribed date, an audit statement of all receipts and expenditure for the year ending.

(13) to send a statement of the assets and liabilities of the society on or before such date and to such authority as may be prescribed.

(14) The General Body of the Society shall prepare guidelines and decide the procedures for its day to day functioning, in conformity with the byelaws of the society and conditions of the agreement to be entered into with the Executive Engineer concerned.

(15) The Society shall furnish such information as may be prescribed to the prescribed authority.

(16) to encourage avenue plantation in its area of operation.

(17) to do such other acts as may be prescribed.

62B. Functions of Water Users Distributory Level Federation

The Water Users Distributory Level Federation shall perform the following functions, namely:

(1) to prepare an operational plan based on its entitlement, area, soil and cropping pattern at the beginning of each irrigation season consistent with the operational plan prepared by the project level federation;
(2) to identify the critical maintenance work that are to be carried out and prioritise these works;
(3) to monitor the maintenance works being executed and ensure that they conform to prescribed standards;
(4) to monitor and regulate the use of water among various societies in its area of operation;
(5) to abide by the decisions of the Water Users Project Level Federation;
(6) to undertake periodical social audit;
(7) to promote economy in the use of water;
(8) to encourage avenue plantation in its area of operation;
(9) to prepare water budgets and crop plans.

62C. Functions of Water Users Project Level Federation

The Water Users Project Level Federation shall perform the following functions, namely:

(1) to prepare an operational plan based on its entitlement area, soil and cropping pattern at the beginning of each irrigation season;
(2) to identify the critical maintenance works that are to be carried out and prioritise these works;
(3) to monitor the maintenance works being executed and ensure that they conform to prescribed standards;
(4) to prepare water budgets and crop plans;
(5) to undertake periodical social audit;
(6) to promote economy in the use of water;
(7) to encourage avenue plantation in its area of operation.

62D. Water Users Apex Level Federation

(1) The committee of management of the Water Users Apex Level Federation shall consist of the following members, namely;
   (a) the Minister-in-charge of Major and Medium Irrigation, Government of Karnataka who shall be the Chairperson;
   (b) the Minister-in-charge of Minor Irrigation, Government of Karnataka who shall be the Co-chairperson;
   (c) the Secretary of Government, Irrigation Department, Government of Karnataka;

(d) one expert in the field of participatory irrigation management nominated by the State Government;
(e) Chief Executive/Managing Director of the Water Users Apex Level Federation;
(f) Chairpersons of all the Command Area Development Authorities;
(g) such number of persons representing different projects and minor irrigation and lift irrigation works in the State and elected in such manner as may be specified in the bye-laws of the Water Users Apex Level Federation.

(2) Vice-Chairperson shall be elected from amongst the members of the Water Users Apex Level Federation.
(3) The Water Users Apex Level Federation shall make recommendation to the State Government on the policies to be adopted and the guidelines to be formulated regarding construction, maintenance and regulation of irrigation work and supply of water therefrom.
(4) The Waters Users Apex Level Federation may give directions to the Waters Users Project Level Federation, Waters Users Distributory Level Federation and the Water Users Society to carry out the purposes of this Act.

62E. Effect of Other Laws

"Save as otherwise provided in this Act, the provisions of this Act and the rules made thereunder shall have effect notwithstanding anything inconsistent therewith contained in any other law."

62F. Power to Exempt Certain Projects

The State Government having regard to the size and nature of any irrigation project or works, may, by notification, exempt such project or works from the application of the provisions relating to Water Users Distributory Level Federation or Water Users Project Level Federation.

3. Amendment of Karnataka Act 28 of 1957

In the Karnataka Irrigation (Levy of Betterment Contribution and Water Rate)Act, 1957 (Karnataka Act 28 of 1957), in section 10,

(i) in sub-section (1A), the words, figures and brackets or the "Karnataka Societies Registration Act, 1960 (Karnataka Act 17 of 1960)" shall be omitted.
(ii) in sub-section (1B),
 (a) the words, figures and brackets "or the Karnataka Societies Registration Act, 1960 (Karnataka Act 17 or 1960)" shall be omitted.
 (b) for the words "water rate", the words "water charges" shall be substituted.

(iii) in sub-section (3), after the words "water rate" the words "or the water charges, as the case may be" shall be inserted.

4. Repeal and Savings

The Karnataka Irrigation and Certain Other Law (Amendment) Ordinance, 2000 (Karnataka Ordinance No. 3 of 2000) is hereby repealed.

(2) Notwithstanding such repeal anything done or any action taken under the Karnataka Irrigation Act, 1965 (Karnataka Act 16 of 1965) and the Karnataka Irrigation (Levy of Betterment Contribution and Water Rate) Act, 1957 (Karnataka Act 28 of 1957) as amended by the said Ordinance, shall be deemed to have been done or taken under the said Acts as amended by this Act.

By order and in the name of the
Governor of Karnataka

(M.R. Hegde)
Secretary to Government,
Department of Parliamentary Affairs
and Legislations

APPENDIX III

Current Issues and the Need for Monitoring and Evaluation: Visit Note on PIM Programme in Andhra Pradesh

DOUGLAS L. VERMILLION

This note identifies key issues related to the Andhra Pradesh Farmers' Management of Irrigation Systems (APFMIS) Act of 1997 and makes suggestions for monitoring and evaluation of the reforms by the government. Comments are based on the author's observations from field visits, reports reviewed, and discussions held in the INPIM seminar on the Participatory Irrigation Management (PIM) programme in Andhra Pradesh (AP).

This note contains the following sections:

(1) Current status of reforms.
(2) Apparent outcomes and impacts to date.
(3) Key issues related to the reform programme.
(4) Needs for monitoring and evaluation.
(5) Annexed summary of field visits and list of reports reviewed.

Current Status of Reforms[1]

By 1996, the irrigation sector in AP was in a state of crisis and major reform was called for. The AP government "White Paper" of 1996 noted the following key problems (amongst others):

- Serious decline in net irrigated area. Only 2.3 mha of the 4.8 mha of potential irrigated area created was actually under irrigation (a gap of 52% between potential created and actual irrigated area);
- Under-financing of operations and maintenance;
- Poor irrigation system efficiencies and water distribution (including significant wastage of water diverted and inequitable distribution of water within schemes, especially between head and tail ends);

- Low cropping intensities and yields; and
- Low farm income.

After several months of consultations involving the state legislature, various government agencies, political parties, farmer groups, and the media, the APFMIS Act, 1997 was passed by the legislative assembly in April 1997. Water charges were more than tripled in the same month (from Rs. 60 per acre to Rs. 200 per acre). In June 1997, nearly[2] 10,292 Water Users' Associations (WUAs) were created through a state-wide election. In November 1997, 174 Distributory Committees (DCs) were created through elections. Project Committees (PCs) at the scheme level are yet to be created. Numerous consultation, training and feedback meetings have been held throughout the state at different levels from June 1997 to the present.

Key features of the APFMIS Act are that it provides for:

- mandatory constitution of farmer organizations and automatic membership,
- hydraulic area of jurisdiction,
- powers to farmer organizations (FOs) to collect water charges and apply sanctions for rule violations,
- reorientation of Irrigation Department (ID) staff as 'competent authorities' (CAs) to provide technical support to FOs,
- federation of FOs up to the scheme level and creation of an apex body at the state level.

With assistance from the World Bank, Rs. 118.1 crores ($28.1m) was provided in 1998 for minimum rehabilitation works (an investment of approximately Rs. 100 per acre). Previously, the government allocated Rs. 40 per acre for operation and maintenance (O&M), but approximately Rs. 35 of this was used for Irrigation and Command Area Development Department (I&CADD) staff salaries and fixed overhead costs, leaving barely Rs. 5 per acre for actual maintenance work. The Rs. 100 per acre allocation for minimum rehabilitation, from WB funds, was to be used only for actual maintenance works, exclusive of I&CADD salaries. From January 1998 to March 1999, 22,887 works were undertaken, of which 70 percent were implemented by WUAs or DCs.

Training in role reorientation was given to officials in the I& CADD in 1995 and to designated "competent authorities" in 1997. Training workshops were held for WUA presidents in September and October 1997 and for WUA and DC leaders in July and October 1999. In 1998 and 1999, conferences for WUA and DC leaders were held at the district, regional and state levels for additional information dissemination, training and obtaining of feedback from farmer organization leaders. Additionally, the training and information campaign activities under the state-wide Janmabhoomi programme supported awareness about the principles of reform.

Structure for M&E Process

As will be seen below, there are numerous issues concerning the reforms in AP. which require M&E data, analysis by stakeholders and further consensus building and decision-making. Representatives of all relevant stakeholders should be involved in working groups or consultative committees at the state, regional, district and scheme levels. The reform programme should involve in consultative committees representatives of all important stakeholder groups. This may include, amongst others, the following: farmer leaders from canal, tank and lift irrigation schemes; officials of the state government; officials from the I & CADD, revenue department, agriculture department, and finance department; NGO's; Panchayat Raj and other local government officials; and possibly local law enforcement officials.

The APFMIS working groups or consultative committees should meet regularly to review M&E data and request collection of additional information, as needed and as new issues arise. There should be an apex committee at the state level which guides policy and overall strategy. Soon, farmer-elected representatives will sit on scheme-level project committees and committees at higher administrative levels right up to the state level. The reforms continue to evolve and new issues are on the horizon for which M&E date will be needed. Reliable, objective information is an important resource to facilitate the process of problem solving and consensus building.

It is recommended that working groups consisting of multiple stakeholders be formed at scheme, district, regional and state levels to review and facilitate the process of reform. The working group at the state level should adopt a core set of indicators, with the request that the lower levels collect and aggregate the same data. Lower levels should be encouraged to adopt their own additional indicators as local circumstances require.

For more critical and complex issues (such as mal-distribution of water and local resource mobilization for O&M), it may be helpful to establish special issue groups to focus on generating ideas, information and support for how to address these issues.

The government and farmer organization should link M&E indicators to certain desired performance criteria. These should then be linked to certain rewards, such as bonuses, awards or study tours.

Current Needs for M&E Indicators

Performance assessment can be divided into three components of reform: (1) assessment of the process of implementation, (2) assessment of immediate outcomes of implementation, and (3) assessment of ultimate impacts. The author's perception of key issues for the APFMIS reform programme and related needs for M&E are listed below according to these components.

Process Issues

Process issues are about the dynamics of change, procedures and achievement of specific targets. An assessment of process asks, "Are we doing things in the right way?" They tend to require more qualitative indicators than do outcome and impact assessments. In many cases multiple indicators for the same issue may be necessary to get a balanced view. The following are what appear to be the most important current process issues for the APFMIS reform programme, with suggestions for related M&E indicators. Issues are presented as questions to be answered through M&E.

For many of these issues, simple indicators will not be able to tell the full story about local dynamics and perceptions. More qualitative and exploratory methods will be needed to enable understanding of more subtle or sensitive issues. These may include participatory rural appraisal methods and key informant and group interviews.

1. *Were the elections for Territorial Constituency (TC), WUA and DC leaders done in a democratic way? Did they result in selection of popular leaders who represent the interests of common farmers, including tail enders, small holders and landless cultivators?*

   Possible indicators to measure this are:

(a) % of elections which were contested versus unanimous,
(b) WUA Social Awareness Index = % of WUA members who know names of TC and WUA leaders, % of members who know level of water charge, % of members who know what are their membership requirements (O&M obligations),
(c) WUA Social Support Index = % of WUA members who support WUA leaders and their policies,
(d) Socio-economic background of WUA leadership = Total size of landholding of WUA leader, caste of WUA leader, type of off-farm business of WUA leader (including whether they have been construction contractors),
(e) % of WUA presidents and TC leaders which are literate.

Data collection for these indicators should be compared, or if necessary, stratified by head/tail end and by land tenure categories.

2. *To what extent do WUA members participate in the WUA (including women)? For farms cultivated by tenants, how do landowners versus tenants participate in the WUA? Also, do WUA members have the ability to inspect WUA records when they like?*

This is about both the breadth of participation of WUA members, social viability of WUA's, and transparency about WUA business. Possible indicators to measure this are:

(a) % of WUA members (by gender and tenure status) who participated in the election of WUA leaders (or selection by unanimous consent),
(b) % of WUA members (by gender and tenure status) who attended key general assembly meetings for establishment of WUA and approval of by-laws, basic policies and level of water charge,
(c) % of WUA members (by gender and tenure status) who participate in mandatory maintenance works done by communal labor? (if applicable),
(d) % of WUA members who are aware of basic WUA policies and decisions (such as level of water charge, water distribution rule, timing of general assembly meetings),
(e) WUA Social Stability Index =% of WUA members who own or have long-term rights to irrigated land,
(f) Number of time WUA records have been examined by WUA members.

3. *To what extend do TC leaders participate in WUA decisions and activities?*

Some observers question whether the reform process is too much focused on the WUA president, to the exclusion of TC leaders and regular WUA members. The following indicators might be used to obtain information about this issue:

(a) % of TC leaders present in meeting when a key WUA decision was made (such as final selection of maintenance works, determination of new water distribution method, approval of water charges, recruitment of ditch tenders or gate keepers, etc.)
(b) % of TC leaders who feel adequately involved in key WUA decisions,
(c) % of TC leaders who participated in walkthroughs for maintenance works.

4. *What is the quality of the working relationship between WUA and DC levels(and later, the PC level)? Are there any conflicts or communication problems between levels?*

Suggested indicators are:

(a) % of WUA presidents who attend DC meetings, % of DC leaders who attend PC meetings (after PC's are formed),
(b) Frequency of disputes between any WUA and the DC over water distribution,
(c) Frequency of disputes between any WUA and the DC over selection of maintenance works,
(d) % of WUA presidents who are aware of the water distribution arrangement along the distributory canal during rabi and kharif seasons.

5. *What records are being kept by WUA leaders and how adequate are they? What records or data have been copied or shown to the WUA Presidents and/ or TC leaders by Village Administrative Officers (VAO).*

Suggested indicators are percentage of WUA's which have:

- a full list of members,
- accounts records which are complete and up-to-date,
- a complete record of maintenance works,
- command area maps, and
- a record of members who have paid or defaulted payments of water charges, and attempted to obtain above information from the revenue department and have been unsuccessful.

6. *Are WUA leaders managing funds in an appropriate manner? To what extent are WUA presidents benefiting personally from maintenance contracts?*
   Suggested indicators are:
(a) % of WUA presidents which have been recalled due to financial improprieties,
(b) % of maintenance works handled completely by WUA presidents,
(c) % of WUA members or TC leaders which trust the WUA president in financial matters.

These simple indicators will not be sufficient. It will be necessary to have financial audits conducted by independent auditors not under the influence of WUA presidents. Results of the audits should be reported to DCs and PCs as well as to the I & CADD and revenue department.

7. *To what extent have disputes or problems arisen over the delineation of boundaries and powers for the WUAs and DCs? Is there any opposition to WUAs or DCs by Panchayat Raj governments or other local bodies?*
   Suggested indicators are:
(a) Frequency of disputes over WUA boundaries,
(b) Frequency of disputes between WUAs,
(c) Frequency of disputes between WUAs and Panchayat Raj governments.

More incisive assessment of sensitive issues may be required through confidential key informant and small focus group interviews.

8. *What kinds of farmers tend to pay the water charge and what kinds of farmers tend not to pay the water charge? To what extent do TC leaders or WUA presidents collaborate with the VAO in the collection of water charges? Are WUAs applying sanctions against water charge defaulters?*
   Suggested indicators are:
(a) % of head versus tail enders who are in default of water charge payments,
(b) % of large landholders versus small holders or tenants who are in default of water charge payments,
(c) % of TC leaders who assisted VAOs with collection of water charges.

(d) % of WUAs which have applied sanctions against members who default on payment of water charges,
(e) Type of sanctions applied by WUAs against water charge defaulters.

More qualitative analysis, through semi-structured interviews with key informants and small focus groups, may be required to elicit information about motivations of farmers to pay or not pay water charges and about the relationship between TCs and VAO and between WUA presidents and other revenue department officials concerning water charges.

A focus group is a group of individuals which all belong to a same category, such as tail enders, laskars, competent authorities, TC leaders, etc. They are used to discuss sensitive matters.

9. *To what extent did WUA members participate in the selection of maintenance works done in the minimal rehabilitation programme? To what extent are WUA members satisfied with the works done?*

Suggested indicators are:
(a) % of TC leaders participating in scheme walkthroughs for maintenance planning,
(b) Number of non-office-holding WUA members which participated in scheme walkthroughs for maintenance planning,
(c) % of WUA members who are satisfied with the selection of maintenance works,
(d) % of WUA members who are satisfied with the quality of maintenance works done.

10. *Did WUAs and DCs contribute the required 15% matching investment for the minimal rehabilitation programme? Are WUAs making plans for mobilizing resources for routine maintenance activities and for capital reserves?*

This is about the issue of financial sustainability, not just current solvency. It is of critical importance for assessing the viability of management transfer programmes. Suggested indicators are:
(a) % of WUAs and DCs which contributed their required 15 percent matching investment under the minimal rehabilitation programme,
(b) Average percentage matching contribution made by WUAs and DCs,
(c) Frequency of different types of contributions made by WUAs and DCs (in labor, materials and cash),
(d) O&M Financial Viability Index = Revenue available for O&M/cost of O&M,
(e) WUA Financial Self Reliance Index = Amount of funds and other contributions for irrigation from WUA members/Total resource requirement for O&M,
(f) % of WUAs which have formulated plans for mobilising resources for routine maintenance and for capital reserve funds,

(g) % of WUAs which have implemented routine maintenance activities with their own resources,
(h) % of WUAs which have started developing a capital reserve or long-term savings fund.

11. *What new procedures have WUAs and DCs adopted for water distribution and regulation? How satisfied are WUA members (especially tail enders) with any changes made in water distribution?*

This is another very important issue. Unless water distribution is improved, it is unlikely that WUAs will be able to raise the level of water charges and increase the collection rate. Unless more funds are raised locally for irrigation O&M than at present, the schemes will continue to deteriorate after minimal rehabilitation and the gap between potential and actual irrigated area will again widen.

Suggested indicators are:
(a) % of WUAs and DCs which have adopted new methods for water distribution,
(b) % of WUAs and DCs which have adopted new arrangements for implementing and regulating water distribution,
(c) % of WUA members (tail versus head enders) which are satisfied with the reliability and equity of water distribution within the hydraulic unit of the WUA,
(d) % of WUA members which believe that water distribution within the hydraulic unit of the WUA has improved after creating of the WUAs.

12. *To what extent do designated "competent authorities" understand and accept their new roles? To what extent are the laskars willing and able to work with the WUAs? To what extent is the I&CADD taking on the role of facilitator versus director? Is any additional training needed for I&CADD staff?*

It is not a simple matter for a large bureaucracy to reorient itself from the role of director to the role of facilitator and service provider. Officials could perceive the reforms to be a threat to their position of influence, budgets, nature of work, and job security. There are subtle ways whereby lower-level agency officials could resist, try to sabotage, or spread negative propaganda about the reforms. More sensitive methods for assessing staff perceptions, behavior and support for the reforms may be required, such as through participatory appraisal and key informant and small focus group interviewing.

The following are some suggested simple indicators:
(a) Skill Capacity Index = % of competent authorities, works inspectors, and laskars with required respective skills,
(b) Functional Responsiveness Index = % of competent authorities, works inspectors and laskars which have performed all of their required function,

(c) % of WUA presidents and TC leaders who are satisfied with attitudes, responsiveness and services provided by competent authorities, works inspectors and laskars,
(d) Organizational Responsiveness Index = Number of requests for assistance by WUA which have been responded to adequately by competent authority.

13. *How adequate are the skills of WUA and DC leaders? What further training may be required?*

This includes the type of skills available and required, as perceived by the competent authorities and farmer organization leaders. A training needs assessment would be required to identify specific training requirements. But some simple indicators can be used to detect whether additional training may be needed. Some suggested simple indicators are:
(a) Skill Capacity Index = % of WUA and DC leaders with required skills,
(b) Number of WUAs with leaders who are deficient in certain skills (such as, bookkeeping, conducting meetings, preparing O&M plans, dispute resolution, scheme walkthroughs, water distribution methods, measuring water flows, etc.),
(c) Frequency of different skill areas requiring additional training, as identified by WUA and DC leaders,
(d) Preference for certain types of training methods (classroom, theory, in-field, on-job, etc.).

Outcome Issues

Outcome issues are about the immediate, direct effects of a programme, or the achievement of essential objectives. An assessment of outcomes asks, "Are we accomplishing our objectives?" The following are what appear to be the most important programme outcome issues for the APFMIS reform programme, with suggestions for related M&E indicators. Issues are presented as questions to be answered through M&E.

1. *To what extent are WUAs and DCs performing necessary functions?*

Different WUAs and DCs will have different requirements for management activities and levels of intensity. This is especially true between canal, tank and lift irrigation schemes. It is important that not all organizations be evaluated according to the same standard. Hence, ratio-type indicators are recommended which incorporate local management requirements. Such indicators will have to be "situationalized" according to specific local conditions. Such indicators could be structured as follow:
(a) Service Delivery Responsiveness = No. of services delivered/No. of services required locally,

(b) Organizational Responsiveness = No. of functions performed/No. of functions required locally.
(c) Member Satisfaction Index = % of WUA members satisfied with WUA performance

Rapid or participatory appraisals may be required to obtain more in-depth understanding of farmer perceptions about WUA and DC performance.

2. *To what extent has there been an increase in actual irrigated area after the reform? To what extent can we attribute this to institutional reform versus the rehabilitation programme?*

There are three problematic aspects of this outcome measure. The first problem is the inaccuracy of data about irrigated area before the reform programme. It is widely known that the revenue department underestimates irrigated area by 20 to 30 percent, while the I&CADD has a tendency to over-estimate area figures. Widely reported findings of an area irrigated after reforms may largely be only the result of correction of previous misinformation about actual area irrigated. There needs to be a reconciliation between revenue department, I&CADD and agriculture department data, perhaps done through the coordination of the TCs and WUAs. Part of the reconciliation process, a close estimate of irrigated area before the APFMIS could be made.

The second problem is the difficulty of attributing increase in irrigated area to the APFMIS reform versus the rehabilitation. It is likely that most of immediate increase in area have been caused by the rehabilitation programme. Increase due to the institutional reforms will only become apparent, at the regional or state levels, after a few years of rehabilitation, or in comparison with schemes which did not have rehabilitation (if there are any which are comparable to the others).

The third problem is the definition of area irrigated. One senses that there are different definitions and some are not clearly defined and some may be misleading. One definition is the area which has been "touched" at least once during the season (or year) by irrigation water (and this could include only canal water or could also include lift irrigation). This can be a misleading definition because a single release of water to a field may not be enough to produce a crop. Another more meaningful definition might be area irrigated and harvested. Another problem is that the potential area created may have been exaggerated originally or may no longer be at all functional due to land filling of canals or other deterioration. Failure to irrigate the original design area may be due more to past errors than present O&M performance. Whatever definition is used, it should be a clear and standard one which enables assessment of current performance of O&M.

After the above problems have been addressed, the following simple indicators could be used:

(a) Irrigation Potential Achievement Index = Actual area irrigated/Potential area created,
(b) Irrigated Area Utilization Efficiency = Actual area irrigated/Current functional service area.

By current functional service area we mean the area which could physically be served by irrigation because irrigation structures exist and are functional to deliver water to the designated area. In schemes where there is as shortage of water to serve the entire functional service area, a multiplier can be used to discount functional service area in accordance with water availability.

3. *Has there been an improvement in efficiency, equity, or reliability of water distribution after implementation of the reform programme?*

Careful measurement of water delivery or distribution performance can require measuring devices and measurement capacity which do not exist locally. More simple indicators include perceptions of farmers, occurrence or frequency of water reaching tail end of canals, etc. As with the measure of area irrigated, it may be difficult in the short-term to determine whether changes are due to rehabilitation or institutional reforms. Additional qualitative assessment about new procedures implemented may be needed to strengthen evidence about effects of institutional reforms.

Some possible indicators are:
(a) Delivery Performance Ratio (DPR) = Actual delivery/ Target delivery (at different locations),
(b) Relative Water Supply = Irrigation + rainfall/ ET+seepage & percolation,
(c) Water Volume Performance = Actual water volume/ Target water volume (for a given area),
(d) Irrigation Scheme Efficiency = Crop water requirement/ Total inflow to scheme,
(e) Irrigation Distribution Efficiency = Amount delivered to turnouts/Total inflow to scheme,
(f) Field Application Efficiency = Crop water requirement/ Amount of water delivery to field (in some areas a surrogate measure for the same concept could be the total time taken for land preparation, as determined by water availabilities),
(g) Head/Tail Equity Ratio = Average DPR of upper-end turnouts/Average DPR of lower-end turnouts
(h) Delivery Reliability Index = Total number of irrigations/Target number of irrigations (for selected head and tail areas).

4. *Has there been an improvement in the quality of maintenance and functional condition of irritation scheme structures after the reform programme?*

It is important to measure this over time so that trends can be detected. Physical sustainability of irrigation schemes is an important objective of management transfer programmes. It is important to have early warning information available so that corrective actions can be taken if needed before schemes deteriorate as before the reforms. Suggested indicators are:

(a) Structural Functionality = Number of functional structures/Total number of structures,

(b) Maintenance Resource Utilization Index = Volume of work/Funds expended,

(c) Canal Flow Capacity = Actual canal flow capacity/Design canal flow capacity.

5. *Has there been an improvement in cropping intensity or shift toward crop diversification after implementation of the reforms?*

Measures for this are well known. Annual cropping is conventionally measured according to assumption of three seasons of a standard length of time. The maximum value is 300 percent (being 100% per season). Where cropping is diversified this may not work well and it may be more informative to base a cropping intensity index upon number of months (out of 12) the crops are in the ground. Cropping intensity is not a suitable indicator for long-growing crops like sugarcane or tree crops.

It may be more important to measure extent to which farmers shift to higher value or non-rice crops after reforms.

Suggested indicators are:

(a) Annual Cropping Intensity = Area cultivated (adding each season or month) /Irrigable, cultivable area (adding each season or month),

(b) % of cultivated area planted in rice,

(c) % of cultivated area planted in non-rice commercial crops.

6. *Have crop yields per unit of land or per unit of water increased after implementation of the reforms?*

The standard indicator of crop production is yield per hectare or acre. Where water is a more constraining or scarce factor than land, yield per unit of water may be a more important indicator. This would require more attention to measurement of irrigation water diverted and delivered. So many factors effect crop yields that it is difficult to attribute changes in yields to an institutional reform programme. Analysis of long-term changes in yield trend lines, weather and agricultural input application rates can help control for other factors which effect yields. Suggested indicators are:

(a) Crop yield per unit of land = metric tons or kilograms per hectare,

(b) Crop yield per unit of water delivered = Metric tons or kilograms/100 cubic meter or mm delivered to field turnout.

7. *What has happened to water charge collection rates after the reforms? How adequate are financial resources raised by the WUA for irrigation O&M requirements?*

Water charge collection rates are a useful, widely-used indicator of financial viability. But they do not tell us about the extent to which funds collected are sufficient to ensure physical sustainability of irrigation infrastructure. For this, additional measures are needed. Suggested indicators are:

(a) Water Charge Collection Rate = Amount of funds collected/Amount of funds assessed,
(b) Water User Payment Rate = Number of water users who paid water charge in full/Total number of water users required to pay,
(c) Maintenance Resource Sufficiency Index = Amount of funds allocated for maintenance/Funds required for routine preventive maintenance + fund required to repair backlog of deferred maintenance.

8. *How prevalent are irrigation-related disputes among farmers after the APFMIS Act, compared with before and to what extent can these be settled within the farmer organizations? How frequent are farmers taking irrigation-related complaints to the I & CADD?*

Suggested indicators are:

(a) Frequency of irrigation-related disputes before and after the reform,
(b) % of irrigation-related disputes arising which are settled within the WUA or DC,
(c) Frequency of farmer complaints about irrigation which are taken to the I & CADD, before and after reforms.

9. *What impacts is the APFMIS having on I&CADD staff and budgets?*

Suggested indicators are:

(a) Number of I & CADD staff at scheme and district levels, before and after reforms,
(b) Number of I & CADD staff reassigned as competent authorities,
(c) Number of I & CADD staff reassigned to WUAs and DCs,
(d) Size of I & CADD budget for establishment and operational budgets at scheme and district levels, before and after reforms (not including rehabilitation programme funds).

Impact Issues

Impact issues are about the ultimate, indirect effects of a programme, or the realization of the basic goals and purposes of the programme. An assessment of impacts asks, "Have we done the right thing?" If reforms are implemented as intended, it is important to then know whether they are having the intended ultimate effects on people and the environment. Normally, this takes longer to assess than outcomes. Impact assessment is more closely associated with evaluation than monitoring. The following are what appear to be the most important programme impact issues for the APFMIS reform programme, with suggestions for related M&E indicators. Issues are presented as questions to be answered through M&E.

1. *What has been the effect of the reform programme on farm income and economic returns to irrigation water? What has been the impact on land values within irrigation commands?*

It can be expected that irrigation management transfer can lead to higher agricultural productivity and, through improved yields and crop diversification, to more profitable and labor-intensive agriculture. Some suggested indicators for economic productivity of irrigated agriculture are:

(a) Economic return to irrigated land = Gross or net value of output/hectare
(b) Economic return to irrigation water = Gross or net value of output/100 cubic meter or mm water delivered,
(c) Season or annual farm income (gross or net) per irrigated hectare (by location and farm size),
(d) Land sale values per hectare in upper and tail end areas.

The above indicators should be measured in head and tail areas and for different farm size categories.

2. *What has been the effect of the reforms on agricultural wages, employment and poverty?*

Suggested indicators are:

(a) Irrigation Wage Generation = Annual average income in irrigated area before and after reform,
(b) Irrigation Employment Generation = Annual person days of labor per hectare in scheme before and after reform,
(c) Relative Poverty Indicator = % of population below poverty line in scheme before and after reforms,
(d) Regional Relative Poverty Index = % of population below poverty line in the region.

10. *What has been the effect of the reforms on the environment?*

It is possible that reforms could have some effect on the environment. One example of a possible effect is what farmers in the Godavari Delta reported to this consultant. They said that levels of salinity in the soil and water had declined after the reform programme due to improvements in water distribution. More water has been reaching the tail-end areas and was used for flushing salts from soils. Also, due to more canal water being available, farmers in tail ends had stopped or decreased their use of borewell water (which has high levels of salinity). Such findings, if they could be documented and attributed to the reform, could be information of potential important global significance. Assessment of environmental impacts of reform should include analysis of causal linkages between institutional reforms and observed impacts.

Two possible indicators are:

(a) Irrigated Area Sustainability Index = Current irrigable area/Initial irrigable area,

(b) Resource Degradation Index = % of potential irrigation service area lost due to environmental problems (including waterlogging, salinity, sodicity, erosion, depletion of soil fertility, siltation of canals, etc.).

Important Emerging Issues

The following are this consultant's perceptions about what are the five most important topics emerging in the APFMIS reform programme that deserve top priority attention. Working groups or consultative committees should address these topics through extensive discussions with stakeholders, participatory appraisals, M&E and action research where needed.

Shift to Member-oriented WUAs

Some observers and participants in the reform process have expressed concerns that until the present, the process of developing WUAs have been mainly top-down and too focused on the WUA president. This may be partly exaggerated. In the next phase of reform, consolidating trust among members in the WUA will be crucial. There should be an emphasis on involving the membership in WUA activities in more meaningful ways and in ensuring that WUA policies and decisions are primarily oriented toward the interests of common members.

The following are some ways whereby membership participation in WUAs could be enhanced:

(1) A general formalized azamoish, or joint inter-departmental reassessment of irrigated area, should be held state-wide, following a standard procedure. This should involve WUA presidents and TC leaders. Before this is done, all parties concerned should be aware that in the future, the amount of funds available for irrigation O&M will be dependent upon the number of members and amount of funds raised. This will help provide an incentive to WUA leaders to ensure that the long-standing under-estimation of area is overcome.

(2) Formalize the rights of share-croppers as full members with the right to vote. This may require legislation which permits landowners or delegate WUA "acting member" status to tenants without landowners loosing their land ownership status.

(3) Formalize co-membership status of married women, which allows them to vote in WUA meetings when the husband is not present and to be elected to WUA positions.
(4) Requirements for members to contribute labor (or pay for it) for seasonal or annual maintenance activities (such as canal cleaning and desilting).
(5) Requirements that WUA records and accounts be regularly open for inspection by all members at a certain place and time.
(6) Establishment of internal and external audit committees for technical, financial and social audits. Members would serve on audit committees for auditing their own WUA and some would participate in auditing of other WUAs.
(7) Formalize a process for recalling WUA presidents and TC leaders which is not too onerous to be done quickly when needed.
(8) Encourage creation of sideline businesses related to irrigated agriculture which could be taken up by WUA members, but organized under a separate unit. Leaders of the business units should not be related to WUA office bearers.

"Moral Equivalent of War" Against Maldistribution of Water

The single most important thing the WUAs, DCs and PCs could do to establish trust among members and to create necessary incentives for members to raise sufficient resources to pay for O&M is to overcome the longstanding maldistribution of water with irrigation schemes. This should be *the critical next step* in the reform process. If this problem is not solved effectively, the reform process will fail. Consultative committees at all levels and special issue groups should begin to focus on solving this problem. Competent authorities, NGOs and law enforcement agencies should be involved in tackling this problem.

WUAs and DCs should hold consultations with their competent authorities and others from the I&CADD to identify new water distribution policies, rules and sanctions. The I&CADD should provide the necessary technical support to ensure that the proposed new water distribution arrangements are technically sound. Then the WUAs and DCs should hold general assembly meetings to obtain the approval of members. Meetings should then be held between WUA leaders, officials of district government, Panchayat Raj, I&CADD, revenue department and the law enforcement agency to inform them about the new water distribution rules and request the agencies to prepare a jointly-signed letter to support and legitimize the new policies, rules and sanctions. Technical audit committees should be concerned with water distribution as well as maintenance. They should include not only members of WUAs but also representatives from relevant government agencies. Findings of the technical audit committee should be made public and should influence allocation of government subsidies and perhaps other sanctions which government agencies could apply in a supportive manner.

While strong enforceable sanctions need to be established and enforced, with back-up from the government, mechanisms of consultation and persuasion with offenders and applying social pressure should always be the first line of response. It may be helpful for letters to be issued by the state and district governments and law enforcement agencies to WUAs and DCs which clearly acknowledge the rules for water distribution adopted by WUAs and DCs and state the extent of their powers to settle disputes and apply sanctions against errant members and others who may break rules regarding water distribution and damage to structures.

Getting the Incentives Right for Local Resource Mobilisation

After the necessary improvements have been made in water distribution, farmers will be in a far better position to support increasing water charges and making other contributions toward the local financial sustainability of their irrigation systems. The following issues will then need to be dealt with.

What is the basic structure of incentives which will be required to motivate farmers to invest in routine O&M and long-term infrastructure improvement at adequate levels to ensure the physical sustainability and productivity of their irrigation systems? The following are examples of what some of these incentives might be:

(1) WUAs take over some or all of the functions of collecting and transmitting water charges. It may be preferable for the VAO to continue to receive the money and keep the accounts while the WUA presidents and TC leaders serve to mobilize member to pay the fee. It is vital that VAO open their books to the WUAs. They should be instructed to do so by the revenue department hierarchy;

(2) Project Committees should be allowed to set their own water charge rates above the level of the state recommended level (which should be considered only as a minimum);

(3) WUAs retain their share of the water charge collected for routine O&M and the VAO transmits the government's share to the government. Shortfalls in collection could be borne by both the WUA and the government, but shortfalls would reduce the amount of funds available to WUAs, DCs and PCs;

(4) The amount of revenue for O&M or rehabilitation allocated by the government to WUA (or DC or PC) is strictly linked to and based on the amount of funds and other resources (including labor) which are mobilized by the WUA. This will create an incentive for WUAs to expand their service

areas so as to expand their revenue base (this has happened in a number of cases of IMT in other countries);

(5) Government subsidies should not be considered as a form of government welfare or largess, but as a kind of government investment to stimulate local investment and self-reliance, economic development and a gradual decline in the amount of subsidies required. Subsidies should always be contingent upon local investments and certain performance criteria (which could be assessed by M&E or audit committees). Also, funds should not be disbursed to WUAs until after the WUA has made its share of the investment;

(6) Government funds would be disbursed to WUAs only after approval is also given by the Project Committee, so that the PC can provide a review, check and balance function;

(7) WUAs should be required to raise capital reserve funds. These would be required as collateral for WUAs to obtain loans and a prerequisite for them to obtain government subsidies.

(8) Different levels of subsidies and terms and conditions for local contributions should be developed for tank and lift irrigation schemes, suitable to their special constraint and needs.

Reorienting the I&CADD

It is not enough to simply have policy statements which declare that the I&CADD will take on the role of facilitator and provider of technical services to the farmer organizations. Certain additional structural changes are needed to support this. Some of these might be as listed below:

(1) Bonuses, promotions and favorable postings could be given to staff who are effective in the role of facilitating development of strong and well-performing WUAs, DCs and PCs.

(2) The current dispute with laskars being transferred to WUAs needs to be resolved as soon as possible. Compromise may be necessary, but they should still be under the supervision of TC leaders and WUA presidents. Perhaps, an appeal process (to I&CADD) or a joint supervision arrangement between the WUA and I&CADD could be set up. Laskars who adamantly reject the change could be replaced by traditional neeradi or others recruited by the WUAs.

(3) Federating WUAs upto scheme level will probably require transfer of many I&CADD works inspectors to the farmer organizations. Eventually farmer organizations may need to recruit their own works inspectors. Young graduates with Industrial Training Institute certificates, Draftsman Civil Course are becoming widely available in rural areas.

(4) The I&CADD may need to conduct some strategic planning, in collaboration with other government bodies, to revise its mandate. This may

involve some withdrawal form its former role in irrigation O&M and adoption of new tasks in water basin management, environmental monitoring and regulation, water rights registration, higher-level water dispute resolution, etc.

(5) The I&CADD (and perhaps revenue department) should participate in technical, financial and social audits of WUAs, DCs, and PCs, at least for the next several years. This would provide a necessary check and legitimizing function and would help "give teeth" to the audit committees.

Empowering and Enhancing the Relevance of WUAs Through Scaling Up

Scaling up farmer organization from the WUA level to scheme level and higher, can open up new opportunities to empower and enhance the value of the farmer organizations to farmers. Actions need to be taken to protect the integrity of the federation process and to exploit the new opportunities for farmers. Some suggestions follow:

(1) Social audit committees, which consist of both farmers and government officials (and possibly NGO staff), could be established prior to elections at the PC level to monitor elections and report any irregularities observed.

(2) Technical and financial audit committees, which consist of both farmers and government officials (and possibly NGO staff), should be established at each level of federation.

(3) Important government communications, regulations and subsidies should go through the Project Committee first, and then trickled own to the DC and WUA levels, so as to strengthen the PC and support the principle of "one scheme, one system of management".

(4) Pilot or demonstration schemes should be selected in each region where full management transfer is implemented right up to the scheme level. This will provide a learning experience and a location where others can visit for training purposes. This will encourage farmer-to-farmer training.

(5) The training programme should identify training modules based on requests from farmer leaders. More emphasis is needed on practical task-oriented, field-based training (as reported by farmers).

(6) More emphasis is needed on agricultural extension, especially on crop diversification for high value, non-rice crops. This will serve the need to reduce water consumption and increase the economic value of output per unit of water. Federation to DC and PC levels will create a sufficiently large scale of organization for farmer organizations to begin hiring their own agricultural extension agents. Alternatively, they could hire existing extension agents from the agriculture department on a contract basis – not having to pay their full cost of salaries, but more their operational costs. Extension agents could be charged not only with the task of extending

information, but also searching out and setting-up marketing opportunities, conducting soil analyses, and obtaining information about new crops and best cultivation and irrigation practices.

(7) Farmer organization federations above the PC level could create information networks with newsletters about crop price information, marketing channels, political issues requiring lobbying form farmers, etc.

(8) Farmer organization federations will need to be linked up to water basin management bodies and be given a role in coordinating water use and environmental management at this level. The rising inter-sectional competition for water requires that farmer irrigation organizations be involved in these coordinating functions.

Summary of Field Visits

Dr. Douglas Vermillion and Dr. Denise Beaulieu visited the Vamshadhara and Narayanapuram schemes in Srigagulam district on December 9 and 10 and the Central and East Godavari Delta on 11-12 December. We met with WUA presidents, TC leaders, regular WUA members, DC presidents, competent authorities and other engineers from the I&CADD, officials of the revenue department and the press. We tended to split into two groups, one of us would interview farmer representatives and the other would interview government officials.

In the Vamshadhara scheme, Srikakulam district, farmers reported that after the reforms they had handled making repairs by themselves and that water distribution had become more reliable and equitable and yields had increased (from about 20 bags/acre to 25 bags/acre; 1 bag = 80kgs). They said, after the Act farmers got united and persuaded farmers not to steal or take extra water. We saw WUAs can handle repairs up to Rs. 50 per acre, but above that level they expect the government to pay. They said that if the government would fully rehabilitate their scheme, the farmer would then be willing to take over the cost of maintenance thereafter.

WUA members want to grow fruit trees along canals and have fisheries to raise extra money to put into their bank accounts for a reserve fund to be used for emergency repairs or special maintenance. A key concern of farmers is that they need more agricultural extension, in order to reap the benefits of improved water management.

They reported that the government has an arrangement to provide assistance to WUAs in the form of matching grants, usually on a 50:50 or 30:70 share basis, subject to negotiation. This is allocated amount WUAs at the district level.

A WUA president suggested that the recall provision in the Act should be changed from requiring a simple majority to requiring a two-third vote, so things can be more stable (I wondered whether this could not be done by the WUA changing its own by-laws). WUA leaders complained that training was more about what the government thought they needed to know and do, than their own practical needs. They said every training just adds to their responsibilities. They recommended that judicial power be given to the I&CADD to enable them to prosecute those who damage irrigation structures. Farmers said that the election process was satisfactory but some factions try to promote their candidates. In general, farmers do not know what the water charge collection rates are. Farmers want to have more water control structures to exploit benefits of improved water supply. Head-end farmers in large schemes still seem indifferent to water problems of tail enders.

In the Narayanapuram Anicut System, I interviewed a competent authority and works inspector and discussed the minimal rehabilitation programme. Before the Act, the I&CADD would get Rs. 60 per acre per year for O&M. Of this, Rs 20 was used for staff costs. They identified annual works according to two criteria: What works have not been done for three or more years and any special problems arising during the previous season. The tendering and contract process was discussed and compared with the new process of maintenance under the minimum rehabilitation programme.

Now, the competent authority, WUA president, TC members and interested farmers walk through all canals in the WUA area (walking about 5kms per day). The WUA leaders present the list of proposed works to the general body of members, wherein the list is shortened and prioritized. Then the competent authority and his staff prepare cost estimates and obtain necessary technical approvals from superiors and the WUA president signs an agreement for the maintenance programme. The government disburses 40 percent of the funds required in advance. No outside contractors are used. Virtually all labor is provided by WUA members. It was reported that final disbursements of funds for maintenance works had been delayed for unknown reasons.

This process is better than before because farmers prioritize the works, use their own labor, avoid using outside contractors and the quality and volume of work is much better. After the Act the Assistant Engineers and staff have more work (with meetings, supervision, walkthroughs, etc.). But "actually we are doing it the way it is supposed to be done." The Assistant Engineer interviewed suggested that the reform should be decentralized more so that the competent authority can disburse funds without having the pay and accounts officer involved.

We also interviewed farmers and I&CADD staff at a tank scheme in the district (I have lost the name of it). We were told that the first priority for maintenance works in tanks is tank sluices and surplus weirs, then division dams and canal regulators, and then branch canals (desilting and repairing breaches). In 1998-99 they received Rs. 90 per acre for maintenance works, of which Rs. 60 was used for head works. WUA contributes a 15 percent matching amount in labor, bullocks and materials. Farmers reported that with less rainfall than some seasons before the Act, the water is still enough to reach the entire area.

We also met with a revenue department official. He explained the revenue collection process for the land tax and water charge. He said that the Recovery Act empowers the department to "attach property" of defaulters, but so far his superiors had instructed him not to enforce this. His collection rate is about 80 percent and those who do not pay tend to be the more wealthy and powerful.

He estimated that the actual cost of irrigation is about Rs. 1,200 per acre, so even with the recent increases in the water charge to Rs. 200 per acre the government is still subsidizing most of the cost of irrigation. A betterment levy is needed to rehabilitate tanks.

He is against having the WUA take over collection of water charge and against direct allocation of funds collected to WUAs. He said there are so many checks and balances and inspections to protect appropriate collection and use of funds and these would be eliminated by the proposed reform. However, he said the WUA could assist the VAO to raise the collection rate. He felt that it would require a new legislative Act to change water charge collection to the WUAs. The government often lets water charge collection be dropped in an election year. He said politicians should not interfere with revenue collection.

On December 11, we met with farmer leaders in the Central Godavari Delta. A WUA president said that they could collect the water charge if given staff and property attachment powers. They could better persuade farmers to pay, even powerful ones. They would say, "If you don't pay, we will all refuse to pay and will ensure that you get no water". They would not allow politicians to interfere with water charge collection. WUA presidents do not want to get paid for their services. WUA and DC leaders expressed willingness to collaborate with the revenue department to collect water charges, but were not so inclined to do it by themselves. They expect to keep the WUA share and pass on only the government's share. This would enable them to obtain the funds in advance and use them as needed.

We were told that if a landholder is absent, a registered tenant cultivator may vote. The government provided an incentive of Rs. 20,000 per WUA if the elections were done uncontested. Open meetings were held to discuss potential candidates and reach unanimous decisions about WUA president and TC members. Pamphlets about the APFMIS Act and duties of WUA presidents and TC members were given to all farmers. TC members are supervising laskars and reported no difficulties working with them.

Some WUAs are already raising funds from fruit trees and sale of grass on canal embankments. All reported that funds used for minimal rehabilitation were well used with no wastage. After the Act, water is delivered simultaneously to tail and head ends. Transplanting time had been reduced from 25 days. Rice crop yields had increased from 20-25 bags per acre to 30-40 bags per acre after the Act. Farmers want agricultural extension help to do land classification and soil testing to determine what non-rice crops can be grown in different places. They need support to diversify into commercial crops and want help to develop a marketing strategy to export crops, especially bananas.

Farmers reported that in tail areas near the sea, farmers are using less fertilizer now because they don't need to use the infertile borewell water. Canal water conveys more soil nutrients. Also, there is less salinity in the soil be because more water going to tail ends is flushing the salt (If true, believe that this is an important environmental impact which deserves international attention. This deserves a research study to document this).

In the East Godavari Delta farmer leaders told us that in addition to laskars, WUAs also needed to have works inspectors be transferred to them (at least one inspector for two WUAs). Out of 15 WUAs in the DC, four had uncontested elections and 11 had contested elections. They were sure that in the next election, all elections would be contested. They said that one competent authority and one works inspector was needed for every five WUAs. They noted that the so-called social audit is really only a WUA general body meeting where farmer members inquire about WUA activities (this does not seem to be adequate, as noted above).

List of Reports Reviewed While Preparing Report

(1) Oblitas, Keith and Raymond Peter, in association with Gautam Pingle, Halla M. Qaddumi, and Jayantha Perera. 1999. *Transferring Irrigation Management to Farmers in Andhra Pradesh, India*. World Bank Technical Paper No. 449. Washington, D.C.: The World Bank.

(2) K.V. Raju. 1999. Participatory Irrigation Management in Andhra Pradesh: A Way Forward. Draft report commissioned by the Indian Network on Participatory Irrigation Management. Banglore: Institute of Social and Economic Change (included in the present volume).

(3) Davuluri Venkateswarlu. 1999. Politics of Irrigation Management Reforms in Andhra pradesh. Paper commissioned by the India Network on Participatory Irrigation Management and presented at the International Researchers' Conference on Participatory Irrigation Management, 11-14 December, Hyderabad, India (included in the present volume).

(4) B. Guru Prem. 1999. Capacity Building for PIM in Andhra Pradesh (India). Paper presented at the Fifth International Seminar on Participatory Irrigation Management, 14-21 December, Hyderabad, India (included in the present volume).

(5) Jasveen Jairath. 1999. Participatory Irrigation Management (PIM) in Andhra Pradesh: Contradictions of a Supply Side Approach. Draft Manuscript of paper commissioned by Indian Network on Participatory Irrigation

Management for the International Researchers' Conference on Participatory Irrigation Management, 11-14 December, Hyderabad, India (included in the present volume).

(6) M.K. Rahaman. 1999. Farmers Participation: Successful Stories in Andhra Pradesh. Draft manuscript of paper presented at the Fifth International Seminar on Participatory Irrigation Management, 14-21 December, Hyderabad, India.

NOTES

1. The main point of this section is to expose the author's basic perceptions and assumptions to scrutiny. This note is based on a visit to Andhra Pradesh in December, 1999 and was presented and discussed at the Fifth International Seminar on Participatory Irrigation Management, 14-21 December, Hyderabad, India.
2. Elections were not held in June for 492 WUAs due to court and government stay orders and other problems.

Index

Process issues

Process issues are about the dynamics of change, procedures and achievement of specific targets. An assessment of process asks: "Are we doing things in the right way?" They tend to require more quantitative indicators than do outcome and impact assessments. In many cases multiple indicators for the same issue may be necessary to generate [illegible]. The following are what appear to be the most important current process issues for the APFMIS reform programme with suggestions for related indicators. Issues are presented as questions to be answered through M&E.

For many of these issues, simple indicators will not be able to tell the full story about local dynamics and perceptions. More qualitative and exploratory methods will be needed to enable understanding of more subtle [illegible] issues. These may include participatory rural appraisal methods, [illegible] informant and group interviews.